# ABELIAN VARIETIES, THETA FUNCTIONS AND THE FOURIER TRANSFORM

This book is a modern introduction to the theory of abelian varieties and theta functions. Here the Fourier transform techniques play a central role, appearing in several different contexts. In transcendental theory, the usual Fourier transform plays a major role in the representation theory of the Heisenberg group, the main building block for the theory of theta functions. Also, the Fourier transform appears in the discussion of mirror symmetry for complex and symplectic tori, which are used to compute cohomology of holomorphic line bundles on complex tori. When developing the algebraic theory (in arbitrary characteristic), emphasis is placed on the importance of the Fourier–Mukai transform for coherent sheaves on abelian varieties. In particular, it is used in the computation of cohomology of line bundles, classification of vector bundles on elliptic curves, and proofs of the Riemann and Torelli theorems for Jacobians of algebraic curves.

Another subject discussed in the book is the construction of equivalences between derived categories of coherent sheaves on abelian varieties, which follows the same pattern as the construction of intertwining operators between different realizations of the unique irreducible representation of the Heisenberg group.

CAMBRIDGE TRACTS IN MATHEMATICS



# 153 Abelian Varieties, Theta Functions and the Fourier Transform

Alexander Polishchuk
*Boston University*

# Abelian Varieties, Theta Functions and the Fourier Transform

CAMBRIDGE
UNIVERSITY PRESS

PUBLISHED BY THE PRESS SYNDICATE OF THE UNIVERSITY OF CAMBRIDGE
The Pitt Building, Trumpington Street, Cambridge, United Kingdom

CAMBRIDGE UNIVERSITY PRESS
The Edinburgh Building, Cambridge CB2 2RU, UK
40 West 20th Street, New York, NY 10011-4211, USA
477 Williamstown Road, Port Melbourne, VIC 3207, Australia
Ruiz de Alarcón 13, 28014 Madrid, Spain
Dock House, The Waterfront, Cape Town 8001, South Africa

http://www.cambridge.org



First published 2003

Printed in the United Kingdom at the University Press, Cambridge

*Typeface* Times Roman 10.25/13 pt. *System* LaTeX $2_\varepsilon$ [TB]

*A catalog record for this book is available from the British Library.*

*Library of Congress Cataloging in Publication Data*
Polishchuk, Alexander, 1971–
Abelian varieties, Theta functions and the Fourier transform / Alexander Polishchuk.
p. cm. – (Cambridge tracts in mathematics ; 153)
Includes bibliographical references and index.
ISBN 0-521-80804-9
1. Abelian varieties. 2. Fourier transformations.
I. Title. II. Series.
QA564 .P64 2002
516.3′5 – dc21 2002024457

ISBN 0 521 80804 9 hardback

*In memory of Andrei Nikolaevich Tyurin*

# Contents

# Preface

In 1981, S. Mukai discovered a nontrivial algebro-geometric analogue of the Fourier transform in the context of abelian varieties, which is now called the *Fourier–Mukai transform* (see [7]). One of the main goals of this book is to present an introduction to the algebraic theory of abelian varieties in which this transform takes its proper place. In our opinion, the use of this transform gives a fresh point of view on this important theory. On the one hand, it allows one to give more conceptual proofs of the known theorems. On the other, the analogy with the usual Fourier analysis leads one to new directions in the study of abelian varieties. By coincidence, the standard Fourier transform usually appears in the proof of functional equation for theta functions; thus, it is relevant for analytic theory of complex abelian varieties. In references [6] and [9], this fact is developed into a deep relationship between theta functions and representation theory. In the first part of this book we present the basics of this theory and its connection with the geometry of complex abelian varieties. The algebraic theory of abelian varieties and of the Fourier–Mukai transform is developed in the second part. The third part is devoted to Jacobians of algebraic curves. These three parts are tied together by the theory of theta functions: They are introduced in Part I and then used in Parts II and III to illustrate abstract algebraic theorems. Part II depends also on Part I in a more informal way: An important role in the algebraic theory of abelian varieties is played by the theory of Heisenberg groups, which is an algebraic analogue of the corresponding theory in Part I.

Another motivation for our presentation of the theory of abelian varieties is the renewed interest on the Fourier–Mukai transform and its generalizations because of the recently discovered connections with the string theory, in particular, with mirror symmetry. Kontsevich's homological mirror conjecture predicts that for mirror dual pairs of Calabi–Yau manifolds there exists an equivalence of the derived category of coherent sheaves on one of these manifolds with certain category defined by using the symplectic structure on another manifold. This implies that the derived category of coherent sheaves on a Calabi–Yau manifold possessing a mirror dual has many autoequivalences. Such autoequivalences indeed often can be constructed, and the Fourier–Mukai transform is a typical example. On the other hand, it seems that the correspondence between coherent sheaves on a Calabi–Yau manifold and Lagrangian submanifolds in a mirror dual manifold predicted by the homological mirror conjecture should be given by an appropriate real analytic analogue of the Fourier–Mukai transform. At the end of Part I we sketch the simplest example of such a transform in the case of complex and symplectic tori.

We would like to stress that the important idea that influenced the structure of this book is the idea of *categorification* (see [1]). Roughly speaking, this is the process of finding category-theoretic analogues of set-theoretic concepts by replacing sets with categories, functions with functors, etc. The nontriviality of this (nonunique) procedure comes from the fact that axioms formulated as equalities should be replaced by isomorphisms, so one should in addition formulate compatibilities between these isomorphisms. Many concepts in the theory of abelian varieties turn out to be categorifications. For example, the category of line bundles and their isomorphisms on an abelian variety can be thought of as a categorification of the set of quadratic functions on an abelian group, the derived category of coherent sheaves on an abelian variety is a categorification of the space of functions on an abelian group, etc. Of course, the Fourier–Mukai transform is a categorification of the usual Fourier transform. In fact, the reader will notice that most of the structures discussed in Part II are categorifications of the structures from Part I. It is worth mentioning here that the idea of categorification was applied in some other areas of mathematics as well. The most spectacular example is the recent work of Khovanov [5] on the categorification of the Jones polynomial of knots.

Perhaps we need to emphasize that this book does not claim to provide an improvement of the existing accounts of the theory of abelian varieties and theta functions. Rather, its purpose is to enhance this classical theory with more recent ideas and to consider it in a slightly different perspective. For example, in our exposition of the algebraic theory of abelian varieties in Part II we did not try to include all the material contained in Mumford's book [8], which remains an unsurpassed textbook on the subject. Our choice of topics was influenced partly by their relevance for the theory of theta functions, which is a unifying theme for all three parts of the book, and partly by the idea of categorification. In Part I we were strongly influenced by the fundamental works of Lion and Vergne [6] and Mumford et al. [9]. However, our exposition is much more concise: We have chosen the bare minimum of ingredients that allow us to define theta series and to prove the functional equation for them. Our account of the theory of Jacobians in Part III is also far from complete, because our main idea was to stress the role of the Fourier–Mukai transform. Nevertheless, we believe that all main features of this theory are present in our exposition.

This book is based on the lectures given by the author at Harvard University in the fall of 1998 and Boston University in the spring of 2001. It is primarily intended for graduate students and postgraduate researchers interested in algebraic geometry. Prerequisites for Part I are basic complex and differential geometry as presented in chapter 0 of [3], basic Fourier analysis, and familiarity with main concepts of representation theory. Parts II and III are much more technical. For example, the definition of the Fourier–Mukai transform requires working with derived categories of coherent sheaves; to understand it, the reader should be familiar with the language of derived categories. References [2] and [10] can serve as a nice introduction to this language. We also assume the knowledge of algebraic geometry in the scope of the first four chapters of Hartshorne's book [4]. Occasionally, we use more complicated facts from algebraic geometry for which we give references. Some of these facts are collected in Appendix C. Each chapter ends with a collection of exercises. The results of some of these exercises are used in the main text.

Now let us describe the content of the book in more details. Chapters 1–7, which constitute Part I of the book, are devoted to the transcendental theory of abelian varieties. In Chapter 1 we classify holomorphic line bundles on complex tori. In Chapters 2–5 our main focus is the theory of theta functions. We show that they appear naturally as sections of holomorphic line bundles over complex tori. However, the most efficient tool for their study comes not from geometry but from representation theory. The relevant group is the Heisenberg group, which is a central extension of a vector space by $U(1)$, such that the commutator pairing is given by the exponent of a symplectic form. Theta functions arise when one compares different models for the unique irreducible representation of the Heisenberg group on which $U(1)$ acts in the standard way. The main result of this study is the functional equation for theta functions proved in Chapter 5. As another by-product of the study of the Heisenberg group representations, we prove in Appendix B some formulas for Gauss sums discovered by Van der Blij and Turaev.

In Chapters 6 and 7 we discuss mirror symmetry between symplectic tori and complex tori. The main idea is that for every symplectic torus equipped with a Lagrangian tori fibration, there is a natural complex structure on the dual tori fibration. Furthermore, there is a correspondence between Lagrangian submanifolds in a symplectic torus and holomorphic vector bundles on the mirror dual complex torus. The construction of this correspondence can be considered as a (toy) real version of the Fourier–Mukai transform. We apply these ideas to compute cohomology of holomorphic line bundles on complex tori.

Part II (Chapters 8–15) is devoted to algebraic theory of abelian varieties over an algebraically closed field of arbitrary characteristic. In Chapters 8–10 we study line bundles on abelian varieties and the construction of the dual abelian variety. Some of the material is a condensed review of the results from [8], chapter III, sections 10–15. However, the proof of the main theorem about duality of abelian varieties is postponed until Chapter 11, where we introduce the Fourier–Mukai transform. Another result proven in Chapter 11 is that line bundles on abelian varieties satisfying certain nondegeneracy condition have cohomology concentrated in one degree. The Fourier–Mukai transform is also applied to construct an action of a central extension of $SL_2(\mathbb{Z})$ on the derived category of a principally polarized abelian variety. Then in Chapter 12 we develop an algebraic analogue of the representation theory of Heisenberg groups and apply it to the proof of Riemann's quartic theta identity. In Chapter 13 we revisit line bundles on abelian varieties and develop algebraic analogues of some structures associated to holomorphic line bundles on complex tori. Chapter 14 is devoted to the study of vector bundles on elliptical curves. The main idea is to combine the action of a central extension of $SL_2(\mathbb{Z})$ on the derived category of sheaves on elliptic curve with the notion of stability of vector bundles. As a result, we recover Atiyah's classification of bundles on elliptical curves. In Chapter 15 we develop a categorification of the theory of representations of Heisenberg groups, in which the role of the usual Fourier transform is played by the Fourier–Mukai transform. The main result is a construction of equivalences between derived categories of coherent sheaves on abelian varieties, which "categorifies" the construction of intertwining operators between different models of the unique representation of the Heisenberg group given in Chapter 4.

In Part III (Chapters 16–22) we present some topics related to Jacobians of algebraic curves. Chapter 16 is devoted to the construction of the Jacobian of a curve by gluing open pieces of its $g$th symmetric power, where $g$ is the genus of the curve. We also present some basic results on symmetric powers of curves with proofs that work in arbitrary characteristic. In Chapter 17 we define the principal polarization on the Jacobian and give a modern treatment to some classical topics related to the geometry of the embedding of a curve in its Jacobian. In particular, we prove Riemann's theorem describing intersections of the curve with theta divisors in the Jacobian. Chapter 18 is devoted to the proof of Fay's trisecant identity, which is a special identity satisfied by theta functions on the Jacobian. The proof is a combination of the theory developed in Chapter 17 with the residue theorem for rational differentials on a curve. In Chapter 19 we present a more detailed study of the symmetric powers of a curve. The main results are the calculation of the Picard groups and the vanishing theorem for cohomology of some natural vector bundles. We also study Chern classes of the vector bundles over the Jacobian whose projectivizations are isomorphic to symmetric powers of the curve. Chapter 20 is devoted to the varieties of special divisors. Its main results are estimates on dimensions of these varieties and an explicit description of tangent cones to their singular points. In Chapter 21 we prove the Torelli theorem stating that a curve can be recovered from its Jacobian and the theta divisor in it. The idea of the proof is to use the fact that the Fourier–Mukai transform exchanges some coherent sheaves supported on the curve embedded into its Jacobian with coherent sheaves supported on the theta divisor. Finally, in Chapter 22 we discuss Deligne's symbol for a pair of line bundles on a relative curve and its relation to the principal polarization of the Jacobian. We also take a look at the strange duality conjecture, which involves generalization of theta functions to moduli space of vector bundles on curves. The main result is a reformulation of this conjecture in a symmetric way by using the Fourier–Mukai transform.

## Acknowledgments

First, I would like to express my gratitude to A. Beilinson who taught me the Fourier–Mukai transform. It is a pleasure to thank D. Arinkin, B. Conrad, R. Donagi, K. Fukaya, B. Gross, J. Harris, S. Kleiman, J. Lipman, and D. Orlov for helpful discussions of various topics in this book. I am especially grateful to D. Kazhdan, T. Pantev, and A. Vaintrob for the valuable feedback on the first draft of the book. Thanks are also due to everyone who attended the courses I taught at Harvard and Boston universities, on which this book is based, for their remarks. I am indebted to I. Dolgachev, who suggested the relation between the Fourier–Mukai transform and the strange duality that is the subject of the second part of Chapter 22. Finally, I would like to thank Steve Rosenberg for his interest in publishing this book.

## References

[1] J. C. Baez, J. Dolan, Categorification, in *Higher Category Theory* (Evanston, 1997), 1–36. E. Getzler and M. Kapranav, eds. American Mathematical Society, Providence, RI, 1998.

[2] S. Gelfand, Yu. Manin, *Methods of Homological Algebra*. Springer-Verlag, Berlin, 1996.
[3] P. Griffiths, J. Harris, *Principles of Algebraic Geometry*. Wiley-Interscience, 1978.
[4] R. Hartshorne, *Algebraic Geometry*. Springer-Verlag, Berlin, 1977.
[5] M. Khovanov, A categorification of the Jones polynomial, *Duke Math. J.* 101 (2000), 359–426.
[6] G. Lion, M. Vergne, *The Weil Representation, Maslov Index, and Theta Series*. Birkhäuser, 1980.
[7] S. Mukai, Duality between $D(X)$ and $D(\hat{X})$ with its application to Picard sheaves. *Nagoya Math. J.* 81 (1981), 153–175.
[8] D. Mumford, *Abelian Varieties.* Oxford University Press, London, 1974.
[9] D. Mumford, et al. *Tata Lectures on Theta I-III*. Birkhauser, Boston, 1982–1991.
[10] C. Weibel, *An Introduction to Homological Algebra*. Cambridge University Press, Cambridge, 1994.

# Part I

## Analytic Theory

# 1

# Line Bundles on Complex Tori

In this chapter we study holomorphic line bundles on complex tori, i.e., quotients of complex vector spaces by integral lattices. The main result is an explicit description of the group of isomorphism classes of holomorphic line bundles on a complex torus $T$. The topological type of a complex line bundle $L$ on $T$ is determined by its first Chern class $c_1(L) \in H^2(T, \mathbb{Z})$. This cohomology class can be interpreted as a skew-symmetric bilinear form $E : \Gamma \times \Gamma \to \mathbb{Z}$, where $\Gamma = H_1(T, \mathbb{Z})$ is the lattice corresponding to $T$. The existence of a holomorphic structure on $L$ is equivalent to the compatibility of $E$ with the complex structure on $\Gamma \otimes \mathbb{R}$ by which we mean the identity $E(iv, iv') = E(v, v')$. On the other hand, the group of isomorphism classes of topologically trivial holomorphic line bundles on $T$ can be easily identified with the dual torus $T^\vee = \mathrm{Hom}(\Gamma, U(1))$. Now the set of isomorphism classes of holomorphic line bundles on $T$ with the fixed first Chern class $E$ is a $T^\vee$-torsor[1]. It can be identified with the $T^\vee$-torsor of quadratic maps $\alpha : \Gamma \to U(1)$ whose associated bilinear map $\Gamma \times \Gamma \to U(1)$ is equal to $\exp(\pi i E)$. These results provide a crucial link between the theory of theta functions and geometry that will play an important role throughout the first part of this book.

The holomorphic line bundle on $T$ corresponding to a skew-symmetric form $E$ and a quadratic map $\alpha$ as above, is constructed explicitly by equipping the trivial line bundle over a complex vector space with an action of an integral lattice. We show that as a result, every holomorphic line bundle on $T$ has a canonical Hermitian metric and a Hermitian connection. We also show that the dual torus, $T^\vee$, has a natural complex structure and the universal family $\mathcal{P}$ of line bundles on $T$ parametrized by $T^\vee$ (called the *Poincaré bundle*) has a natural holomorphic structure that we describe. In Chapter 9 we will study a purely algebraic version of this duality for abelian varieties.

[1] Following Grothendieck, we will use the term *G-torsor* when referring to a principal homogeneous space for a group $G$.

## 1.1. Cohomology of the Structure Sheaf

Let $V$ be a finite-dimensional complex vector space, $\Gamma$ be a lattice in $V$ (i.e., $\Gamma$ is a finitely generated $\mathbb{Z}$-submodule of $V$ such that the natural map $\mathbb{R} \otimes_{\mathbb{Z}} \Gamma \to V$ is an isomorphism).

**Definition.** The complex manifold $T = V/\Gamma$ is called a *complex torus.*

As a topological space $T$ is just a product of circles, so the cohomology ring $H^*(T, \mathbb{Z}) = \oplus_r H^r(T, \mathbb{Z})$ (resp., $H^*(T, \mathbb{R})$) can be identified naturally with the exterior algebra $\bigwedge^* H^1(T, \mathbb{Z})$ (resp., $\bigwedge^* H^1(T, \mathbb{R})$). Furthermore, we have a natural isomorphism $\Gamma \xrightarrow{\sim} H_1(T, \mathbb{Z})$ sending $\gamma \in \Gamma$ to the cycle $\mathbb{R}/\mathbb{Z} \to T : t \mapsto t\gamma$. Therefore, we get canonical isomorphisms $H^*(T, \mathbb{Z}) \simeq \bigwedge^* \Gamma^\vee$ and $H^*(T, \mathbb{R}) \simeq \bigwedge^* \operatorname{Hom}_{\mathbb{R}}(V, \mathbb{R})$, where $\Gamma^\vee = \operatorname{Hom}(\Gamma, \mathbb{Z})$ is the lattice dual to $\Gamma$.

Recall that one has the direct sum decomposition

$$V \otimes_{\mathbb{R}} \mathbb{C} = V \oplus \overline{V},$$

where $V$ is identified with the subset of $V \otimes_{\mathbb{R}} \mathbb{C}$ consisting of vectors of the form $v \otimes 1 - iv \otimes i$, $\overline{V}$ is the complex conjugate subspace consisting of vectors $v \otimes 1 + iv \otimes i$. We also have the corresponding decomposition

$$\operatorname{Hom}_{\mathbb{R}}(V, \mathbb{C}) = V^\vee \oplus \overline{V}^\vee,$$

where $V^\vee = \operatorname{Hom}_{\mathbb{C}}(V, \mathbb{C})$ is the dual complex vector space to $V$, $\overline{V}^\vee$ is the space of $\mathbb{C}$-antilinear functionals on $V$. Since $T$ is a Lie group, the tangent bundle to $T$ is trivial and the above decomposition is compatible with the decomposition of the bundle of complex valued 1-forms on $T$ according to types $(1, 0)$ and $(0, 1)$. Hence, we have canonical isomorphisms

$$\mathcal{E}^{p,q} \simeq \bigwedge^p V^\vee \otimes_{\mathbb{C}} \bigwedge^q \overline{V}^\vee \otimes_{\mathbb{C}} \mathcal{E}^{0,0},$$

where $\mathcal{E}^{p,q}$ is the sheaf of smooth $(p, q)$-forms on $T$.

The first basic result about $T$ as a complex manifold is the calculation of cohomology of the structure sheaf $\mathcal{O}$, i.e., the sheaf of holomorphic functions.

**Proposition 1.1.** *One has a canonical isomorphism* $H^r(T, \mathcal{O}) \simeq \bigwedge^r \overline{V}^\vee$.

*Proof.* To calculate cohomology of $\mathcal{O}$ one can use the Dolbeault resolution:

$$0 \to \mathcal{O} \to \mathcal{E}^{0,0} \xrightarrow{\overline{\partial}} \mathcal{E}^{0,1} \xrightarrow{\overline{\partial}} \mathcal{E}^{0,2} \to \cdots.$$

We can consider elements of $\bigwedge^p \overline{V}^\vee$ as translation-invariant $(0, p)$-forms on $T$. Note that translation-invariant forms are automatically closed. We claim that this gives an embedding

$$i : \bigwedge^p \overline{V}^\vee \hookrightarrow H^p(T, \mathcal{O}).$$

Indeed, let $\int : \mathcal{E}^{0,0} \to \mathbb{C}$ be the integration map (with respect to some translation-invariant volume form on $T$) normalized by the condition $\int 1 = 1$. Then we can extend $\int$ to the map $\int : \mathcal{E}^{0,p} \to \bigwedge^p \overline{V}^\vee$. It is easy to see that $\int \circ \overline{\partial} = 0$, so $\int$ induces the map on cohomology

$$\int : H^p(T, \mathcal{O}) \to \bigwedge^p \overline{V}^\vee$$

such that $\int \circ i = \mathrm{id}$. Hence, $i$ is an embedding. Let $\Omega^q$ be the sheaf of holomorphic $q$-forms on $T$. Since $\Omega^q \simeq \bigwedge^q V^\vee \otimes \mathcal{O}$, there is an induced embedding

$$i : \oplus_{p,q} \bigwedge^q V^\vee \otimes \bigwedge^p \overline{V}^\vee \to \oplus_{p,q} H^p(T, \Omega^q).$$

Notice that the source of this embedding can be identified with $H^*(T, \mathbb{C}) \simeq \bigwedge^*(V^\vee \oplus \overline{V}^\vee)$. Recall that for every Kähler complex compact manifold $X$ one has Hodge decomposition $H^*(X, \mathbb{C}) \simeq \oplus_{p,q} H^p(X, \Omega^q)$ (e.g., [52], Chapter 0, Section 7). Since any translation-invariant Hermitian metric on $T$ is Kähler, it follows that $\dim H^*(T, \mathbb{C}) = \dim \oplus_{p,q} H^p(T, \Omega^q)$. Therefore, the embedding $i$ is an isomorphism. □

## 1.2. Appell–Humbert Theorem

It is well known that all holomorphic line bundles on $\mathbb{C}^n$ are trivial. Indeed, from the exponential exact sequence

$$0 \to \mathbb{Z} \to \mathcal{O} \overset{\exp}{\to} \mathcal{O}^* \to 0 \tag{1.2.1}$$

we see that it suffices to prove triviality of $H^1(\mathbb{C}^n, \mathcal{O})$. But $H^{>0}(\mathbb{C}^n, \mathcal{O}) = 0$ by Poincaré $\overline{\partial}$-lemma ([52], Chapter 0, Section 2.)

For every complex manifold $X$ we denote by $\mathrm{Pic}(X)$ the Picard group of $X$, i.e., the group of isomorphism classes of holomorphic line bundles on $X$. Triviality of $\mathrm{Pic}(\mathbb{C}^n)$ leads to the following computation of $\mathrm{Pic}(T)$ in terms of group cohomology of the lattice $\Gamma$.

**Proposition 1.2.** *Every holomorphic line bundle $L$ on $T$ is a quotient of the trivial bundle over $V$ by the action of $\Gamma$ of the form $\gamma(z, v) = (e_\gamma(v)z, v+\gamma)$,* *where $\gamma \in \Gamma$, $z \in \mathbb{C}$, $v \in V$, for some 1-cocycle $\gamma \mapsto e_\gamma$ of $\Gamma$ with values in the group $\mathcal{O}^*(V)$ of invertible holomorphic functions on $V$. Here the action of $\Gamma$ on $\mathcal{O}^*(V)$ is induced by its action on $V$. This correspondence extends to an isomorphism of groups*

$$\mathrm{Pic}(T) \simeq H^1(\Gamma, \mathcal{O}^*(V)).$$

*Proof.* Let $\pi : V \to T$ be the canonical projection. Since $\mathrm{Pic}(V)$ is trivial, for every holomorphic line bundle $L$ on $T$ the line bundle $\pi^*L$ on $V$ is trivial. Choose a trivialization $\pi^*L \simeq \mathcal{O}_V$. Then the natural action of $\Gamma$ on $\pi^*L$ becomes an action on the trivial bundle, which should be of the form stated in formulation for some collection $(e_\gamma(v), \gamma \in \Gamma)$ of invertible holomorphic functions on $V$. Unravelling the definition of the action we get the following condition on these functions:

$$e_{\gamma+\gamma'}(v) = e_\gamma(v + \gamma')e_{\gamma'}(v)$$

for every $\gamma, \gamma' \in \Gamma$. This is precisely the cocycle equation for the map $\Gamma \to \mathcal{O}^*(V) : \gamma \mapsto e_\gamma$. If we change the trivialization by another one, the function $e_\gamma(v)$ gets replaced by $e_\gamma(v)f(v+\gamma)f(v)^{-1}$ where $f$ is an invertible holomorphic function on $V$. In other words, the cocycle $\gamma \mapsto e_\gamma$ changes by a coboundary. Thus, we get an isomorphism of $\mathrm{Pic}(T)$ with $H^1(\Gamma, \mathcal{O}^*(V))$. □

**Definition.** We will call 1-cocycles $\Gamma \to \mathcal{O}^*(V) : \gamma \mapsto e_\gamma$ *multiplicators* defining a line bundle on $T$.

From the exponential sequence (1.2.1) we get the long exact sequence

$$\begin{aligned} 0 &\to H^1(T, \mathbb{Z}) \to H^1(T, \mathcal{O}) \to H^1(T, \mathcal{O}^*) \\ &\xrightarrow{\delta} H^2(T, \mathbb{Z}) \to H^2(T, \mathcal{O}) \to \cdots. \end{aligned}$$

Recall that the first Chern class $c_1(L) \in H^2(T, \mathbb{Z})$ of a line bundle $L$ on $T$ is defined as the image of the isomorphism class $[L] \in H^1(T, \mathcal{O}^*)$ under the boundary homomorphism $\delta$. We can consider $c_1(L)$ as a skew-symmetric bilinear form $\Gamma \times \Gamma \to \mathbb{Z}$. Note that $c_1(L)$ determines $L$ as a topological (or $C^\infty$) line bundle. Indeed, this follows immediately from the exponential sequence for continuous (resp., $C^\infty$) functions and from the fact that the sheaf of continuous (resp., $C^\infty$) functions is flabby.

The following natural problems arise.

1. Find out which topological line bundles admit a holomorphic structure, that is, describe the image of $\delta$.

2. For every topological type of holomorphic line bundles find convenient multiplicators producing it.
3. Describe the group of topologically trivial holomorphic line bundles on $T$.

The solution of these problems is given in Theorem 1.3. The main ingredient of the answer is the following construction of multiplicators. Let $H$ be a Hermitian form[2] on $V$, $E = \operatorname{Im} H$ be the corresponding skew-symmetric $\mathbb{R}$-bilinear form on $V$. Assume that $E$ takes integer values on $\Gamma \times \Gamma$. Let $\alpha : \Gamma \to U(1) = \{z \in \mathbb{C} : |z| = 1\}$ be a map such that

$$\alpha(\gamma_1 + \gamma_2) = \exp(\pi i E(\gamma_1, \gamma_2))\alpha(\gamma_1)\alpha(\gamma_2) \tag{1.2.2}$$

(such $\alpha$ always exists; see Exercise 7). Set

$$e_\gamma(v) = \alpha(\gamma) \exp\left(\pi H(v, \gamma) + \frac{\pi}{2} H(\gamma, \gamma)\right),$$

where $\gamma \in \Gamma$, $v \in V$. It is easy to check that $\gamma \mapsto e_\gamma$ is a 1-cocycle. We denote by $L(H, \alpha)$ the corresponding holomorphic line bundle on $T$.

It is easy to see that

$$L(H_1, \alpha_1) \otimes L(H_2, \alpha_2) \simeq L(H_1 + H_2, \alpha_1 \alpha_2),$$
$$[-1]^* L(H, \alpha) = L(H, \alpha^{-1}),$$

where $[-1] : T \to T$ is the involution of $T$ sending $v$ to $-v$.

**Definition.** Let $E$ be a skew-symmetric $\mathbb{R}$-bilinear form on $V$. We say that $E$ is *compatible with the complex structure* if $E(iv, iw) = E(v, w)$. We will say that $E$ is *strictly compatible* with the complex structure if in addition $E(iv, v) > 0$ for $v \neq 0$.

**Remark.** In some books the definition of compatibility of $E$ with the complex structure is equivalent to the strict compatibility in our definition. Note that strict compatibility implies that $E$ is nondegenerate.

**Theorem 1.3.**

*(1) A skew-symmetric bilinear form $E : \Gamma \times \Gamma \to \mathbb{Z}$ is the first Chern class of some holomorphic line bundle on $T$ if and only if $E$ (extended to an $\mathbb{R}$-bilinear form on $V$) is compatible with the complex structure on $V$.*

[2] By a Hermitian form we mean an $\mathbb{R}$-bilinear form, which is $\mathbb{C}$-linear in the first argument and $\mathbb{C}$-antilinear in the second argument (no positivity condition is imposed).

*(2) A skew-symmetric $\mathbb{R}$-bilinear form $E$ on $V$ is compatible with complex structure if and only if there exists a Hermitian form $H$ on $V$ such that $E = \operatorname{Im} H$ (then such $H$ is unique). Assume in addition that $E$ takes integer values on $\Gamma \times \Gamma$. Then there exists a map $\alpha : \Gamma \to U(1)$ satisfying (1.2.2), and for every such $\alpha$ one has $c_1(L(H, \alpha)) = -E$.*

*(3) The map $\alpha \mapsto L(0, \alpha)$ defines an isomorphism from $\operatorname{Hom}(\Gamma, U(1))$ to the group of isomorphism classes of topologically trivial holomorphic line bundles on $T$.*

*Proof.* 1. Consider the canonical map

$$H^r(T, \mathbb{C}) \to H^r(T, \mathcal{O})$$

We can identify $H^r(T, \mathbb{C})$ with $\bigwedge^r (V \otimes_{\mathbb{R}} \mathbb{C})^{\vee}$. We have a decomposition $V \otimes_{\mathbb{R}} \mathbb{C} \simeq V \oplus \overline{V}$, and it is easy to see that the above map is given by restricting an alternating $r$-form from $V \otimes_{\mathbb{R}} \mathbb{C}$ to $\overline{V}$. Now consider the composed map

$$H^2(T, \mathbb{R}) \to H^2(T, \mathbb{C}) \to H^2(T, \mathcal{O}).$$

An element in $H^2(T, \mathbb{R})$ corresponds to a skew-symmetric real bilinear form $E$ on $V$. The above map sends it to a $\mathbb{C}$-bilinear form on $\overline{V}$ obtained by extending scalars to $\mathbb{C}$ and restricting the form to the subspace $\overline{V} \subset V \otimes_{\mathbb{R}} \mathbb{C}$. The latter subspace consists of elements of the form $v \otimes 1 + iv \otimes i \in V \otimes_{\mathbb{R}} \mathbb{C}$. Thus, the condition that $E$ maps to zero in $H^2(T, \mathcal{O})$ means that

$$(E \otimes \mathbb{C})(v \otimes 1 + iv \otimes i, w \otimes 1 + iw \otimes i) = 0$$

for any $v, w \in V$. It is easy to see that this condition is equivalent to $E(iv, iw) = E(v, w)$. Thus, a skew-symmetric bilinear form $E : \Gamma \times \Gamma \to \mathbb{Z}$ comes from a holomorphic line bundle if and only if it is compatible with a complex structure.

2. The Hermitian form $H$ is constructed from $E$ by the formula $H(v, w) = E(iv, w) + iE(v, w)$. It is easy to see that in this way we get a bijective correspondence between Hermitian forms and skew-symmetric $\mathbb{R}$-bilinear forms compatible with complex structure.

Now assume that $E$ takes integer values on $\Gamma \times \Gamma$. The proof of existence of a map $\alpha$ satisfying (1.2.2) is sketched in Exercise 7. It remains to check that the class $c_1(L(H, \alpha)) \in H^2(T, \mathbb{Z})$ corresponds to the skew-symmetric form $-E$. By general nonsense (see Exercise 5) the coboundary map $H^1(T, \mathcal{O}^*) \to H^2(T, \mathbb{Z})$ can be identified with the coboundary map

$$\delta : H^1(\Gamma, \mathcal{O}^*(V)) \to H^2(\Gamma, \mathbb{Z}) \simeq \bigwedge^2 \Gamma.$$

The value of the latter map on a 1-cocycle $\gamma \mapsto e_\gamma(v)$ can be computed as follows. For every $\gamma \in \Gamma$ choose a holomorphic function $f_\gamma$ on $V$ such that $e_\gamma(v) = \exp(2\pi i f_\gamma(v))$. Then the 2-cocycle

$$F(\gamma_1, \gamma_2) = t_{\gamma_1}^* f_{\gamma_2} - f_{\gamma_1+\gamma_2} + f_{\gamma_1}$$

takes values in $\mathbb{Z}$ and represents $\delta(e_\gamma)$. Under the natural isomorphism $H^2(\Gamma, \mathbb{Z}) \simeq \mathrm{Hom}(\bigwedge^2 \Gamma, \mathbb{Z})$ the class of the 2-cocycle $F(\gamma_1, \gamma_2)$ corresponds to the skew-symmetric form

$$F(\gamma_2, \gamma_1) - F(\gamma_1, \gamma_2)$$

(see Exercise 6). It follows that the first Chern class of the line bundle associated with a 1-cocycle $\gamma \mapsto e_\gamma$ is represented by the skew-symmetric form

$$f_{\gamma_2}(v + \gamma_1) - f_{\gamma_1}(v + \gamma_2) + f_{\gamma_1}(v) - f_{\gamma_2}(v).$$

In our case we can take

$$f_\gamma(v) = \delta(\gamma) + \frac{1}{2i} H(v, \gamma) + \frac{1}{4i} H(\gamma, \gamma),$$

where $\alpha(\gamma) = \exp(2\pi i \delta(\gamma))$, which implies that $c_1(L(H, \alpha))$ corresponds to the form $-E$.

3. Consider the following exact sequence

$$0 \to H^1(T, \mathbb{Z}) \to H^1(T, \mathbb{R}) \to H^1(T, U(1)) \to H^2(T, \mathbb{Z}) \to H^2(T, \mathbb{R}).$$

The last arrow is injective, therefore, the map $H^1(T, \mathbb{R}) \to H^1(T, U(1))$ is surjective. On the other hand, the map $H^1(T, \mathbb{R}) \to H^1(T, \mathcal{O})$ is an isomorphism, so from the commutative diagram

$$\begin{array}{ccccc} H^1(\mathbb{Z}) & \longrightarrow & H^1(T, \mathbb{R}) & \longrightarrow & H^1(T, U(1)) \\ \downarrow & & \downarrow & & \downarrow \\ H^1(\mathbb{Z}) & \longrightarrow & H^1(T, \mathcal{O}) & \longrightarrow & H^1(T, \mathcal{O}^*) \end{array} \tag{1.2.3}$$

we deduce that the image of $H^1(T, \mathcal{O}) \to H^1(T, \mathcal{O}^*)$ coincides with the image of the injective map $H^1(T, U(1)) \to H^1(T, \mathcal{O}^*)$. Note that we have a natural isomorphism $H^1(T, U(1)) \simeq \mathrm{Hom}(\Gamma, U(1))$. It is easy to check that in terms of this isomorphism the embedding $H^1(T, U(1)) \to H^1(T, \mathcal{O}^*) = \mathrm{Pic}(T)$ is given by $\alpha \mapsto L(0, \alpha)$. □

As a corollary, we get the following description of Pic($T$) due to Appell and Humbert.

**Corollary 1.4.** *The group* Pic($T$) *is isomorphic to the group of pairs* $(H, \alpha)$, *where $H$ is a Hermitian form on $V$ such that $E = \mathrm{Im}\, H$ takes integer values on $\Gamma$, $\alpha$ is a map from $\Gamma$ to $U(1)$ such that (1.2.2) is satisfied. The group law on the set of pairs is given by* $(H_1, \alpha_1)(H_2, \alpha_2) = (H_1 + H_2, \alpha_1\alpha_2)$.

The only nonobvious part of the above argument is the invention of line bundles $L(H, \alpha)$. We will see in Section 2.5 that in the case of positive definite $H$ their construction is quite natural from the point of view of the Heisenberg group.

## 1.3. Metrics and Connections

The line bundle $L(H, \alpha)$ constructed in Section 1.2 comes equipped with a natural Hermitian metric. To construct it, first we define a metric on the trivial line bundle on $V$ by setting

$$h(v) = \exp(-\pi H(v, v)).$$

**Proposition 1.5.** *The metric $h$ descends to a metric on $L(H, \alpha)$. There is a unique connection on $L(H, \alpha)$ that is compatible with this metric and with the complex structure on $L(H, \alpha)$. Its curvature is equal to $\pi i E$ considered as a translation-invariant 2-form on $T$, where $E = \mathrm{Im}\, H$.*

*Proof.* It is easy to check that the metric $h$ is invariant with respect to the action of $\Gamma$ on the trivial bundle, which we used to define $L(H, \alpha)$. Therefore, it descends to a metric on $L(H, \alpha)$. It is well known that for every Hermitian metric on a holomorphic line bundle there exists a unique connection compatible with this metric and the complex structure ([52], Chapter 0, Section 5). To describe this connection on $L(H, \alpha)$ we are going to write the corresponding $\Gamma$-invariant connection $\nabla$ on the trivial line bundle on $V$. The section

$$s = \exp\left(\frac{\pi}{2} H(v, v)\right)$$

of the trivial bundle on $V$ is orthonormal with respect to our metric $h$. Hence, we should have

$$\langle \nabla s, s\rangle + \langle s, \nabla s\rangle = 0. \tag{1.3.1}$$

We can write $\nabla s = ds + s\omega$ for some $(1,0)$-form $\omega$, where

$$ds = \frac{\pi}{2}(H(dv, v) + H(v, dv))s.$$

Here the notation $H(dv, v)$ and $H(v, dv)$ should be understood as follows. Let us identify $V$ with $\mathbb{C}^n$ in such a way that $H(z, z') = \sum_{i=1}^{r} z_i \overline{z_i'}$, where $r$ is the rank of $H$. Then we have $H(dv, v) = \sum_{i=1}^{r} \overline{z_i} dz_i$, etc. Now we can rewrite equation (1.3.1) as

$$\pi(H(dv, v) + H(v, dv)) + \omega + \overline{\omega} = 0.$$

This implies that $\omega = -\pi H(dv, v)$. Thus, we obtain

$$\nabla = d - \pi H(dv, v).$$

The curvature of this connection is equal to $\pi H(dv, dv)$. If we identify $V$ with $\mathbb{C}^n$ as above then $H(dv, dv) = \sum_{i=1}^{r} dz_i \wedge d\overline{z_i}$. Note that $\overline{H(dv, dv)} = -H(dv, dv)$. Hence, the curvature is equal to

$$\pi H(dv, dv) = \frac{\pi}{2}(H - \overline{H})(dv, dv) = \pi i E(dv, dv). \qquad \square$$

In the case $H = 0$ we obtain that the line bundle $L(0, \alpha)$, where $\alpha \in \mathrm{Hom}(\Gamma, U(1))$, can be equipped with a flat unitary connection compatible with the complex structure. It is not difficult to check that the corresponding 1-dimensional representation of the fundamental group $\pi_1(T) = \Gamma$ is given by the character $\alpha$.

## 1.4. Poincaré Line Bundle

According to Theorem 1.3, topologically trivial holomorphic line bundles on $T$ are parametrized (up to an isomorphism) by the group $T^\vee = \mathrm{Hom}(\Gamma, U(1))$. Note that we have the following isomorphisms:

$$\mathrm{Hom}(\Gamma, U(1)) = \mathrm{Hom}(\Gamma, \mathbb{R})/\mathrm{Hom}(\Gamma, \mathbb{Z}) = \mathrm{Hom}_{\mathbb{R}}(V, \mathbb{R})/\Gamma^\vee.$$

Also, one has a canonical isomorphism $\overline{V}^\vee \stackrel{\sim}{\rightarrow} \mathrm{Hom}_{\mathbb{R}}(V, \mathbb{R})$ sending a $\mathbb{C}$-antilinear map $\phi : V \to \mathbb{C}$ to $\mathrm{Im}\,\phi$. Hence, we can identify $T^\vee = \mathrm{Hom}_{\mathbb{R}}(V, \mathbb{R})/\Gamma^\vee$ with the complex torus $\overline{V}^\vee/\Gamma^\vee$. It is easy to see that $\Gamma^\vee \subset \overline{V}^\vee$ coincides with the set of all $\mathbb{C}$-antilinear maps $\phi : V \to \mathbb{C}$ such that $\mathrm{Im}\,\phi(\Gamma) \subset \mathbb{Z}$.

**Definition.** $T^\vee$ is called the *dual complex torus to* $T$.

Since $T^\vee$ parametrizes all topologically trivial line bundles on $T$, it is natural to expect that there is a universal line bundle on $T \times T^\vee$. Such a line bundle is constructed in the following definition.

**Definition.** The *Poincaré line bundle* is the holomorphic line bundle $\mathcal{P}$ on $T \times T^\vee = (V \oplus \overline{V}^\vee)/(\Gamma \oplus \Gamma^\vee)$ obtained as $L(H_{\text{univ}}, \alpha_{\text{univ}})$, where $H_{\text{univ}}$ is the natural Hermitian form on $V \oplus \overline{V}^\vee$:

$$H_{\text{univ}}((v, \phi), (v', \phi')) = \phi(v') + \overline{\phi'(v)},$$
$$\alpha_{\text{univ}}(\gamma, \gamma^\vee) = \exp(\pi i \langle \gamma^\vee, \gamma \rangle).$$

For every $\alpha \in T^\vee$ we have a natural isomorphism of holomorphic bundles on $T$

$$\mathcal{P}|_{T \times \{\alpha\}} \simeq L(0, \alpha).$$

Furthermore, every (holomorphic) family of topologically trivial line bundles on $T$ parametrized by a complex manifold $S$ is induced by $\mathcal{P}$ via a holomorphic map $S \to T^\vee$.

In Part II we will consider an algebraic analogue of duality between complex tori. The corresponding algebraic Poincaré line bundle will be the main ingredient in the definition of the Fourier–Mukai transform in Chapter 11.

## Exercises

1. Let $f : V \to V'$ be a $\mathbb{C}$-linear map of complex vector spaces mapping a lattice $\Gamma \subset V$ into a lattice $\Gamma' \subset V'$. Then $f$ induces the holomorphic map $f : T = V/\Gamma \to T' = V'/\Gamma'$ of the corresponding complex tori. Show that for a line bundle $L(H, \alpha)$ on $T'$ associated with a Hermitian form $H$ on $V'$ and a map $\alpha : \Gamma' \to U(1)$ as in Section 1.2 one has
$$f^* L(H, \alpha) \simeq L(f^* H, f^* \alpha).$$
2. Let $t_v : V \to V : x \mapsto x + v$ be a translation. Prove that
$$t_v^* L(H, \alpha) \simeq L(H, \alpha \cdot \nu_v),$$
where $\nu_v(\gamma) = \exp(2\pi i E(v, \gamma))$. Check that this isomorphism is compatible with metrics introduced in Section 1.3 up to a constant factor.

3. Let $E : \Gamma \times \Gamma \to \mathbb{Z}$ be a skew-symmetric form compatible with the complex structure on $V$. Let $\Gamma_0 \subset \Gamma$ be the kernel of $E$. Prove that $V_0 = \mathbb{R}\Gamma_0$ is a complex subspace of $V$.
4. Let $L(H, \alpha)$ be a holomorphic line bundle on $T$ corresponding to some data $(H, \alpha)$ as in Section 1.2. Let $V_0 \subset V$ be the kernel of $E = \operatorname{Im} H$, $\Gamma_0 = V_0 \cap \Gamma$. Assume that $\alpha|_{\Gamma_0} \equiv 1$. Prove that $L$ is a pull-back of a holomorphic line bundle on $T' = V/V_0 + \Gamma$ under the natural projection $T \to T'$.
5. In this exercise *a sheaf* always means a sheaf of abelian groups. A $\Gamma$-equivariant sheaf on $V$ is a sheaf $\mathcal{F}$ on $V$ equipped with the system of isomorphisms $i_\gamma : t_\gamma^*\mathcal{F} \simeq \mathcal{F}$, where $t_\gamma : V \to V$ is the translation by $\gamma$. These isomorphisms should satisfy the following cocycle condition:

$$i_{\gamma+\gamma'} = i_\gamma \circ t_\gamma^*(i_{\gamma'}).$$

We denote by $\Gamma - \mathrm{Sh}_V$ the category of $\Gamma$-equivariant sheaves on $V$ and by $\mathrm{Sh}_T$ the category of sheaves on $T$.

(a) Show that the functor $\pi^\cdot$ establishes an equivalence of categories $\mathrm{Sh}_T \widetilde{\to} \Gamma - \mathrm{Sh}_V$. Deduce that if $\mathcal{F}$ is an injective sheaf on $T$ then $\pi^\cdot\mathcal{F}$ is an injective object in the category $\Gamma - \mathrm{Sh}_V$.

(b) Let $\mathcal{F}$ be a sheaf on $T$. Construct a functorial isomorphism $H^0(T, \mathcal{F}) \to H^0(V, \pi^\cdot\mathcal{F})^\Gamma$.

(c) Let $\mathbb{Z}_V$ denotes the constant sheaf on $V$ corresponding to $\mathbb{Z}$. Then for every $\Gamma$-module $M$ the constant sheaf $M \otimes \mathbb{Z}_V$ on $V$ is equipped with a natural $\Gamma$-action. Show that for every $\Gamma$-equivariant sheaf $\mathcal{G}$ on $V$ there is a functorial isomorphism

$$\mathrm{Hom}_\Gamma(M, H^0(V, \mathcal{G})) \simeq \mathrm{Hom}_{\Gamma-\mathrm{Sh}_V}(M \otimes \mathbb{Z}_V, \mathcal{G}).$$

Deduce from this that if $\mathcal{G}$ is an injective object of the category $\Gamma - \mathrm{Sh}_V$ then $H^0(V, \mathcal{G})$ is an injective $\Gamma$-module.

(d) Let $\mathcal{F}$ be an injective sheaf on $T$. Show that $H^{>0}(V, \pi^\cdot\mathcal{F}) = 0$. [*Hint*: Use the fact that $\pi$ is a local homeomorphism to show that $\pi^\cdot\mathcal{F}$ is flabby.]

(e) Let $\mathcal{F}$ be a sheaf on $T$ such that $H^{>0}(V, \pi^\cdot\mathcal{F}) = 0$. Choose an injective resolution $\mathcal{F}_\bullet$ of $\mathcal{F}$. Prove that cohomology of the complex $H^0(V, \pi^\cdot\mathcal{F}_\bullet)^\Gamma$ can be identified with $H^*(\Gamma, H^0(V, \pi^\cdot\mathcal{F}))$. Now using (b) construct isomorphisms

$$H^i(T, \mathcal{F}) \to H^i(\Gamma, H^0(V, \pi^\cdot\mathcal{F})).$$

Show that if $0 \to \mathcal{F}' \to \mathcal{F} \to \mathcal{F}'' \to 0$ is an exact sequence of sheaves such that $\pi^{\cdot}\mathcal{F}'$, $\pi^{\cdot}\mathcal{F}$ and $\pi^{\cdot}\mathcal{F}''$ are acyclic, then the above maps fit into a morphism of long exact sequences.

(f) Show that the sheaf-theoretic pull-back of the exponential exact sequence on $T$ gives the exponential exact sequence on $V$.

(g) Prove that the global exponential sequence on $V$

$$0 \to \mathbb{Z} \to \mathcal{O}(V) \to \mathcal{O}^*(V) \to 0$$

is exact.

(h) Identify the connecting homomorphism $H^1(T, \mathcal{O}^*) \to H^2(T, \mathbb{Z})$ with the connecting homomorphism in group cohomology $H^1(\Gamma, \mathcal{O}^*(V)) \to H^2(\Gamma, \mathbb{Z})$.

6. The goal of this exercise is to identify the isomorphism $i : H^2(\Gamma, \mathbb{R}) \simeq H^2(T, \mathbb{R})$ obtained in the previous exercise with the natural map $H^2(\Gamma, \mathbb{R}) \to \mathrm{Hom}(\bigwedge^2 \Gamma, \mathbb{R})$ sending a 2-cocycle $c : \Gamma \times \Gamma \to \mathbb{R}$ to the skew-symmetric bilinear form $c(\gamma_2, \gamma_1) - c(\gamma_1, \gamma_2)$.

(a) Show that the (real) de Rham complex on $V$: $\mathcal{E}^0(V) \to \mathcal{E}^1(V) \to \cdots$ is a resolution of $\mathbb{R}$ by acyclic $\Gamma$-modules. Derive from this the following description of the isomorphism $i$. Start with a 2-cocycle $c : \Gamma \times \Gamma \to \mathbb{R}$ of $\Gamma$ with coefficients in $\mathbb{R}$. Then there exists a collection of smooth functions $f_\gamma$ on $V$ such that

$$c(\gamma_1, \gamma_2) = t_{\gamma_1}^* f_{\gamma_2} - f_{\gamma_1+\gamma_2} + f_{\gamma_1},$$

where $c(\gamma_1, \gamma_2)$ is considered as a constant function on $V$. Next, there exists a 1-form $\omega$ on $V$ such that $df_\gamma = t_\gamma^*\omega - \omega$ for every $\gamma$. This implies that the 2-form $d\omega$ is $\Gamma$-invariant. Hence, it descends to a closed 2-form on $T$. Its cohomology class is $i(c)$.

(b) Recall that the isomorphism $H^2(T, \mathbb{R}) \to \mathrm{Hom}(\bigwedge^2 \Gamma, \mathbb{R}) = \mathrm{Hom}_{\mathbb{R}}(\bigwedge_{\mathbb{R}}^2 V, \mathbb{R})$ sends the cohomology class of a closed 2-form $\eta$ on $T$ to $\int \eta$, where the map

$$\int : \mathcal{E}^2(T) \to \mathrm{Hom}_{\mathbb{R}}\left(\bigwedge\nolimits_{\mathbb{R}}^2 V, \mathbb{R}\right)$$

is obtained from the isomorphism

$$\mathcal{E}^2(T) \simeq \mathrm{Hom}_{\mathbb{R}}\left(\bigwedge\nolimits_{\mathbb{R}}^2 V, \mathbb{R}\right) \otimes \mathcal{E}^0(T)$$

via the integration map $\int : \mathcal{E}^0(T) \to \mathbb{R}$. Choosing real coordinates on $V$ associated with a basis of $\Gamma$ show that this map sends the 2-form on $T$ representing $i(c)$ to the skew-symmetric bilinear form $c(\gamma_2, \gamma_1) - c(\gamma_1, \gamma_2)$.

7. Let $\overline{E} : \Gamma \times \Gamma \to \mathbb{Z}/2\mathbb{Z}$ be a skew-symmetric bilinear form modulo 2 (*skew-symmetry* means that $\overline{E}(\gamma, \gamma) = 0$ for every $\gamma \in \Gamma$). Prove that there exists a map $f : \Gamma \to \mathbb{Z}/2\mathbb{Z}$, such that

$$\overline{E}(\gamma_1, \gamma_2) = f(\gamma_1 + \gamma_2) + f(\gamma_1) + f(\gamma_2).$$

Deduce that for every skew-symmetric bilinear form $E : \bigwedge^2 \Gamma \to \mathbb{Z}$ there exists a map $\alpha : \Gamma \to \{\pm 1\}$ satisfying (1.2.2).

8. Let $T$ be a complex torus, $e_1, \ldots, e_{2n}$ be the basis of the lattice $H^1(T, \mathbb{Z})$, $e_1^*, \ldots, e_{2n}^*$ be the dual basis of $H^1(T^\vee, \mathbb{Z})$, where $T^\vee$ is the dual torus. Show that the first Chern class of the Poincaré bundle on $T \times T^\vee$ is given by

$$c_1(\mathcal{P}) = \sum_{i=1}^{2n} e_i \wedge e_i^*.$$

# 2

# Representations of Heisenberg Groups I

This chapter in an introduction to the representation theory of Heisenberg groups. This theory will be our principal tool in the study of theta functions.

Throughout this chapter $V$ is a real vector space with a fixed symplectic form $E$. The main object of our study will be the Heisenberg group $\mathcal{H}(V)$ associated with $V$. By the definition, $\mathcal{H}(V)$ is the central extension of $V$ by $U(1)$ corresponding to the 2-cocycle $\exp(\pi i E)$. According to the theorem of Stone and von Neumann, there is a unique (up to an equivalence) irreducible unitary representation of $\mathcal{H}(V)$ on which $U(1)$ acts naturally. We consider three different types of structures on $V$ compatible with the symplectic structure, namely, Lagrangian subspaces, maximal isotropic lattices, and complex structures. Each of these structures gives rise to a model for the above representation of $\mathcal{H}(V)$. The model relevant for the geometry is the one associated to a complex structure on $V$. It is called the *Fock representation* (the corresponding representation space is the space of holomorphic functions on $V$). For an isotropic lattice $\Gamma \subset V$ equipped with a lifting to a subgroup in $\mathcal{H}$, the space of $\Gamma$-invariants in Fock representation can be identified with the space of global sections of some holomorphic line bundle on the torus $V/\Gamma$. The elements of this space are called *canonical theta functions*.[3] They will be studied in Chapters 3 and 5.

For some parts of the theory of Heisenberg groups it is convenient to work with the category of locally compact abelian groups. However, we will mostly care about two classes of such groups: real vector spaces and finite abelian groups, so the reader may choose to think only about locally compact abelian groups out of these two classes.

[3] Canonical theta functions differ from classical theta functions by some exponential factor (see Section 5.6).

## 2.1. Heisenberg Groups

Recall that for a locally compact abelian group $K$, its *Pontryagin dual* group $\widehat{K}$ is defined as the group of continuous homomorphisms from $K$ to $U(1)$. The important fact about this duality is that the functor $K \mapsto \widehat{K}$ on the category of locally compact abelian groups is exact. On the subcategory of real vector spaces the Pontryagin duality becomes the usual duality of vector spaces.

**Definition.** (i) A *Heisenberg group* $H$ is a central extension

$$0 \to U(1) \to H \xrightarrow{p} K \to 0,$$

where $U(1) = \{z \in \mathbb{C}^* \mid |z| = 1\}$, $K$ is a locally compact abelian group, such that the commutator pairing

$$e : K \times K \to U(1) : (k, k') \mapsto [\widetilde{k}, \widetilde{k'}],$$

where $\widetilde{k}, \widetilde{k'} \in H$ are arbitrary lifts of $k, k' \in K$, identifies $K$ with its Pontryagin dual $\widehat{K}$.

(ii) We call a closed subgroup $\widetilde{L} \subset H$ *isotropic* if $\widetilde{L} \cap U(1) = 1$.

(iii) A closed subgroup $L \subset K$ is called *isotropic* if $e|_{L\times L} \equiv 1$.

Since $[H, H] \subset U(1)$, every isotropic subgroup $\widetilde{L} \subset H$ is commutative. Therefore, its image in $H/U(1) = K$ is an isotropic subgroup $L \subset K$. Thus, one can specify an isotropic subgroup $\widetilde{L} \subset H$ by giving an isotropic subgroup $L \subset K$, equipped with a continuous lifting homomorphism $\sigma : L \to H$ (such that $\widetilde{L} = \sigma(L)$).

**Proposition 2.1.** *For every isotropic subgroup $L \subset K$ there exists a continuous lifting homomorphism $\sigma : L \to H$.*

*Proof.* We have to prove that the central extension

$$1 \to U(1) \to p^{-1}(L) \to L \to 1$$

in the category of commutative locally compact groups splits. But this follows from the Pontryagin duality: the dual exact sequence gives an extension of $\widehat{U(1)} = \mathbb{Z}$, hence it splits. □

**Definition.** We say that an isotropic subgroup $L \subset K$ is *Lagrangian* if it is maximal isotropic, or equivalently, if $L^{\perp} = L$, where $L^{\perp}$ denotes the orthogonal complement with respect to $e$.

Henceforward, whenever we talk about an isotropic subgroup $L$ in $K$, we assume that its lifting $\sigma : L \to H$ is chosen. By abuse of notation we will denote the corresponding isotropic subgroup $\sigma(L) \subset H$ simply by $L \subset H$.

**Proposition 2.2.** *Let $L \subset K$ be an isotropic subgroup, $N_H(L) \subset H$ be its normalizer in $H$. Then $N_H(L) = p^{-1}(L^{\perp})$, so $N_H(L)/L$ is a central extension of $L^{\perp}/L$ by $U(1)$. Furthermore, $N_H(L)/L$ a Heisenberg group.*

The proof is left to the reader.

## 2.2. Representation Associated with an Isotropic Subgroup

For an isotropic subgroup $L \subset K$ let us consider the space $C(L)$ consisting of continuous functions $\phi : H \to \mathbb{C}$ such that

$$\phi(\lambda h) = \lambda\phi(h),\ \lambda \in U(1),$$
$$\phi(\sigma(l)h) = \phi(h),\ l \in L$$

(recall that $\sigma : L \to H$ is a lifting homomorphism). There is a natural linear action of $H$ on this space given by $(h\phi)(h') = \phi(h'h)$. To get a unitary representation of $H$ we modify the space $C(L)$ as follows. Note that for every $\phi_1, \phi_2 \in C(L)$ the function $\phi_1\overline{\phi_2}$ is $U(1)$-invariant, hence it descends to a function on $K/L$. In particular, for every $\phi \in C(L)$ we can consider $|\phi|^2$ as a function on $K(L)$. Now we set

$$\mathcal{F}(L) = \widehat{\left\{\phi \in C(L) \mid \int_{K/L} |\phi|^2 dk < \infty\right\}},$$

where $dk$ is a Haar measure[4] on $K/L$, $\widehat{}$ denotes the completion with respect to the hermitian metric given by

$$\langle\phi_1, \phi_2\rangle = \int_{K/L} \phi_1\overline{\phi_2} dk.$$

**Remark.** The representation $\mathcal{F}(L)$ is equivalent to the induced representation $\mathrm{Ind}_{p^{-1}(L)}^{H} \chi_L$, where $\chi_L$ is the 1-dimensional representation of $p^{-1}(L) \subset H$

[4] By the definition, Haar measure is a translation-invariant measure. Such measures exist on general locally compact abelian groups. In the cases that we will be considering the relevant group is either a commutative Lie group or a finite group. In the former case, one can integrate using an invariant volume form; in the second case, we replace the integral by the sum.

corresponding to the projection

$$\chi_L : p^{-1}(L) \simeq U(1) \times \sigma(L) \to U(1).$$

Indeed, for a function $\phi : H \to \mathbb{C}$ the condition $\phi \in \mathcal{F}(L)$ can be rewritten as

$$\phi(h_1 h) = \chi_L(h_1)\phi(h),$$

where $h \in H, h_1 \in p^{-1}(L)$ (plus integrability of $|\phi|^2$ on $K/L$).

The following theorem due to Stone and von Neumann plays the principal role in the representation theory of $H$.

**Theorem 2.3.** *If the subgroup $L \subset K$ is Lagrangian then the representation of $H$ on $\mathcal{F}(L)$ is irreducible. There is a unique (up to an equivalence) unitary irreducible representation of $H$ on which $U(1)$ acts in the standard way.*

Note that this theorem implies that all representations $\mathcal{F}(L)$ for different Lagrangian subgroups are equivalent to each other. The unique irreducible representation of $H$ (on which $U(1)$ acts in the standard way) is called *Schrödinger representation* of $H$. We will not prove this theorem (the particular case when $K$ is finite is considered in Exercises 3 and 4, the complete proof can be found in [97] or [84]). However, in Chapter 4 we will construct explicitly intertwining operators between representations $\mathcal{F}(L)$ for some pairs of Lagrangian subgroups in $K$.

## 2.3. Real Heisenberg Group

We are mainly interested in the classical real Heisenberg group $\mathcal{H}(V)$ of a symplectic vector space $(V, E)$. By the definition, $\mathcal{H}(V) = U(1) \times V$ and the group law is given by

$$(\lambda, v) \cdot (\lambda', v') = (\exp(\pi i E(v, v'))\lambda\lambda', v + v').$$

Note that for a real Lagrangian subspace $L \subset V$ there is a canonical lifting homomorphism $L \to \mathcal{H}(V) : l \mapsto (1, l)$. So by Theorem 2.3, we have an irreducible representation $\mathcal{F}(L)$ of $\mathcal{H}(V)$. A function $\phi \in \mathcal{F}(L)$ is determined by its restriction to $\{1\} \times V \subset \mathcal{H}(V)$, so we can identify $\mathcal{F}(L)$ with the completion (with respect to $L^2$-norm on $V/L$) of the space of functions $\phi$ on $V$ such that $\phi(v + l) = \exp(\pi i E(v, l))\phi(v)$, where $v \in V, l \in L$. The action of $\mathcal{H}(V)$ on such functions is given by

$$(\lambda, v) \cdot \phi(v') = \lambda \exp(\pi i E(v', v))\phi(v + v').$$

Real Lagrangian subspaces form one class of isotropic subgroups in $V$. Another important example of an isotropic subgroup in $\mathcal{H}(V)$ is the following. Let $\Gamma \subset V$ be an isotropic lattice, i.e., a lattice such that $E|_{\Gamma\times\Gamma}$ is integer-valued. Then a lifting homomorphism $\sigma : \Gamma \to \mathcal{H}(V)$ has form

$$\sigma(\gamma) = \sigma_\alpha(\gamma) = (\alpha(\gamma), \gamma), \tag{2.3.1}$$

where a map $\alpha : \Gamma \to U(1)$ should satisfy

$$\alpha(\gamma_1 + \gamma_2) = \exp(\pi i E(\gamma_1, \gamma_2))\alpha(\gamma_1)\alpha(\gamma_2).$$

Note that this is precisely the equation (1.2.2) on the data used in the construction of a line bundle on $V/\Gamma$. The corresponding space $\mathcal{F}(\Gamma)$ can be identified with the space of $L^2$-sections of certain line bundle on $V/\Gamma$ (see Exercise 1). Henceforward, referring to the above situation, we will say that a lifting homomorphism $\sigma_\alpha$ is given by the quadratic map $\alpha$ and will freely use the correspondence $\alpha \mapsto \sigma_\alpha$ when discussing liftings of a lattice to the Heisenberg group. A lattice $\Gamma \subset V$ is maximal isotropic if and only if $\Gamma = \Gamma^\perp = \{v \in V \mid E(v, \Gamma) \subset \mathbb{Z}\}$, i.e., if and only if the skew-symmetric form $E|_{\Gamma\times\Gamma}$ is unimodular. We will refer to such lattices as *self-dual.*

## 2.4. Representations Associated with Complex Structures

One can consider a complex structure $J$ on $V$ compatible with $E$ as an infinitesimal isotropic subgroup in $\mathcal{H}(V)$, i.e., as a *Lie subalgebra* in $\mathrm{Lie}(\mathcal{H}(V)) \otimes_{\mathbb{R}} \mathbb{C}$. Indeed, a complex structure on $V$ can be specified by a complex subspace $P \subset V \otimes_{\mathbb{R}} \mathbb{C}$ of dimension $\frac{1}{2}\dim_{\mathbb{R}} V$ such that $P \cap V = 0$. The correspondence goes as follows: given a complex structure $J$, the subspace $P_J \subset V \otimes_{\mathbb{R}} \mathbb{C}$ consists of elements $v \otimes 1 + Jv \otimes i$, $v \in V$ (this is precisely the subspace $\overline{V}$ from the canonical decomposition $V \otimes_{\mathbb{R}} \mathbb{C} = V \oplus \overline{V}$). The complex structure on $V$ can be recovered from $P_J$ via the isomorphism $V \simeq V \otimes_{\mathbb{R}} \mathbb{C}/P_J$. By the definition, a function $f$ on $V$ is $J$-holomorphic if and only if $df(P_J) = 0$, where $P_J$ is extended to a translation-invariant complex distribution of subspaces on $V$. Now it is easy to check that $E$ is compatible with a complex structure $J$ if and only if the corresponding subspace $P_J \subset V \otimes_{\mathbb{R}} \mathbb{C}$ is isotropic with respect to $E$ (extended to a $\mathbb{C}$-bilinear form). Therefore, $0 \oplus P_J$ is a Lie subalgebra in $\mathrm{Lie}(\mathcal{H}(V)) \otimes_{\mathbb{R}} \mathbb{C} = \mathbb{C} \oplus V \otimes_{\mathbb{R}} \mathbb{C}$.

Thus, one can expect that a choice of $J$ also gives a realization of the unique irreducible unitary representation of $\mathcal{H}$. We are going to show that this is indeed the case provided that $J$ satisfies the condition $E(Jx, x) > 0$ for $x \in V \setminus 0$, i.e., $E$ and $J$ are strictly compatible. The latter condition is equivalent to the condition that the Hermitian form $H$ on $V$, such that $\operatorname{Im} H = E$, is positive definite.

We are going to construct several equivalent models for the unitary representation of $\mathcal{H}(V)$ associated with a complex structure $J$. The first model is the space

$$\mathcal{F}^{-}(J) = \Big\{\phi : \mathcal{H}(V) \to \mathbb{C} \mid \phi(\lambda h) = \lambda^{-1}\phi(h), \lambda \in U(1);$$

$$d\phi\left(P_J^r\right) = 0;\ \int_V |\phi|^2 dv < \infty\Big\},$$

where the action of $\mathcal{H}(V)$ is given by $(h\phi)(h') = \phi(h'h)$. Here $P_J^r$ denotes the right-invariant distribution of subspaces on $\mathcal{H}(V)$, which is equal to $0 \oplus P_J$ at the point $(1, 0)$. The unitary structure on $\mathcal{F}^{-}(J)$ is given by

$$\langle \phi_1, \phi_2 \rangle = \int_V \phi_1 \overline{\phi_2} dv$$

($\mathcal{F}^{-}(J)$ is complete with respect to this metric).

The second model is a slight modification of the first one. Its importance is due to the fact that it uses holomorphic functions on $V$ with respect to the complex structure $J$.

**Definition.** The *Fock representation* is the unitary representation of $\mathcal{H}(V)$ on the space

$$\operatorname{Fock}(V, J) = \left\{ f \text{ holomorphic on } V,\ \int_V |f(v)|^2 \exp(-\pi H(v, v)) dv < \infty \right\},$$

given by

$$U_{(\lambda, v)} f(v') = \lambda^{-1} \exp\left(-\pi H(v', v) - \frac{\pi}{2} H(v, v)\right) f(v + v'), \quad (2.4.1)$$

where $(\lambda, v) \in \mathcal{H}(V)$, $v' \in V$. The Hermitian form on $\operatorname{Fock}(V, J)$ is given by

$$\langle f, g \rangle = \int_V f(v)\overline{g(v)} \exp(-\pi H(v, v)) dv$$

(the space $\operatorname{Fock}(V, J)$ is complete with respect to the corresponding norm).

The equivalence $\mathcal{F}^{-}(F) \to \operatorname{Fock}(V, J)$ is given by the map $\phi \mapsto f$, where

$$f(v) = \exp\left(\frac{\pi}{2} H(v, v)\right) \phi(1, v).$$

Indeed, the invariance of $\phi$ with respect to the distribution $P_J^r$ is equivalent to the fact that $f$ is $J$-holomorphic (see Exercise 6).

**Remarks.** 1. Note that $\lambda \in U(1)$ acts on $\mathcal{F}^-(F)$ and $\mathrm{Fock}(V, J)$ by $\lambda^{-1}$. In fact, these representations are irreducible, so by Theorem 2.3, they are isomorphic to the *dual* (or complex conjugate) of the representation $\mathcal{F}(L)$ associated with a real Lagrangian subspace $L \subset V$.

2. Unlike representations $\mathcal{F}(L)$ the Fock representation contains a canonical *vacuum vector* $f = 1$. This vector is characterized (up to a constant) by the property that the action of the Lie algebra of $\mathcal{H}(V)$ on $f$ is well defined and the subalgebra $P_J \subset \mathrm{Lie}(\mathcal{H}(V)) \otimes_{\mathbb{R}} \mathbb{C}$ annihilates $f$ (the fact that such vector is unique can be used to prove irreducibility of $\mathrm{Fock}(V, J)$).

3. If the Hermitian form $H$ is not positive then the space $\mathrm{Fock}(V, J)$ defined above, is zero (see Exercise 7).

Yet another model of the representation of $\mathcal{H}(V)$ associated with the complex structure $J$ strictly compatible with $E$, is the space

$$\mathcal{F}^+(J) = \Big\{ \phi : \mathcal{H}(V) \to \mathbb{C} \mid \phi(\lambda h) = \lambda \phi(h), \lambda \in U(1);$$

$$d\phi(P_J^l) = 0; \int_V |\phi|^2 dv < \infty \Big\},$$

where $\mathcal{H}(V)$ acts by $(h\phi)(h') = \phi(h^{-1}h')$, $P_J^l$ denotes the left-invariant distribution of subspaces on $\mathcal{H}(V)$, which is equal to $0 \oplus P_J$ at the point $(1, 0)$. The isomorphism between $\mathcal{F}^+(J)$ and $\mathcal{F}^-(J)$ is given by the correspondence $\phi(h) \mapsto \phi(h^{-1})$.

## 2.5. Canonical Theta Functions

There is an action of $\mathcal{H}(V)$ on the trivial line bundle $\mathbb{C} \times V \to V$, such that an element $(\lambda, v) \in \mathcal{H}(V)$ acts by a holomorphic map

$$(\lambda', v') \mapsto \left(\lambda' \lambda^{-1} \exp\left(\pi H(v', v) + \frac{\pi}{2} H(v, v)\right), v + v'\right).$$

Clearly, the induced action of $\mathcal{H}(V)$ on the space of holomorphic functions on $V$ is precisely the Fock representation. On the other hand, if $\Gamma$ is an isotropic lattice in $V$ equipped with a lifting to $\mathcal{H}(V)$ then there is an induced holomorphic action of $\Gamma$ on $\mathbb{C} \times V$. This is the action we used in Section 1.2 to construct the holomorphic line bundle $L(H, \alpha^{-1})$ on $V/\Gamma$. The space of global holomorphic sections of $L(H, \alpha^{-1})$ can be identified with the space of holomorphic functions on $V$ satisfying the equations

$$f(v+\gamma)=\alpha(\gamma)\exp\left(\pi H(v,\gamma)+\frac{\pi}{2}H(\gamma,\gamma)\right)f(v), \qquad (2.5.1)$$

where $v \in V$, $\gamma \in \Gamma$.

**Definition.** We denote by $T(H,\Gamma,\alpha)$ the space of holomorphic functions on $V$ satisfying (2.5.1). Its elements are called *canonical theta functions* for $(H,\Gamma,\alpha)$.

Note that $T(H,\Gamma,\alpha)$ is equipped with the natural Hermitian metric

$$\langle f,g\rangle_\Gamma=\int_{V/\Gamma} f(v)\overline{g(v)}\exp(-\pi H(v,v))dv \qquad (2.5.2)$$

coming from the Hermitian metric on $L(H,\alpha^{-1})$.

Comparing the equations (2.5.1) with the definition of the Fock representation we would like interpret the condition $f \in T(H,\Gamma,\alpha)$ for a holomorphic function $f$ on $V$ as the invariance of $f$ under operators $U_{\alpha(\gamma),\gamma}$ for all $\gamma \in \Gamma$ (where $\Gamma$ is lifted to $H$ using $\alpha$). Since nonzero elements of $T(H,\Gamma,\alpha)$ are not square-integrable, to achieve such an interpretation we have to extend operators $U_{\lambda,v}$ to a larger space of holomorphic functions. The correct way to enlarge Fock$(V,J)$ is to consider the space Fock$_{-\infty}(V,J)$ consisting of holomorphic functions $f$ on $V$ such that

$$f(x)=O\left(\|x\|^n\cdot\exp\left(\frac{\pi}{2}H(x,x)\right)\right) \quad \text{for some } n.$$

The action of $\mathcal{H}(V)$ on Fock$(V,J)$ extends naturally to this larger space.

**Proposition 2.4.** *One has* $H^0(T,L(H,\alpha^{-1})) \simeq T(H,\Gamma,\alpha) = (\text{Fock}_{-\infty}(V,J))^\Gamma$.

*Proof.* The only fact one has to check is the inclusion $T(H,\Gamma,\alpha) \subset$ Fock$_{-\infty}(V,J)$. We leave this as an exercise for the reader. □

The space Fock$_{-\infty}(V,J)$ is dual to the subspace Fock$_\infty(V,J) \subset$ Fock$(V,J)$ consisting of $f \in$ Fock$(V,J)$ such that

$$f(x)=O\left(\|x\|^{-n}\exp\left(\frac{\pi}{2}H(x,x)\right)\right) \quad \text{for all } n.$$

The subspace Fock$_\infty(V,J)$ has the following representation-theoretic meaning: it is the space of $C^\infty$*-vectors* in the Fock representation. By the definition, these are vectors on which the action of the universal enveloping algebra of the Lie algebra of $\mathcal{H}(V)$ is well defined. We do not elaborate on the details of this construction (see [84]).

The spaces $\mathrm{Fock}_{\infty}(V, J)$ and $\mathrm{Fock}_{-\infty}(V, J)$ are isomorphic to the complex conjugates of similar spaces $\mathcal{F}_{\infty}(L)$ and $\mathcal{F}_{-\infty}(L)$ constructed for the representation $\mathcal{F}(L)$ associated with a Lagrangian subgroup of $\mathcal{H}(V)$.

**Proposition 2.5.** *Let $\Gamma \subset V$ be a self-dual lattice. Then the space $\mathcal{F}_{-\infty}(\Gamma)^{\Gamma}$ is* 1*-dimensional.*

*Proof.* The space $\mathcal{F}_{\infty}(\Gamma)$ consists of $C^{\infty}$-functions contained in $\mathcal{F}(\Gamma)$ (since the Lie algebra of $\mathcal{H}(V)$ acts on $\mathcal{F}(\Gamma)$ by infinitesimal translations, i.e., by differentiations). Therefore, the space $\mathcal{F}_{-\infty}(\Gamma)$ consists of all distributions $\phi$ on $V$ satisfying

$$\phi(x+\gamma) = \alpha(\gamma)^{-1} \exp(\pi i E(x, \gamma))\phi(x)$$

for any $\gamma \in \Gamma$. The value of such a distribution on $f \in \mathcal{F}_{\infty}(\Gamma)$ is $\int_{V/\Gamma} \overline{\phi} f$. On the other hand, the condition of $\Gamma$-invariance for $\phi \in \mathcal{F}_{-\infty}(\Gamma)$ reads as

$$\alpha(\gamma) \exp(\pi i E(x, \gamma))\phi(x+\gamma) = \phi(x).$$

Comparing these two conditions we get that $\phi$ is proportional to the following linear combination of delta-functions at lattice points:

$$\sum_{\gamma \in \Gamma} \alpha(\gamma)^{-1} \delta(\gamma).$$

□

This proposition implies the following result for which we will give an independent proof in Chapter 7.

**Corollary 2.6.** *If the lattice $\Gamma$ is self-dual (and $H$ is positive-definite), then $T(H, \Gamma, \alpha)$ is* 1*-dimensional.*

**Remark.** We obtained above a description of the (projectivization of the) spaces of global sections of line bundles $L(H, \alpha)$ in terms that are independent of complex structure. This description relied heavily on the fact that the Hermitian form $H$ is positive. It is easy to see that if $H$ is nondegenerate but not positive then $L(H, \alpha)$ has no holomorphic sections. As we will show in Chapter 7, in this case one has $H^j(T, L(H, \alpha)) = 0$ for $j \neq i$, while $H^i(T, L(H, \alpha)) \neq 0$, where $i$ is the number of negative eigenvalues of $H$. Moreover, in the case when $\Gamma$ is self-dual, the space $H^i(T, L(H, \alpha))$

is 1-dimensional. An interesting problem is to find an analogue of the isomorphism of Proposition 2.4 for $H^i(T, L(H, \alpha))$. The first step in this direction was recently done by I. Zharkov (see [138]) who constructed a canonical cohomology class in $H^i(\Gamma, H^0(V, \mathcal{O}))$, where $H^0(V, \mathcal{O})$ is the space of holomorphic functions on $V$. There remains a question, whether there exists an $\mathcal{H}(V)$-submodule $\mathcal{F}_{-\infty} \subset H^0(V, \mathcal{O})$ such that the above class lies in $H^i(\Gamma, \mathcal{F}_{-\infty})$ and such that the projectivization of $\mathcal{F}_{-\infty}$ does not depend on a choice of complex structure.

## Exercises

1. Let $(V, E)$ be a symplectic vector space and $\Gamma \subset V$ be an isotropic lattice equipped with a lifting to the Heisenberg group $\mathcal{H}(V)$. Let $T = V/\Gamma$ be the corresponding torus. We can consider the natural morphism

$$\Gamma \backslash \mathcal{H}(V) \to T$$

as an $\mathcal{H}(V)$-equivariant principal $U(1)$-bundle over $T$. Let $\mathcal{L}$ be the $\mathcal{H}(V)$-equivariant complex line bundle over $T$ associated with the opposite $U(1)$-bundle (i.e., the fibers of $\mathcal{L}$ are $U(1)$-equivariant maps from the fibers of our $U(1)$-bundle to $\mathbb{C}$). By construction the line bundle $\mathcal{L}$ is equipped with a Hermitian metric. Show that the representation $\mathcal{F}(\Gamma)$ can be identified with the natural representation of $\mathcal{H}(V)$ on the space of square-integrable sections of $\mathcal{L}$.
2. In the situation of the previous exercise show that $c_1(\mathcal{L}) \in H^2(T, \mathbb{Z})$ is given by the skew-symmetric form $E|_{\Gamma\times\Gamma}$. [*Hint*: Use the trivialization of the pull-back of $\mathcal{L}$ to $V$ and Exercises 5 and 6 of Chapter 1.]
3. Let $H$ be a finite Heisenberg group (i.e., such that the corresponding group $K$ is finite).
   (a) Prove that there exists a Lagrangian subgroup $I \subset K$.
   (b) Let $W$ be a representation of $H$ such that $U(1)$ acts by the identity character. Consider the decomposition

$$W = \oplus_{\chi\in\hat{I}} W_\chi$$

   in isotipic components with respect to the $I$-action. Show that an element $h \in H$ sends $W_\chi$ to $W_{\chi-\phi_h}$ where $\phi_h$ is a character of $I$ defined by $\phi_h(i) = [h, i]$.
4. In the situation of the previous exercise show that the natural map

$$W \to \mathcal{F}(I) \otimes W_1 : w \mapsto (h \mapsto (hw)_1)$$

is an isomorphism of $H$-modules, where for $w \in W$ we denote by $w_1$ its projection to $W_1$. Show that $W$ is irreducible if and only if $W_1 = W^I$ is one-dimensional. Deduce from this the Stone-von Neumann theorem for $H$ and the irreducibility of $\mathcal{F}(I)$.

5. Let $H$ be a finite Heisenberg group, $W$ be its Schrödinger representation. Show that $W^* \otimes W$ is isomorphic as $H \times H$-representation to the space of functions $\phi$ on $H$ such that $\phi(zh) = z\phi(h)$ for $z \in U(1)$ with the $H \times H$-action given by $(h_1, h_2)\phi(h) = \phi(h_1^{-1}hh_2)$.
6. (a) Let us identify the Lie algebra of $U(1)$ with $\mathbb{R}$ in such a way that the exponential map $\mathrm{Lie}(U(1)) \to U(1)$ is given by $x \mapsto \exp(2\pi i x)$ and consider the induced identification of $\mathrm{Lie}(\mathcal{H}(V))$ with $\mathbb{R} \oplus V$ (as vector spaces). Show that the distribution $P_J^r$ on $\mathcal{H}(V)$ can be described explicitly as follows:

$$\left(P_J^r\right)_{(\lambda_0, v_0)} = \left\{\left(\frac{i}{2}H(v_0, v), v \otimes 1 + Jv \otimes i\right), v \in V\right\} \subset \mathbb{C} \oplus V \otimes_{\mathbb{R}} \mathbb{C}.$$

   (b) Let $\phi$ be a function on $V$. We extend $\phi$ to $\mathcal{H}(V)$ by setting $\widetilde{\phi}(\lambda, v) = \lambda^{-1}\phi(v)$. Show that $\widetilde{\phi}$ is invariant with respect to the distribution $P_J^r$ if and only if the function $v \mapsto \exp(\frac{\pi}{2}H(v, v))\phi(v)$ is $J$-holomorphic.
7. Show that if the Hermitian form $H$ on $V$ is not positive then a holomorphic function $f$ on $V$ satisfying $\int_V |f(v)|^2 \exp(-\pi H(v, v))dv < \infty$, is zero. [*Hint*: Reduce to the 1-dimensional case.]

# 3

# Theta Functions I

In this chapter we construct *theta series*, special elements in the spaces of canonical theta functions $T(H, \Gamma, \alpha)$ defined in the previous chapter. Recall that this space depends on the following data: a complex vector space $V$ equipped with a positive-definite Hermitian form $H$, a lattice $\Gamma \subset V$, such that the symplectic form $E = \operatorname{Im} H$ takes integer values on $\Gamma$, and a lifting of $\Gamma$ to a subgroup of the Heisenberg group $\mathcal{H}(V)$ given by a quadratic map $\alpha : \Gamma \to U(1)$ such that $\alpha|_{\Gamma \cap L} \equiv 1$. The theta series $\theta^{\alpha}_{H,\Gamma,L}$ is a canonical theta function depending on one additional datum $L \subset V$, a real Lagrangian subspace compatible with $\Gamma$. In the case when the lattice $\Gamma$ is maximal isotropic, according to Corollary 2.6, the space of canonical theta functions is 1-dimensional, so for different choices of a Lagrangian subspace $L$ the elements $\theta^{\alpha}_{H,\Gamma,L}$ should be proportional. In Chapter 5 we will compute these proportionality coefficients and deduce from this the classical functional equation for theta series. In the case when $\Gamma$ is not necessarily maximal isotropic, we equip the space of canonical theta functions with the structure of a representation of a finite Heisenberg group associated with $\Gamma$, and show that it is irreducible.

As a geometric application of theta functions we prove the theorem of Lefschetz stating that a complex torus $V/\Gamma$ can be embedded holomorphically into a projective space if and only if there exists a positive-definite Hermitian form $H$ on $V$ such that $\operatorname{Im} H$ takes integer values on $\Gamma$.

In Appendix A we describe a relation between 1-dimensional theta series and Weierstrass sigma function.

Throughout this chapter $V$ denotes a complex vector space, $H$ is a positive-definite Hermitian form on $V$, $\Gamma \subset V$ is a lattice such that the symplectic form $E = \operatorname{Im} H$ takes integer values on $\Gamma \times \Gamma$, $\alpha : \Gamma \to U(1)$ is a map satisfying equation (1.2.2).

## 3.1. Action of the Finite Heisenberg Group

Let $\sigma_\alpha : \Gamma \to \mathcal{H}(V)$ be a lifting homomorphism (2.3.1) given by $\alpha$. Recall that the space $T(H, \Gamma, \alpha)$ of canonical theta functions consists of holomorphic functions $f$ on $V$, invariant under the action of $\Gamma$ in (an extension of) Fock representation, where $\Gamma$ is lifted to $\mathcal{H}(V)$ using $\sigma_\alpha$. Let $N(\Gamma)$ be the normalizer of $\Gamma$ in $\mathcal{H}(V)$. Then the group $G(E, \Gamma, \alpha) := N(\Gamma)/\Gamma$ acts naturally on the space $T(H, \Gamma, \alpha)$. As we have seen in Section 2.1, $G(E, \Gamma, \alpha)$ is a Heisenberg group. More precisely, it is a central extension of the finite abelian group $\Gamma^\perp/\Gamma$ by $U(1)$, where $\Gamma^\perp = \{v \in V : E(v, \Gamma) \subset \mathbb{Z}\}$. Equivalent way to define an action of $G(E, \Gamma, \alpha)$ on $T(H, \Gamma, \alpha)$ is the following. Recall that we have a natural holomorphic action of $\mathcal{H}(V)$ on the trivial line bundle over $V$ (see Section 2.5). The restriction of this action to $N(\Gamma)$ descends to an action of $G(E, \Gamma, \alpha)$ on the line bundle $L(H, \alpha^{-1})$ over $V/\Gamma$, compatible with the action of $\Gamma^\perp/\Gamma$ on $V/\Gamma$ by translations. Hence, we get an action of $G(E, \Gamma, \alpha)$ on all the cohomology groups of $L(H, \alpha^{-1})$. It remains to use the identification $T(H, \Gamma, \alpha) \simeq H^0(V/\Gamma, L(H, \alpha^{-1}))$.

**Proposition 3.1.** *$T(H, \Gamma, \alpha)$ is an irreducible representation of $G(E, \Gamma, \alpha)$ of dimension $\sqrt{[\Gamma^\perp : \Gamma]}$.*

*Proof.* According to Exercise 4 of Chapter 2 it suffices to prove that if $I \subset \Gamma^\perp/\Gamma$ is a Lagrangian subgroup equipped with a lifting $\widetilde{I} \subset G(E, \Gamma, \alpha)$, then the invariants of $I$ in $T(H, \Gamma, \alpha)$ is a 1-dimensional subspace. Let $\Gamma_1 \subset \Gamma^\perp$ be the preimage of $I$ under the homomorphism $\Gamma^\perp \to \Gamma^\perp/\Gamma$, and let $\widetilde{\Gamma}_1$ be the preimage of $\widetilde{I}$ under the homomorphism $N(\Gamma) \to G(E, \Gamma, \alpha)$. Then $\widetilde{\Gamma}_1$ is a lifting of $\Gamma_1$ to $\mathcal{H}(V)$ and $\Gamma_1$ is maximal isotropic. By Corollary 2.6, this implies that the space of $\Gamma_1$-invariant holomorphic functions on $V$ is 1-dimensional. But

$$H^0(V, \mathcal{O})^{\Gamma_1} = (H^0(V, \mathcal{O})^{\Gamma})^I = T(H, \Gamma, \alpha)^I,$$

so the latter space is 1-dimensional.

To prove the last assertion let us consider any finite Heisenberg group $H_f$, i.e., a Heisenberg group such that $K = H_f/U(1)$ is finite. Then we can realize the Schrödinger representation of $H_f$ in the space of functions on $K/I$ where $I \subset K$ is a Lagrangian subgroup, so its dimension is equal to $|K/I| = \sqrt{|K|}$ (cf. Exercise 4 of Chapter 2). □

**Remark.** Recall that $T(H, \Gamma, \alpha)$ is identified with global sections of a holomorphic bundle $L = L(H, \alpha^{-1})$ on $T = V/\Gamma$. One of the goals of the algebraic theory of theta-functions is to define the relevant finite Heisenberg group and its action on $H^0(T, L)$ algebraically (see Chapter 12). It is easy to see that if some of eigenvalues of $H$ are negative then $T(H, \Gamma, \alpha) = 0$. Indeed, restricting an element $f \in T(H, \Gamma, \alpha)$ to an affine subspace on which $H(v, v)$

is bounded (and tends to $-\infty$ when $v \to \infty$) and using the quasi-periodicity conditions one can derive from the maximum principle that $f = 0$. When $H$ is not necessarily positive (but nondegenerate), as we will show in Chapter 7, there is a unique nontrivial cohomology space $H^i(T, L)$. This space is still an irreducible representation of the Heisenberg group $G(E, \Gamma, \alpha)$.

## 3.2. A Choice of Lifting

Recall that liftings of $\Gamma$ to a subgroup in the Heisenberg group correspond to maps $\alpha : \Gamma \to U(1)$ satisfying the equation (1.2.2). An important construction of such a map is described in the following proposition.

**Proposition 3.2.** *Let $\Gamma = \Gamma_1 \oplus \Gamma_2$ be a decomposition of $\Gamma$ into a direct sum of subgroups such that $E|_{\Gamma_i \times \Gamma_i} = 0$ for $i = 1, 2$. Let us define the map $\alpha_0 = \alpha_0(\Gamma_1, \Gamma_2) : \Gamma \to \{\pm 1\}$ by the formula*

$$\alpha_0(\gamma) = \exp(\pi i E(\gamma_1, \gamma_2)), \tag{3.2.1}$$

*where $\gamma = \gamma_1 + \gamma_2$, $\gamma_i \in \Gamma_i$. Then $\alpha_0$ satisfies the equation (1.2.2).*

The proof is straightforward and is left to the reader. We will call a decomposition $\Gamma = \Gamma_1 \oplus \Gamma_2$ of the type described in the above proposition, an *isotropic decomposition* of $\Gamma$.

Any two maps $\alpha$ and $\alpha'$ satisfying equation (1.2.2) are related by the formula

$$\alpha'(\gamma) = \alpha(\gamma) \exp(2\pi i E(c, \gamma)) \tag{3.2.2}$$

for some $c \in V$ which is uniquely determined modulo $\Gamma^{\perp}$. It is easy to see that the corresponding homomorphisms $\sigma_{\alpha'}$ and $\sigma_\alpha$ are related as follows:

$$\sigma_{\alpha'}(\gamma) = (1, c)\sigma_\alpha(\gamma)(1, c)^{-1}. \tag{3.2.3}$$

Therefore, we can define an isomorphism of the corresponding finite Heisenberg groups

$$i_c : G(E, \Gamma, \alpha) \to G(E, \Gamma, \alpha') : g \mapsto (1, c)g(1, c)^{-1}. \tag{3.2.4}$$

Now the operator $U_{(1,c)}$ (corresponding to the action of $(1, c)$ on Fock representation) restricts to an isomorphism between $T(H, \Gamma, \alpha)$ and $T(H, \Gamma, \alpha')$ compatible with the actions of $G(E, \Gamma, \alpha)$ and $G(E, \Gamma, \alpha')$ via $i_c$.

### 3.3. Theta Series

Let $L \subset V$ be a real Lagrangian subspace, such that $L$ is generated by $\Gamma \cap L$ over $\mathbb{R}$ and $\alpha|_{\Gamma\cap L} \equiv 1$. In this situation we are going to give a definition of the theta series. In Chapter 5 we will explain how one can discover these series more naturally.

**Lemma 3.3.** *$L$ generates $V$ over $\mathbb{C}$.*

*Proof.* The condition that the Hermitian form $H$ is positive-definite, is equivalent to the condition $E(iv, v) > 0$ for every non-zero $v \in V$. Since $E|_{L\times L} \equiv 0$, this implies that for $v \in L \setminus 0$ we cannot have $iv \in L$. Therefore, $iL \cap L = 0$ and hence $V = iL + L$. □

Since $H|_{L\times L}$ is a symmetric form, according to Lemma 3.3, it extends uniquely to a $\mathbb{C}$-bilinear symmetric form $S : V \times V \to \mathbb{C}$.

**Proposition–Definition 3.4.** *(i) The theta series*

$$\theta^{\alpha}_{H,\Gamma,L}(v) = \exp\left(\frac{\pi}{2}S(v,v)\right) \times \sum_{\gamma\in\Gamma/\Gamma\cap L} \alpha(\gamma)\cdot\exp\left(\pi(H-S)(v,\gamma) - \frac{\pi}{2}(H-S)(\gamma,\gamma)\right). \tag{3.3.1}$$

*is well defined and is absolutely convergent (uniformly on compacts in $V$).*

*(ii) $\theta^{\alpha}_{H,\Gamma,L}$ is a nonzero element of the space $T(H,\Gamma,\alpha)$ of canonical theta functions.*

*Proof.* (i) First we have to check that each term of the series (3.3.1) depends only on $\gamma \bmod \Gamma \cap L$. Indeed, if we change $\gamma$ to $\gamma + \gamma_1$ where $\gamma_1 \in \Gamma \cap L$, then this expression gets multiplied by

$$\alpha(\gamma)^{-1}\alpha(\gamma+\gamma_1)\exp\left(-\frac{\pi}{2}(H-S)(\gamma_1,\gamma)\right).$$

But we have

$$\begin{aligned}(H-S)(\gamma_1,\gamma) &= H(\gamma_1,\gamma) - S(\gamma,\gamma_1) = H(\gamma_1,\gamma) - H(\gamma,\gamma_1)\\ &= 2\pi i E(\gamma_1,\gamma).\end{aligned}$$

On the other hand,

$$\alpha(\gamma)^{-1}\alpha(\gamma+\gamma_1) = \alpha(\gamma_1)\exp(\pi i E\gamma_1,\gamma)) = \exp(\pi i E(\gamma_1,\gamma)),$$

so the correctness follows. Convergence of the series (3.3.1) follows from Exercise 2.

(ii) The proof is left to the reader. □

The following proposition shows that the study of theta series can always be reduced to the case of a self-dual lattice.

**Proposition 3.5.** *(i) For every Lagrangian subspace* $L \subset V$ *as above the subgroup* $\Gamma_L = \Gamma + \Gamma^\perp \cap L \subset V$ *is a self-dual lattice.*

*(ii) The map* $\alpha$ *has a unique extension to a map* $\alpha : \Gamma_L \to U(1)$ *satisfying (1.2.2), such that* $\alpha|_{\Gamma^\perp \cap L} \equiv 1$.

*(iii) One has*

$$\theta^\alpha_{H,\Gamma,L} = \theta^\alpha_{H,\Gamma_L,L}.$$

*Proof.* (i) Since $\Gamma \subset \Gamma_L \subset \Gamma^\perp$, $\Gamma_L$ is a lattice. One has

$$(\Gamma + \Gamma^\perp \cap L)^\perp = \Gamma^\perp \cap (\Gamma + L) = \Gamma + \Gamma^\perp \cap L.$$

(ii) For an element $\gamma + l \in \Gamma_L$, where $\gamma \in \Gamma, l \in \Gamma^\perp \cap L$, set $\alpha(\gamma + l) = \alpha(\gamma)$. We leave for the reader to check that this map is well defined and satisfies equation (1.2.2).

(iii) This follows immediately from the definition, since $\Gamma_L/\Gamma_L \cap L = \Gamma/\Gamma \cap L$. □

Note that the 1-dimensional subspace $T(H, \Gamma_L, \alpha) \subset T(H, \Gamma, \alpha)$ coincides with the space of $I$-invariants in $T(H, \Gamma, \alpha)$, where $I = \Gamma^\perp \cap L/\Gamma \cap L$ is a maximal isotropic subgroup in $\Gamma^\perp/\Gamma$, lifted to the finite Heisenberg group $G(E, \Gamma, \alpha)$ trivially. Thus, we obtain the following corollary.

**Corollary 3.6.** *The theta series* $\theta^\alpha_{H,\Gamma,L}$ *is a generator of the* 1*-dimensional subspace* $T(H, \Gamma, \alpha)^I \subset T(H, \Gamma, \alpha)$.

Certain compatibility of the definition of theta series with operators of the Fock representation (see Section 2.4) is described in the following proposition.

**Proposition 3.7.** *(i) For any* $c \in L$ *one has*

$$\theta^{\alpha'}_{H,\Gamma,L} = U_{(1,c)}\theta^\alpha_{H,\Gamma,L},$$

*where* $\alpha'$ *and* $\alpha$ *are related by (3.2.2).*

*(ii) Let* $\Gamma' \subset \Gamma$ *be a sublattice. Then*

$$\theta^\alpha_{H,\Gamma,L} = \sum_{\gamma \in \Gamma/(\Gamma' + \Gamma \cap L)} \alpha(\gamma)^{-1} U_{(1,\gamma)} \theta^\alpha_{H,\Gamma',L} = \sum_{\gamma \in \Gamma/(\Gamma' + \Gamma \cap L)} U_{(\alpha(\gamma),\gamma)} \theta^\alpha_{H,\Gamma',L}.$$

The proof is left to the reader. Note that part (ii) of this proposition is sometimes referred to as the "isogeny theorem."

**Remarks.** 1. When one has an isotropic decomposition $\Gamma = \Gamma_1 \oplus \Gamma_2$ such that $\Gamma \cap L = \Gamma_2$ and $\alpha = \alpha_0(\Gamma_1, \Gamma_2)$, the function $\theta^{\alpha}_{H,\Gamma,L}$ coincides with the function $\vartheta^0$ defined in Chapter 3, Section 2.3 of [18].

2. For given data $(H, \Gamma, \alpha)$, a Lagrangian subspace $L$ compatible with $(\Gamma, \alpha)$ does not always exist. We will discuss the conditions of existence of such $L$ in Section 5.2. We will also show there that for every given $(H, \Gamma)$ as above, there exists $\alpha$ and a Lagrangian subspace $L$ compatible with $(\Gamma, \alpha)$.

### 3.4. Lefschetz Theorem

The standard application of the theory of theta functions is the following theorem of Lefschetz.

**Theorem 3.8.** *Let $L$ be a holomorphic line bundle on the complex torus $T = V/\Gamma$ with $c_1(L) = E$; then for $n \geq 3$ global holomorphic sections of $L^n$ define an embedding of $T$ as a complex submanifold into $\mathbb{P}^N$.*

*Proof.* Let us assume that $n = 3$ (the case $n > 3$ can be proven similarly). The line bundle $L$ has form $L(H, \alpha^{-1})$ for some $\alpha : \Gamma \to U(1)$ satisfying equation (1.2.2), so that the space of global sections of $L$ is identified with the space of canonical theta functions $T(H, \Gamma, \alpha)$ (see Proposition 2.4). The main driving force of the proof is the observation that for every $\theta \in T(H, \Gamma, \alpha)$ and for every $a, b \in V$ the function

$$\theta(v - a)\theta(v - b)\theta(v + a + b)$$

belongs to $T(3H, \Gamma, \alpha^3)$, i.e., defines a section of $L^3$ (this follows easily from Exercise 2 in Chapter 1). This immediately implies that $L^3$ is generated by its global sections, i.e., for every $x \in T$ there is a section of $L^3$ that does not vanish at $x$. Hence, we have the morphism $\Phi : T \to \mathbb{P}(H^0(L^3)^*)$. If $\Phi$ were not injective, we would have a pair of points $v_1, v_2 \in V$ such that $v_2 - v_1 \notin \Gamma$, and a constant $\lambda \in \mathbb{C}^*$ such that for every $f \in T(3H, \Gamma, \alpha^3)$ one has $f(v_2) = \lambda f(v_1)$. To get a contradiction it is enough to consider $f(v) = \theta(v - a)\theta(v - b)\theta(v + a + b)$ for various $\theta \in T(H, \Gamma, \alpha) \setminus 0$. Then the trick is to consider both sides of the identity $f(v_2) = \lambda f(v_1)$ as functions of $a$ and to take the logarithmic derivative of the identity. As a result we get the invariance of the meromorphic differential $d \log \frac{\theta(v_2+v)}{\theta(v_1+v)}$ under translations. It means that $\log \frac{\theta(v_2+v)}{\theta(v_1+v)}$ is a linear function (with constant term), so we get an identity of the form

$$\theta(v+\delta) = A\exp(l(v))\theta(v), \tag{3.4.1}$$

where $\delta = v_2 - v_1$, $A \in \mathbb{C}^*$, $l$ is a $\mathbb{C}$-linear form on $V$. In particular, both parts should have the same quasi-periodicity factors for $\Gamma$. It follows that

$$\exp(\pi H(\delta, \gamma)) = \exp(l(\gamma))$$

for all $\gamma \in \Gamma$. We claim that this can happen only if $\delta \in \Gamma^{\perp}$ and $l(v) = \pi H(v, \delta)$. Indeed, we have

$$\pi H(\delta, \gamma) = l(\gamma) + 2\pi i m(\gamma)$$

for some homomorphism $m : \Gamma \to \mathbb{Z}$. Extending $m$ to an $\mathbb{R}$-linear map $m : V \to \mathbb{R}$ we get

$$\pi H(\delta, v) - l(v) = 2\pi i m(v)$$

for every $v \in V$. It follows that

$$\pi H(v, \delta) - l(v) = 2\pi i E(v, \delta) + 2\pi i m(v).$$

But the LHS is $\mathbb{C}$-linear and the RHS takes values in $i\mathbb{R}$. It follows that both sides are zero which implies our claim. Thus, we can rewrite equation (3.4.1) as follows:

$$\theta(v+\delta) = A' \exp\left(\pi H(v, \delta) + \frac{\pi}{2} H(\delta, \delta)\right) \theta(v). \tag{3.4.2}$$

Set $\Gamma' = \Gamma + \mathbb{Z}\delta$. We have seen that $\Gamma' \subset \Gamma^{\perp}$ and the assumption $\delta \notin \Gamma$ implies that $\Gamma'$ is stricly bigger than $\Gamma$. Now it is easy to derive from equation (3.4.2) and from the condition $\theta \in T(H, \Gamma, \alpha)$ that in fact $\theta$ belongs to the space $T(H, \Gamma', \alpha')$ for some $\alpha'$ extending $\alpha$ (see Exercise 6). However, there are only finitely many $\Gamma'$ between $\Gamma$ and $\Gamma^{\perp}$ and the dimension of $T(H, \Gamma', \alpha')$ is strictly smaller than that of $T(H, \Gamma, \alpha)$ as follows from Proposition 3.1. Hence, for a generic $\theta$ this situation does not occur.

The proof of the fact that $\Phi : T \to \mathbb{P}^N$ induces an embedding of tangent spaces goes along similar lines. Assume that a holomorphic tangent vector $D_{v_0}$ at the point $v_0 \in V/\Gamma$ maps to zero under $\Phi$. Then we have $D_{v_0} f = c \cdot f$ for all $f \in T(3H, \Gamma, \alpha^3)$, where $c \in \mathbb{C}$ is a constant. Let us extend $D_{v_0}$ to a constant holomorphic vector field $D$ on $V$. Considering the above condition for the subset of $f$ of the form $\theta(v-a)\theta(v-b)\theta(v+a+b)$, we deduce that $D(\log \theta)$ is a linear function (with constant term) for every non-zero $\theta \in T(H, \Gamma, \alpha)$. Considering $D$ as a vector in $V$ and integrating the equation

$D(\log\theta)(v) = a(v) + b$ (where $a \in \mathrm{Hom}_{\mathbb{C}}(V, \mathbb{C})$, $b \in \mathbb{C}$) we get that for all $t \in C$ and for all $v$ such that $\theta(v) \neq 0$, one has

$$\theta(v + tD) = \exp(t^2 a(D)/2 + t(a(v) + b))\theta(v).$$

For a fixed $t$ this condition has the same form as (3.4.1), so we can bring it to contradiction using the fact that $tD \notin \Gamma^{\perp}$ for $t$ small enough. □

**Definition.** A positive-definite Hermitian form $H$ on $V$ such that $E = \mathrm{Im}\, H$ takes integer values on $\Gamma$ is called a *polarization* of the complex torus $T = V/\Gamma$. A complex torus admitting a polarization is called a *complex abelian variety*.

**Remarks.** 1. Since by Chow theorem (see [52]; Chapter 1, Section 3) analytic subvarieties of $\mathbb{P}^N$ are algebraic, Lefschetz theorem implies that $\Phi(T)$ is an algebraic subvariety of $\mathbb{P}^N$. Thus, a choice of polarization on a complex abelian variety gives an algebraic structure on it. The GAGA principle (see [124]) implies that every holomorphic bundle on $T$ is algebraic with respect to this structure. Hence, the algebraic structure does not depend on a choice of polarization.

Writing explicitly equations defining the subvariety $\Phi(T) \subset \mathbb{P}^N$ is a beautiful chapter in the theory of theta functions. The main idea is to use the group structure on $T$ (plus the fact that line bundles behave like quadratic forms with respect to addition on $T$) to derive some universal identities between theta functions. We will consider some of these relations in Chapter 12 (see also [64], [97], [98]).

2. It is easy to see that conversely, if a complex torus $T = V/\Gamma$ is projective then there exists a positive definite Hermitian form $H$ on $V$, such that $\mathrm{Im}(H)$ is integer-valued on $\Gamma$. Indeed, an embedding of $T$ into $\mathbb{P}^N$ is always given by global sections of some line bundle $L(H, \alpha)$ on $T$. As we have seen earlier (see remark after Proposition 3.1) if $L(H, \alpha)$ has nonzero global sections then all eigenvalues of $H$ should be nonnegative. Let $V_0 \subset V$ be the kernel of $H$. Then $\Gamma \cap V_0$ is a lattice in $V_0$, so we have the corresponding complex subtorus $T_0 = V_0/\Gamma \cap V_0 \subset T$. The restriction of $L(H, \alpha)$ to $T_0$ has form $L(0, \alpha_0)$ where $\alpha_0 = \alpha|_{\Gamma \cap V_0}$. It is easy to see that if $\alpha_0 \neq 1$ then $L(0, \alpha_0)$ has no global sections, otherwise it is trivial. In both cases global sections do not distinguish points in $T_0$, hence $V_0 = 0$.

3. In the above proof we only used elements in $T(3H, \Gamma, \alpha^3)$ of the form $\theta(v - a)\theta(v - b)\theta(v + a + b)$. It is natural to guess that in fact these elements span the whole space. This is indeed true (see Exercise 4).

Let us consider some examples of complex abelian varieties.

**Examples.** 1. If $T = \mathbb{C}/\Gamma$ is a 1-dimensional complex torus (*complex elliptic curve*) then for every nondegenerate $\mathbb{Z}$-valued symplectic form $E$ on $\Gamma$ the corresponding Hermitian form $H$ on $\mathbb{C}$ is nondegenerate. Changing the sign of $E$ if necessary we can achieve that $H$ is positive. Hence, every complex elliptic curve is projective.

2. Let $T = V/\Gamma$ be a complex torus and $T' = V/\Gamma' \to T$ be a finite unramified covering of $T$ corresponding to a sublattice $\Gamma' \subset \Gamma$ of finite index. Then $T$ is a complex abelian variety if and only if $T'$ is. Indeed, any polarization of $T$ is also a polarization of $T'$. Conversely, if $H$ is a polarization of $T'$ then $nH$ will be a polarization of $T$ for some $n > 0$.

3. If $T = V/\Gamma$ is a complex abelian variety, then a polarization $H$ defines a finite morphism from $T$ to the dual complex torus $T^\vee$. It follows that $T^\vee$ is also an abelian variety, called the *dual abelian variety* to $T$. The algebraic construction of $T^\vee$ from $T$ will be given in Chapter 9.

## Exercises

1. (a) Show that the action of $G(E, \Gamma, \alpha)$ on $T(H, \Gamma, \alpha)$ is unitary with respect to the metric (2.5.2).
   (b) Let $I \subset \Gamma^\perp/\Gamma$ be a Lagrangian subgroup, $\widetilde{I} \subset G(E, \Gamma, \alpha)$ be its lifting to a subgroup in $G(E, \Gamma, \alpha)$. Let $f \in T(H, \Gamma, \alpha)^{\widetilde{I}}$ be a generator of length 1. Show that the functions $(U_{(1,c)}f)$ form an orthonormal basis in $T(H, \Gamma, \alpha)$, where $c \in \Gamma^\perp$ runs through the complete system of representatives modulo $I$.
2. (a) Show that the quadratic form $\mathrm{Re}(H - S)$ descends to $V/L$.
   (b) Show that for $l \in L$ one has $(H - S)(il, il) = 2H(l, l)$. Hence, the form $\mathrm{Re}(H - S)$ on $V/L$ is positive-definite.
3. Show that $\theta^{\alpha}_{H,\Gamma,L}(-v) = \theta^{\alpha^{-1}}_{H,\Gamma,L}(v)$.
4. (a) Prove that the space $T(2H, \Gamma, \alpha_1\alpha_2)$ is spanned by the functions of the form $f(v) = \theta_1(v - a)\theta_2(v + a)$ where $a \in V$, $\theta_1 \in T(H, \Gamma, \alpha_1)$, $\theta_2 \in T(H, \Gamma, \alpha_2)$. [*Hint*: It suffices to prove that the subspace spanned by these functions is invariant under the action of the Heisenberg group $G(2H, \Gamma, \alpha_1\alpha_2)$.]
   (b) Generalize the proof of part (a) to show that for every $n > 1$ the space $T(nH, \Gamma, \alpha_1 \cdots \alpha_n)$ is spanned by the functions of the form $f(v) = \theta_1(v - a_1) \cdots \theta_n(v - a_n)$, where $a_i \in V$, $\sum_i a_i = 0$, $\theta_i \in T(H, \Gamma, \alpha_i)$.
5. The classical (1-dimensional) theta series (with zero characteristics) is given by

$$\theta(z, \tau) = \sum_{n\in\mathbb{Z}} \exp(\pi i n^2 \tau + 2\pi i n z),$$

where $z, \tau \in \mathbb{C}$, $\mathrm{Im}(\tau) > 0$. Let us consider the lattice $\Gamma = \Gamma_\tau = \mathbb{Z} + \mathbb{Z}\tau \subset \mathbb{C}$. Define the Hermitian form $H = H_\tau$ on $\mathbb{C}$ by $H(z_1, z_2) = [z_1\overline{z_2}]/[\mathrm{Im}(\tau)]$. The corresponding symplectic form $E = \mathrm{Im}\, H$ takes integer values on $\Gamma$ since $E(\tau, 1) = 1$. Let us also define the map $\alpha_0 : \Gamma \to \{\pm 1\}$ by $\alpha_0(m + n\tau) = (-1)^{mn}$. As a Lagrangian subspace in $\mathbb{C}$ we take $\mathbb{R} \subset \mathbb{C}$.

(a) Show that $\theta^{\alpha_0}_{H_\tau, \Gamma_\tau, \mathbb{R}}(z) = \exp(\frac{\pi z^2}{2\,\mathrm{Im}\,\tau}) \cdot \theta(z, \tau)$.

(b) Show that the line bundle $L = L(H, \alpha_0)$ on $\mathbb{C}/\Gamma$ has degree one.

(c) Prove that for fixed $\tau$ the function $\theta(z, \tau)$ has simple zeroes at $z \in \frac{\tau+1}{2} + \mathbb{Z} + \mathbb{Z}\tau$ and no other zeros.

6. Let $\Gamma \subset V$ be a lattice. Assume that for some nonzero holomorphic function $f$ on $V$ one has

$$f(x + \gamma) = c(\gamma) \cdot \exp\left(\pi H(x, \gamma) + \frac{\pi}{2} H(\gamma, \gamma)\right) f(x)$$

for all $x \in V$, $\gamma \in \Gamma$. Show that $c(\gamma_1 + \gamma_2) = c(\gamma_1)c(\gamma_2) \exp(\pi i E(\gamma_1, \gamma_2))$.

7. Let us fix $\tau$ in the upper half-plane and consider the following function of complex variables $z_1, z_2$:

$$F(z_1, z_2) = \sum_{(\alpha(z_1)+m)(\alpha(z_2)+n)>0} \mathrm{sign}(\alpha(z_1) + m) \exp(2\pi i[\tau mn + nz_1 + mz_2]),$$

where $\alpha(z) = \mathrm{Im}(z)/\,\mathrm{Im}(\tau)$. It is a holomorphic function of $z_1$ and $z_2$ on the open set $\mathrm{Im}\, z_i \notin \mathbb{Z}(\mathrm{Im}\,\tau)$, $i = 1, 2$.

(a) Show that $F$ extends to a meromorphic function on $\mathbb{C}^2$ with simple poles at the lattice points $z_1 \in \Gamma_\tau$ or $z_2 \in \Gamma_\tau$ where $\Gamma_\tau = \mathbb{Z} + \mathbb{Z}\tau$.

(b) Show that $F$ satisfies the following identities:

$$F(z_2, z_1) = F(z_1, z_2),$$
$$F(z_1 + m + n\tau, z_2) = \exp(-2\pi i n z_2) F(z_1, z_2),$$
$$F(z_1, z_2 + m + n\tau) = \exp(-2\pi i n z_1) F(z_1, z_2).$$

(c) Using the quasi-periodicity properties of $F$ prove the following identity discovered by Kronecker:

$$F(z_1, z_2) = \frac{\theta'((\tau + 1)/2)}{2\pi i} \cdot \frac{\theta(z_1 + z_2 - (\tau + 1)/2)}{\theta(z_1 - (\tau + 1)/2)\theta(z_2 - (\tau + 1)/2)},$$

where $\theta(z) = \theta(z, \tau)$ is the classical theta series defined in Exercise 5.

## Appendix A. Theta Series and Weierstrass Sigma Function

In this appendix we will present the relation between theta series and Weierstrass sigma and zeta functions.

Let $\Gamma \subset \mathbb{C}$ be a lattice. Weierstrass sigma function is defined as the following infinite product.

$$\sigma(z) = \sigma(z, \Gamma) = z \prod_{\gamma \in \Gamma \setminus \{0\}} \left(1 - \frac{z}{\gamma}\right) \exp\left(\frac{z}{\gamma} + \frac{z^2}{2\gamma^2}\right).$$

In fact, this product converges absolutely and uniformly on compacts in $\mathbb{C}$, so $\sigma(z)$ is a holomorphic function on $\mathbb{C}$. Its logarithmic derivative is the Weierstrass zeta function

$$\zeta(z) = \zeta(z, \Gamma) = \frac{\sigma'(z)}{\sigma(z)} = \frac{1}{z} + \sum_{\gamma \in \Gamma \setminus \{0\}} \left(\frac{1}{z - \gamma} + \frac{1}{\gamma} + \frac{z}{\gamma^2}\right).$$

The derivative of $\zeta$ is equal to $-\wp(z)$, where

$$\wp(z) = \frac{1}{z^2} + \sum_{\gamma \in \Gamma \setminus \{0\}} \left(\frac{1}{(z - \gamma)^2} - \frac{1}{\gamma^2}\right)$$

is the Weierstrass $\wp$-function. Since $\wp(z)$ is invariant under translations by $\Gamma$, for every $\gamma \in \Gamma$ there is a constant $\eta_\gamma \in \mathbb{C}$ such that

$$\zeta(z + \gamma) = \zeta(z) + \eta_\gamma.$$

Now let $\Gamma = \mathbb{Z} + \mathbb{Z}\tau$, where $\tau$ is in the upper half-plane, $\sigma(z, \tau)$ (resp., $\zeta(z, \tau)$) be the sigma (resp., zeta) function associated with this lattice. The function $\gamma \mapsto \eta_\gamma$ on $\Gamma$ is additive, so it is determined by the numbers $\eta_1$ and $\eta_\tau$ (so-called *quasi-periods*). Furthermore, these numbers satisfy the following *Legendre relation*:

$$\eta_1 \tau - \eta_\tau = 2\pi i.$$

This can be proven by comparing the integral of $\zeta(z)dz$ along the boundary of the parallelogram formed by 1 and $\tau$ (slightly shifted) with the residue of this 1-form at the unique pole inside.

The following theorem gives a relation between $\sigma(z, \tau)$ and the theta series on elliptic curve $\mathbb{C}/\Gamma$ given by

$$\theta_{11}(z, \tau) = \sum_{n \in \mathbb{Z}} \exp\left[\pi i \left(n + \frac{1}{2}\right)^2 \tau + 2\pi i \left(n + \frac{1}{2}\right)\left(z + \frac{1}{2}\right)\right].$$

Note that $\theta_{11}(z, \tau) = \exp(\frac{\pi i \tau}{4} + \pi i(z + \frac{1}{2}))\theta(z + \frac{\tau + 1}{2}, \tau)$, where $\theta(z, \tau)$ is the theta series considered in Exercise 5.

**Theorem 3.9.** *One has*

$$\theta_{11}(z,\tau) = -2\pi\eta(\tau)^3 \exp\left(-\frac{\eta_1 z^2}{2}\right)\sigma(z,\tau), \tag{3.4.3}$$

*where $\eta(\tau)$ is the Dedekind $\eta$-function:*

$$\eta(\tau) = \exp\left(\frac{\pi i \tau}{12}\right)\prod_{n=1}^{\infty}(1-q^n),$$

*where $q = \exp(2\pi i \tau)$.*

*Proof.* First we claim that $\sigma(z)$ satisfies the following quasi-periodicity with respect to the lattice $\mathbb{Z}+\mathbb{Z}\tau$:

$$\sigma(z+\gamma) = \epsilon(\gamma)\exp\left(\eta(\gamma)\left(z+\frac{\gamma}{2}\right)\right)\sigma(z),$$

where $\gamma \in \Gamma$, $\epsilon : \Gamma \to \{\pm 1\}$ is the homomorphism defined by $\epsilon(1)=\epsilon(\tau)= -1$. Indeed, it suffices to check this for $\gamma=1$ and $\gamma=\tau$. Considering logarithmic derivatives we immediately see that

$$\sigma(z+\gamma) = c(\gamma)\exp(\eta(\gamma)z)\sigma(z)$$

for some $c(\gamma) \in \mathbb{C}^*$. Substituting $z = -\gamma/2$ (where $\gamma = 1$ or $\gamma = \tau$) and using the fact that $\sigma$ is odd, we obtain that $c(\gamma) = -\exp(\eta(\gamma)\gamma/2)$. Using the Legendre relation we derive that the right-hand side of (3.4.3) satisfies the same quasi-periodicity in $z$ as $\theta_{11}(z)$:

$$\begin{aligned}\theta_{11}(z+1) &= -\theta_{11}(z),\\ \theta_{11}(z+\tau) &= -\exp(-\pi i\tau - 2\pi i z)\theta_{11}(z).\end{aligned}$$

Hence, we obtain that (3.4.3) holds up to a nonzero constant (depending on $\tau$). To evaluate this constant we will use a trick due to Kronecker. Namely, we claim that the following identity holds.

$$\theta_{11}\left(\frac{\tau}{2}+\frac{1}{4},\tau\right) = \exp\left(\frac{3\pi i}{4}(\tau+1)\right)\theta_{11}(2\tau, 4\tau).$$

Indeed, splitting the series for $\theta_{11}(\frac{\tau}{2}+\frac{1}{4},\tau)$ into two series, the sum over even $n$ and the sum over odd $n$, we get

$$\theta_{11}\left(\frac{\tau}{2}+\frac{1}{4},\tau\right)=\sum_n(-1)^n\exp\left[\pi i\left(4n^2+4n+\frac{3}{4}\right)\tau+\frac{3\pi i}{4}\right]$$
$$+\sum_n\exp\left[\pi i(2n+1)^2\tau+2\pi i\left(n+\frac{1}{2}\right)\left(2\tau+\frac{1}{2}\right)\right.$$
$$\left.+\frac{3\pi i}{4}(\tau+1)\right].$$

It remains to note that the first sum is zero (as seen by substituting $n\mapsto -n-1$).

Now we want to show that similar identity holds for the right-hand side of (3.4.3). To this end we will use the following formula:

$$\sigma(z,\tau)=\frac{1}{2\pi i}\exp\left(\frac{\eta_1 z^2}{2}\right)\left(u^{\frac{1}{2}}-u^{-\frac{1}{2}}\right)\prod_{n=1}^{\infty}\frac{(1-q^n u)(1-q^n u^{-1})}{(1-q^n)^2},\quad(3.4.4)$$

where in the right-hand side we use multiplicative variables $q=\exp(2\pi i\tau)$, $u=\exp(2\pi i z)$ (and where $u^{\frac{1}{2}}=\exp(\pi i z)$). This identity in turn is proven as follows. It is easy to see that ratio of the left-hand and right-hand sides is periodic in $z$ with respect to $\Gamma$. On the other hand, both sides have zeros of first order at all points of $\Gamma$ and nowhere else. This implies that the equality holds up to a constant. It remains to compare derivatives in $z$ at $z=0$.

Let us denote by $g(z,\tau)$ the right-hand side of (3.4.3). Using (3.4.4) one can easily show that

$$g\left(\frac{\tau}{2}+\frac{1}{4},\tau\right)=\exp\left(\frac{3\pi i}{4}(\tau+1)\right)g(2\tau,4\tau).$$

It follows that the ratio of the left- and right-hand sides of (3.4.3), which we know to be a function of $\tau$, is invariant with respect to the substitution $\tau\mapsto 4\tau$ (or $q\mapsto q^4$). On the other hand, as follows from (3.4.4), this ratio is represented by a converging power series in $q$ with initial term 1. Hence, it is equal to 1. □

The beautiful formula of the following corollary is due to Jacobi.

**Corollary 3.10.** *One has*

$$-2\pi\eta(\tau)^3=\theta'_{11}(0,\tau)$$

*where $\theta'_{11}$ denotes the derivative with respect to $z$. Equivalently,*

$$\prod_{n=1}^{\infty}(1-q^n)^3=\sum_{n\in\mathbb{Z}}(-1)^n nq^{\frac{n^2+n}{2}}=\sum_{n\geq 0}(-1)^n(2n+1)q^{\frac{n^2+n}{2}}.$$

# 4

# Representations of Heisenberg Groups II: Intertwining Operators

In this chapter we construct and study canonical intertwining operators between standard models for the irreducible representation of a Heisenberg group $H$ considered in Chapter 2. This theory will play a crucial role in the derivation of the functional equation for theta functions in Chapter 5.

For a pair of compatible Lagrangian subgroups $L_1$, $L_2$ in $H$ (the definition of compatibility will be given in Section 4.2) we construct a canonical intertwining operator $R(L_1, L_2)$ between the corresponding induced representations of $H$. In the particular case when the intersection of $L_1$ and $L_2$ is trivial, this operator is the usual Fourier transform. In general, it is given by some partial Fourier transform. For a triple of pairwise compatible Lagrangian subgroups $L_1, L_2, L_3$ the operator $R(L_3, L_1) \circ R(L_2, L_3) \circ R(L_1, L_2)$ is a scalar multiple of the identity. We compute the corresponding constant $c(L_1, L_2, L_3)$ in two cases: when $(L_1 \cap L_3) + (L_2 \cap L_3)$ has finite index in $(L_1 + L_2) \cap L_3$ and when $L_i$'s are Lagrangian subspaces of a symplectic vector space. In both cases this constant can be expressed in terms of some quadratic function $q$ on a locally compact abelian group $A(L_1, L_2, L_3)$. In the former case the group $A(L_1, L_2, L_3)$ is finite and $c(L_1, L_2, L_3)$ is equal to the Gauss sum associated with $q$. In the latter case $A(L_1, L_2, L_3)$ is a vector space and $c(L_1, L_2, L_3) = \exp(-\frac{\pi i m}{4})$, where $m$ is the Maslov index of the triple $(L_1, L_2, L_3)$, which is equal to the signature of the quadratic form $-q$. Gauss sums associated with quadratic forms on finite abelian groups will appear in the functional equation for theta functions. In this chapter we show that they are always given by 8th roots of unity. On the other hand, the cocycle equation for constants $c(L_1, L_2, L_3)$ (which follows from their definition) can be used to evaluate some Gauss sums explicitly. In the main text we give an example of such computation, while in Appendix B we derive a more general formula for Gauss sums due to Van der Blij and Turaev.

## 4.1. Fourier Transform

Let $K$ be a locally compact abelian group, $\widehat{K}$ be the Pointriagin dual group. As was already mentioned earlier, the reader will not lose much by assuming that $K$ is either a real vector space or a finite abelian group. For every Haar measure $\mu$ on $K$ the Fourier transform is the operator $S_\mu : L^2(K) \to L^2(\widehat{K})$ given by

$$S_\mu(\phi)(\hat{k}) = \int_{k\in K} \hat{k}(k)\phi(k)\mu.$$

In this situation there is a unique *dual measure* $\hat{\mu}$ on $\widehat{K}$ such that $S_\mu$ is unitary with respect to the Hermitian metrics defined using $\mu$ and $\hat{\mu}$. Moreover, it is known that in this case the inverse Fourier transform $S_\mu^{-1}$ coincides with the operator

$$\overline{S}_{\hat{\mu}}(\phi')(k) = \int_{\hat{k}\in\widehat{K}} \hat{k}(k)^{-1}\phi(k)\hat{\mu}.$$

One can rewrite all this without making a choice of measure as follows. Let us denote by $\mathrm{meas}(K)$ the $\mathbb{R}_{>0}$-torsor of Haar measures on $K$ (where $\mathbb{R}_{>0}$ is the multiplicative group of positive real numbers). Let $\mathrm{meas}(K)^{\frac{1}{2}}$ be the $\mathbb{R}_{>0}$-torsor, obtained from $\mathrm{meas}(K)$ by taking the push-out with respect to the homomorphism $\mathbb{R}_{>0} \to \mathbb{R}_{>0} : x \mapsto x^{\frac{1}{2}}$. One can think about elements of $\mathrm{meas}(K)^{\frac{1}{2}}$ as formal square roots of Haar measures. For every real vector space $W$ and an $\mathbb{R}_{>0}$-torsor $T$ we denote by $WT$ the tensor product of $W$ with the 1-dimensional space corresponding to $T$. Now we claim that there is a canonical operator

$$S : L^2(K)\,\mathrm{meas}(K)^{\frac{1}{2}} \to L^2(\widehat{K})\,\mathrm{meas}(\widehat{K})^{\frac{1}{2}}. \tag{4.1.1}$$

Indeed, since the construction of $S_\mu$ is linear in $\mu$, it gives an operator

$$L^2(K)\,\mathrm{meas}(K) \to L^2(\widehat{K}).$$

On the other hand, the passage to the dual measure $\mu \mapsto \hat{\mu}$ gives an isomorphism of $\mathbb{R}_{>0}$-torsors $\mathrm{meas}(K) \simeq \mathrm{meas}(\widehat{K})^{-1}$. Therefore, twisting the source and the target by $\mathrm{meas}(K)^{-\frac{1}{2}} \simeq \mathrm{meas}(\widehat{K})^{\frac{1}{2}}$ we get (4.1.1). By the construction, the space $L^2(K)\,\mathrm{meas}(K)^{\frac{1}{2}}$ (resp., similar space for $\widehat{K}$) has a natural Hermitian metric, and the operator $S$ is unitary.

In the case when $K$ is compact, $\widehat{K}$ is discrete, the natural Haar measure on $K$ such that the total volume of $K$ is 1, is dual to the natural measure on $\widehat{K}$ (so that every point has weight 1).

In the case when $K$ is finite, we can trivialize $\mathrm{meas}(K)$ and $\mathrm{meas}(\widehat{K})$ by choosing the measures such that every point has weight 1. Then the Fourier transform corresponds to the operator

$$S\phi(\hat{k}) = |K|^{-\frac{1}{2}} \cdot \sum_{k \in K} \hat{k}(k)\phi(k).$$

## 4.2. Construction of Intertwining Operators

Let $H$ be a Heisenberg group, $K = H/U(1)$ be the corresponding locally compact abelian group. Let $L \subset K$ be a Lagrangian subgroup. As before, we always assume that a lifting homomorphism $\sigma : L \to H$ is chosen, so $L$ can be also considered as a subgroup in $H$. Recall that we have the corresponding irreducible representation of $H$ on the space $\mathcal{F}(L)$ which is a completion of the space of functions $\phi : H \to \mathbb{C}$ such that $\phi(\lambda h) = \lambda\phi(h)$ for $\lambda \in U(1)$ and $\phi(\sigma(l)h) = \phi(h)$ for $l \in L$. In order to construct canonical intertwining operators, we will change the space $\mathcal{F}(L)$ slightly. Namely, we set

$$\mathcal{F}'(L) = \mathcal{F}(L)\,\mathrm{meas}(K/L)^{\frac{1}{2}}$$

(morally, one should think of elements of $\mathcal{F}'(L)$ as half-measures on $K/L$). Note that the space $\mathcal{F}'(L)$ has a canonical Hermitian metric. Namely, for $\phi_1, \phi_2 \in \mathcal{F}'(L)$ the product $\phi_1\overline{\phi_2}$ descends to a measure on $K/L$, which then can be integrated.

**Definition.** Lagrangian subgroups $L_1, L_2 \subset K$ equipped with liftings to $H$, are called *compatible* if $L_1 + L_2$ is a closed subgroup and their lifting homomorphisms to $H$ agree on $L_1 \cap L_2$.

Note that for compatible subgroups one has $L_1 + L_2 = (L_1 \cap L_2)^{\perp}$.

Now for a pair of compatible Lagrangian subgroups $L_1, L_2$ we are going to define a natural unitary intertwining operator between representations of $H$

$$R(L_1, L_2) : \mathcal{F}'(L_1) \to \mathcal{F}'(L_2).$$

Let us first fix a measure $\mu$ on $L_2/L_1 \cap L_2$. Note that for $\phi \in \mathcal{F}(L_1)$ the function $|\phi|$ on $H$ descends to a function on $H/U(1)L_1 = K/L_1$. Assume that $|\phi|$, considered as a function on $K/L_1$, has compact support. Then we can define

$$R(L_1, L_2)_\mu \phi(h) = \int_{l_2 \in L_2/L_1 \cap L_2} \phi(\sigma(l_2)h)\mu.$$

As in the case of the Fourier transform, one can check that the $L^2$-norm of $R(L_1, L_2)_\mu \phi$ is equal (up to a constant factor) to the $L^2$-norm of $\phi$. Hence, this map extends to an intertwining operator

$$R(L_1, L_2)_\mu : \mathcal{F}(L_1) \to \mathcal{F}(L_2).$$

Since the map $\mu \mapsto R(L_1, L_2)_\mu$ is compatible with rescaling by $\mathbb{R}_{>0}$, we obtain a natural operator

$$\mathcal{F}(L_1)\,\mathrm{meas}(L_2/L_1 \cap L_2) \to \mathcal{F}(L_2).$$

This is essentially $R(L_1, L_2)$. To rewrite it as an operator from $\mathcal{F}'(L_1)$ to $\mathcal{F}'(L_2)$ one has to use the following result.

**Proposition 4.1.** *There is a canonical isomorphism of* $\mathbb{R}_{>0}$*-torsors*

$$\mathrm{meas}(L_2/L_1 \cap L_2) \simeq \mathrm{meas}(K/L_1)^{\frac{1}{2}}\,\mathrm{meas}(K/L_2)^{-\frac{1}{2}}.$$

*Proof.* Since the group $(L_1 + L_2)/(L_1 \cap L_2)$ is self-dual, we have a trivialization of $\mathrm{meas}(L_1 + L_2/L_1 \cap L_2)$ corresponding to the unique self-dual measure. Now from the exact sequence

$$0 \to L_1/L_1 \cap L_2 \to (L_1 + L_2)/L_1 \cap L_2 \to L_2/L_1 \cap L_2 \to 0$$

we get a trivialization of the $\mathbb{R}_{>0}$-torsor

$$\mathrm{meas}(L_1/L_1 \cap L_2)\,\mathrm{meas}(L_2/L_1 \cap L_2) \simeq \mathrm{meas}(L_1)\,\mathrm{meas}(L_2)\,\mathrm{meas}(L_1 \cap L_2)^{-2}.$$

In other words, we have an isomorphism

$$\mathrm{meas}(L_1 \cap L_2) \simeq \mathrm{meas}(L_1)^{\frac{1}{2}}\,\mathrm{meas}(L_2)^{\frac{1}{2}}.$$

Hence,

$$\begin{aligned}\mathrm{meas}(L_2/L_1 \cap L_2) &\simeq \mathrm{meas}(L_2)\,\mathrm{meas}(L_1 \cap L_2)^{-1} \simeq \mathrm{meas}(L_2)^{\frac{1}{2}}\,\mathrm{meas}(L_1)^{-\frac{1}{2}} \\ &\simeq \mathrm{meas}(K/L_1)^{\frac{1}{2}}\,\mathrm{meas}(K/L_2)^{-\frac{1}{2}}\end{aligned}$$

as required. □

In the case when $L_1 \cap L_2 = 0$, the groups $K/L_1$ and $K/L_2$ are naturally dual to each other and the operator $R(L_1, L_2)$ coincides with the Fourier transform (Exercise 1).

In general, we can consider the spaces $K/L_1$ and $K/L_2$ as fibrations over $K/(L_1 + L_2)$ with dual fibers $(L_1 + L_2)/L_1$ and $(L_1 + L_2)/L_2$ and the operator $R(L_1, L_2)$ is essentially the relative Fourier transform in the fibers. In particular, from the inversion formula for the Fourier transform we get that $R(L_2, L_1) = R(L_1, L_2)^{-1}$.

**Definition.** (i) A triple $(L_1, L_2, L_3)$ of Lagrangian subgroups in $K$ is called *admissible* if $L_i$'s are pairwise compatible.
(ii) For an admissible triple $(L_1, L_2, L_3)$ the constant $c(L_1, L_2, L_3) \in U(1)$ is defined by the equation

$$R(L_3, L_1) \circ R(L_2, L_3) \circ R(L_1, L_2) = c(L_1, L_2, L_3)\,\mathrm{id}_{\mathcal{F}(L_1)}, \quad (4.2.1)$$

From the definition we immediately derive the following properties:

$$c(L_1, L_2, L_3) = c(L_2, L_3, L_1) = c(L_3, L_2, L_1)^{-1}, \quad (4.2.2)$$

$$c(L_1, L_2, L_4)c(L_2, L_3, L_4) = c(L_1, L_2, L_3)c(L_1, L_3, L_4). \quad (4.2.3)$$

Below we are going to relate the constant $c(L_1, L_2, L_3)$ to certain quadratic function associated with $(L_1, L_2, L_3)$.

## 4.3. Quadratic Function Associated with an Admissible Triple of Lagrangian Subgroups

Let $A$ be a locally compact abelian group.

**Definition.** A continuous function $q : A \to U(1)$ is called *quadratic* if the function

$$\langle a, a' \rangle = q(a + a')q(a)^{-1}q(a')^{-1} \quad (4.3.1)$$

from $A \times A$ to $U(1)$ is a bihomomorphism. We say that $q$ is *nondegenerate* if (4.3.1) induces an isomorphism $A \simeq \widehat{A}$. We call $q$ a (nondegenerate) *quadratic form* if $q$ is a (nondegenerate) quadratic function on $A$ such that $q(-a) = q(a)$ for any $a \in A$.

Now let $L_1, L_2, L_3$ be an admissible triple of Lagrangian subgroups in $K$. Consider the following complex of locally compact abelian groups:

$$0 \to L_1 \cap L_2 \cap L_3 \xrightarrow{d_1} L_1 \cap L_2 \oplus L_2 \cap L_3 \oplus L_3 \cap L_1 \xrightarrow{d_2} L_1 \oplus L_2 \oplus L_3 \xrightarrow{d_3}$$
$$\to L_1 + L_2 + L_3 \to 0, \quad (4.3.2)$$

where $d_1(x) = (x, x, x)$, $d_2(x_{12}, x_{23}, x_{31}) = (x_{12} - x_{31}, x_{23} - x_{12}, x_{31} - x_{23})$, $d_3(x_1, x_2, x_3) = x_1 + x_2 + x_3$.

**Proposition 4.2.** *The complex (4.3.2) has only one potentially nontrivial cohomology group* $A(L_1, L_2, L_3) = \ker(d_3)/\operatorname{im}(d_2)$. *One has*

$$A(L_1, L_2, L_3) \simeq (L_1 + L_2) \cap L_3/(L_1 \cap L_3 + L_2 \cap L_3).$$

The proof is left to the reader.

Let us define a function $\widetilde{q} : \ker(d_3) \to U(1)$ by setting

$$\widetilde{q}(x_1, x_2, x_3) = \sigma(x_1)\sigma(x_2)\sigma(x_3),$$

where $\sigma : L_i \to H, i = 1, 2, 3$, are lifting homomorphisms. Since $x_1 + x_2 + x_3 = 0$, the expression in the RHS belongs to $U(1) \subset H$.

**Theorem 4.3.** *The function $\widetilde{q}$ is quadratic. It descends to a nondegenerate quadratic function $q = q_{L_1,L_2,L_3} : A(L_1, L_2, L_3) \to U(1)$.*

*Proof.* We denote by $s(\cdot, \cdot)$ the function (4.3.1) on $\ker(d_3) \times \ker(d_3)$ associated with $\widetilde{q}$. Let us compute $s(\cdot, \cdot)$. For arbitrary $(x_1, x_2, x_3), (x_1', x_2', x_3') \in \ker(d_3)$ we have

$$\begin{aligned}&\sigma(x_1 + x_1')\sigma(x_2 + x_2')\sigma(x_3 + x_3')\\ &\quad = [\sigma(x_1'), \sigma(x_2)] \cdot [\sigma(x_1')\sigma(x_2'), \sigma(x_3)] \cdot \sigma(x_1)\sigma(x_2)\sigma(x_3)\sigma(x_1')\sigma(x_2')\sigma(x_3').\end{aligned}$$

Since $\sigma(x_1')\sigma(x_2')$ differs from $\sigma(x_3')^{-1}$ by a central element, it commutes with $\sigma(x_3)$, hence we have

$$s((x_1, x_2, x_3), (x_1', x_2', x_3')) = e(x_1', x_2).$$

Since this is a bihomomorphism, $\widetilde{q}$ is indeed a quadratic function. Using this formula one can also easily check (exercise!) that the kernel of $s$ coincides with the subgroup $\operatorname{im}(d_2) \subset \ker(d_3)$. Furthermore, it is clear that $\widetilde{q}|_{\operatorname{im}(d_2)} \equiv 1$, which finishes the proof. □

In the following proposition we list some simple properties of this construction. Let us say that a Heisenberg group $H$ is *symmetric* if it is equipped with an involution $\tau : H \to H$ such that $\tau|_{U(1)} = \mathrm{id}$ and the induced involution of $K = H/U(1)$ is $k \mapsto -k$. For example, if $H$ is the Heisenberg group $\mathcal{H}(V)$ associated with a symplectic vector space, then there is a natural involution $\tau(\lambda, v) = (\lambda, -v)$. Now for a Lagrangian subgroup $L \subset K$, a lifting homomorphism $\sigma : L \to K$ is called *symmetric* if $\sigma(-l) = \tau(\sigma(l))$.

**Proposition 4.4.** *(i) One has $q_{L_1,L_2,L_3} = q_{L_2,L_3,L_1} = [-1]^* q^{-1}_{L_3,L_2,L_1}$;*

*(ii) if $L_3 = L_1 \cap L_3 + L_2 \cap L_3$ then $A(L_1, L_2, L_3) = 0$;*

*(iii) one has a natural isomorphism of groups compatible with quadratic functions on them*

$$(A(L_1, L_2, L_3), q_{L_1,L_2,L_3}) \simeq (A(\overline{L}_1, \overline{L}_2, \overline{L}_3), q_{\overline{L}_1,\overline{L}_2,\overline{L}_3}),$$

*where* $\overline{L}_1 = L_1/L_1 \cap L_2$, $\overline{L}_2 = L_2/L_1 \cap L_2$, $\overline{L}_3 = L_3 \cap (L_1 + L_2)/L_1 \cap L_2 \cap L_3$ *are the induced Lagrangian subgroups for the reduced Heisenberg group* $\overline{H} = N_H(L_1 \cap L_2)/L_1 \cap L_2$;

*(iv) if the lifting homomorphisms* $\sigma : L_i \to H$, $i = 1, 2, 3$, *are symmetric, then* $q_{L_1,L_2,L_3}$ *is a quadratic form.*

*Proof.* (i) The first equality follows from the fact that for $x_1 + x_2 + x_3 = 0$ one has $[\sigma(x_1), \sigma(x_2)\sigma(x_3)] = 1$. On the other hand,

$$\sigma(x_1)\sigma(x_2)\sigma(x_3) = (\sigma(-x_3)\sigma(-x_2)\sigma(-x_1))^{-1},$$

hence $q_{L_1,L_2,L_3} = [-1]^* q_{L_3,L_2,L_1}$.

The proofs of (ii), (iii) and (iv) are straightforward. □

In the following theorem we show that in the case when $A(L_1, L_2, L_3)$ is finite, the constant $c(L_1, L_2, L_3) \in U(1)$ is completely determined by the quadratic function $q_{L_1,L_2,L_3}$.

**Theorem 4.5.** *Let* $L_1, L_2, L_3$ *be an admissible triple of Lagrangian subgroups in* $K$. *Assume that the group* $A = A(L_1, L_2, L_3)$ *is finite. Then*

$$c(L_1, L_2, L_3) = |A|^{-\frac{1}{2}} \cdot \sum_{a \in A} q_{L_1,L_2,L_3}(a).$$

Let us first look at the RHS of this formula. With every non-degenerate quadratic function $q$ on a finite abelian group $A$ we associate the *Gauss sum*

$$\gamma(q) = |A|^{-\frac{1}{2}} \sum_{a \in A} q(a).$$

Here are some simple properties of $\gamma(q)$.

**Proposition 4.6.** *(i) Let* $A = A_1 \oplus A_2$, $q = q_1 \oplus q_2$, *where* $q_i$ *is a non-degenerate quadratic function on* $A_i$ $(i = 1, 2)$. *By the definition, this means that* $q(a_1, a_2) = q_1(a_1)q_2(a_2)$. *Then*

$$\gamma(q_1 \oplus q_2) = \gamma(q_1) \cdot \gamma(q_2).$$

*(ii) Assume that there is a subgroup* $I \subset A$ *such that* $q|_I = 1$ *and* $|I| = |A|^{\frac{1}{2}}$. *Then* $\gamma(q) = 1$.

*(iii) One has* $\gamma(q^{-1}) = \gamma(q)^{-1}$.

*(iv) One has* $|\gamma(q)| = 1$.

*Proof.* The first assertion is clear. To prove (ii) we observe that the pairing $\langle\ ,\ \rangle$ induces the surjective homomorphism $A/I \to \widehat{I}$. Since the orders of both groups are equal to $|A|^{\frac{1}{2}}$, this map is an isomorphism. Now for every $a \in A$ we have

$$\sum_{i\in I} q(a+i) = \sum_{i\in I} \langle a, i\rangle q(a),$$

which is zero for $a \notin I$. Hence,

$$\gamma(q) = |I|^{-1} \cdot \sum_{i\in I} 1 = 1.$$

To prove (iii) let us consider the form $q \oplus q^{-1}$ on $A \oplus A$. Applying (ii) to the diagonal subgroup $\{(a, a), a \in A\}$ in $A \oplus A$ and using (i) we derive that

$$\gamma(q) \cdot \gamma(q^{-1}) = \gamma(q \oplus q^{-1}) = 1.$$

Finally, we have

$$\gamma(q)\overline{\gamma(q)} = \gamma(q \oplus q^{-1}) = 1,$$

which gives (iv). □

*Proof of Theorem 4.5* . We divide the proof into two steps. First, we will prove the assertion in the case when one of Lagrangian subgroups is contained in the sum of two others. Then we will reduce the general case to this one.

Assume first, that one of $L_i$'s is contained in the sum of two others. Since both constants $c(L_1, L_2, L_3)$ and $\gamma(q_{L_1,L_2,L_3})$ behave in the same way under permutation of our triple (see (4.2.2) and Proposition 4.4) we can assume that $L_2 \subset L_1 + L_3$. We will use the presentation $A = A(L_1, L_2, L_3) = L_2/(L_1 \cap L_2 + L_2 \cap L_3)$, so that $q(x_2) = q_{L_1,L_2,L_3}(x_2) = \sigma(x_1)^{-1}\sigma(x_2)\sigma(x_3)^{-1}$ for $x_2 \in L_2$, where $x_1 \in L_1$ and $x_3 \in L_3$ are chosen in such a way that $x_2 = x_1 + x_3$. Note that since both numbers $c(L_1, L_2, L_3)$ and $\gamma(q_{L_1,L_2,L_3})$ have absolute value 1, it suffices to prove that they differ by a positive constant, so we can be imprecise with our choices of Haar measures when integrating. Now for every $\phi \in \mathcal{F}(L_1)$ and $h \in H$, the expression $R(L_2, L_3)R(L_1, L_2)\phi(h)$ is equal (up to a positive constant) to

$$\int_{x_3\in L_3/L_2\cap L_3} \int_{x_2\in L_2/L_1\cap L_2} \phi(\sigma(x_2)\sigma(x_3)h) = \int_{x\in L_2+L_3/L_1\cap L_2} \phi(\widetilde{\sigma}(x)h),$$

where $\widetilde{\sigma} : L_2 + L_3 \to H$ is the homomorphism defined by $\widetilde{\sigma}(x_2 + x_3) =$

$\sigma(x_2)\sigma(x_3)$. We have an exact sequence

$$0 \to L_3/L_1 \cap L_2 \cap L_3 \to (L_2 + L_3)/L_1 \cap L_2$$
$$\to L_2/(L_1 \cap L_2 + L_2 \cap L_3) = A \to 0.$$

Note also that $L_1 \cap L_2 \cap L_3 = L_1 \cap L_3$ because $L_2 \subset L_1 + L_3$. Thus, we can rewrite the above integral as

$$\sum_{x_2 \in A} \int_{x_3 \in L_3/L_1 \cap L_3} \phi(\widetilde{\sigma}(x_2 + x_3)h).$$

Let $x_2 \in L_2$ be a representative of an element in $A$. Choose a decomposition $x_2 = y_1 + y_3$ where $y_1 \in L_1$, $y_3 \in L_3$. Then

$$\widetilde{\sigma}(y_1) = \sigma(x_2)\sigma(y_3)^{-1} = q(x_2)\sigma(y_1),$$

hence we can rewrite the inner integral as

$$\int_{x_3 \in L_3/L_1 \cap L_3} \phi(\widetilde{\sigma}(x_2 + x_3)h) = \int_{x_3 \in L_3/L_1 \cap L_3} \phi(\widetilde{\sigma}(y_1)\sigma(x_3)h)$$
$$= q(x_2) \int_{x_3 \in L_3/L_1 \cap L_3} \phi(\sigma(x_3)h).$$

Therefore, the above double integral is equal up to a positive constant to $\gamma(q)R(L_1, L_3)\phi(h)$. This implies that $c(L_1, L_2, L_3) = \gamma(q_{L_1,L_2,L_3})$ in this case.

Now let $L_1, L_2, L_3$ be an arbitrary admissible triple of Lagrangian subgroups. We construct another Lagrangian subgroup $L = L(L_1, L_2, L_3)$ by setting

$$L(L_1, L_2, L_3) = L_1 \cap L_2 + (L_1 + L_2) \cap L_3$$

(the lifting $L(L_1, L_2, L_3) \to H$ is given by the formula $l_{12} + l_3 \mapsto \sigma(l_{12})\sigma(l_3)$, where $l_{12} \in L_1 \cap L_2$, $l_3 \in L_3$). It is straighforward to check that $L$ is Lagrangian and is compatible with $L_i$ for $i = 1, 2, 3$. Note also that since $L = L \cap L_1 + L \cap L_3$, we have $c(L_1, L_3, L) = 1$. Similarly, $c(L_2, L_3, L) = 1$. Applying equation (4.2.3) to our quadruple of Lagrangian subgroups we deduce that

$$c(L_1, L_2, L_3) = c(L_1, L_2, L).$$

On the other hand, we have

$$A(L_1, L_2, L) \simeq L/(L \cap L_1 + L \cap L_2) = L/(L_1 \cap L_2 + L_1 \cap L_3 + L_2 \cap L_3)$$
$$\simeq (L_1 + L_2) \cap L_3/(L_1 \cap L_3 + L_2 \cap L_3) \simeq A(L_1, L_2, L_3).$$

It is easy to check that under this isomorphism we have $q_{L_1,L_2,L} = q_{L_1,L_2,L_3}$. Finally, since $L \subset L_1 + L_2$ by the first part of the proof we have

$$c(L_1, L_2, L) = \gamma(q_{L_1,L_2,L}) = \gamma(q_{L_1,L_2,L_3})$$

which finishes the proof. □

Let us point out the following corollary from the second part of the proof.

**Corollary 4.7.** *For an arbitrary admissible triple $L_1, L_2, L_3$ of Lagrangian subgroups in $K$ one has*

$$c(L_1, L_2, L_3) = c(L_1, L_2, L(L_1, L_2, L_3)),$$

*where $L(L_1, L_2, L_3) = L_1 \cap L_2 + (L_1 + L_2) \cap L_3$.*

## 4.4. Maslov Index

Now we specialize to the case of the real Heisenberg group $\mathcal{H}(V)$ associated with a symplectic vector space $(V, E)$ (see Section 2.3). Recall that by the definition we have fixed a splitting $\mathcal{H}(V) = U(1) \times V$, and we always equip a real Lagrangian subspace $L \subset V$ with an obvious lifting homomorphism to $\mathcal{H}(V)$. Then every two real Lagrangian subspaces are compatible.

Applying the construction of the quadratic function to a triple of real Lagrangian subspaces $L_1, L_2, L_3$ in $V$ we get a nondegenerate quadratic form $q_{L_1,L_2,L_3}$ on the vector space $A(L_1, L_2, L_3)$. The signature of the quadratic form $-q_{L_1,L_2,L_3}$ is called the *Maslov index* of our triple and is denoted by $m(L_1, L_2, L_3)$.

**Theorem 4.8.** *For a triple of real Lagrangian subspaces $L_1, L_2, L_3$ one has*

$$c(L_1, L_2, L_3) = \exp\left(-\frac{\pi i}{4} \cdot m(L_1, L_2, L_3)\right)$$

*Proof.* The first step of the proof is to check the statement in the case when $L_1$, $L_2$ and $L_3$ are transversal. In this case, we have a direct sum decomposition $V = L_1 \oplus L_3$, so we can consider the following symmetric form on $L_2$: $S(x, y) = E(x_1, y_3)$ where $x = x_1 + x_3$, $y = y_1 + y_3$ with $x_i, y_i \in L_i$. Now let $L_2 = U \oplus V$ be an $S$-orthogonal decomposition. Let $U_1$ and $V_1$ (resp., $U_3$ and $V_3$) be projections of $U$ and $V$ to $L_1$ (resp., $L_3$) with respect to the decomposition $V = L_1 \oplus L_3$. Then the subspaces $U_1 \oplus U_3$ and $V_1 \oplus V_3$ are

$E$-orthogonal to each other. Thus, the whole picture decomposes into an orthogonal sum and we reduce the problem to the case when $V$ is 2-dimensional with symplectic basis $\{e, f\}$, $L_1$ is generated by $e$, $L_3$ is generated by $f$ and $L_2$ is generated by $e + f$. Then the proof is obtained by a direct computation (see Exercise 2).

In the general case we can assume that $L_1 \cap L_2 \neq 0$ and then use Proposition 4.4(iii) to lower the dimension. Namely, by Corollary 4.7 we have

$$c(L_1, L_2, L_3) = c(L_1, L_2, L(L_1, L_2, L_3)).$$

On the other hand,

$$m(L_1, L_2, L_3) = m(L_1, L_2, L(L_1, L_2, L_3))$$

as follows from Proposition 4.4(iii). Therefore, it suffices to prove the statement with $L_3$ replaced by $L(L_1, L_2, L_3)$. But all three Lagrangians $L_1$, $L_2$ and $L(L_1, L_2, L_3)$ contain the nonzero isotropic subspace $L_1 \cap L_2 \subset V$, so assuming by induction that the theorem holds in lower dimensions we can finish the proof. □

The Maslov index $m(L_1, L_2, L_3)$ has the following nice property: if Lagrangian subspaces $L_i$ vary continuously in such a way that the dimensions of all pairwise intersections are constant, then $m(L_1, L_2, L_3)$ remains constant (see Exercise 3(c)). This property together with Theorem 4.8 implies that the spaces $\mathcal{F}'(L)$ defined in Section 4.2 constitute an infinite-dimensional local system over the Lagrangian Grassmanian $\mathcal{L}(V)$. More precisely, fix a Largangian subspace $L_0 \subset V$ and consider an open subset $D(L_0) \subset \mathcal{L}(V)$ consisting of the Lagrangian subspaces $L \subset V$ such that $L \cap L_0 = 0$. Then the operators $R(L_0, L)$ give a trivialization of the vector bundle formed by $\mathcal{F}'(L)$ over $D(L_0)$, such that the transition functions over $D(L_0, L_0') = D(L_0) \cap D(L_0')$ are locally constant, because the Maslov index $m(L_0, L_0', L)$ is locally constant for $L \in D(L_0, L_0')$.

### 4.5. Lagrangian Subspaces and Lattices

For applications to theta functions we need to study intertwining operators between $\mathcal{H}(V)$-modules $\mathcal{F}(L)$ and $\mathcal{F}(\Gamma)$ defined in Chapter 2, where $L \subset V$ is a real Lagrangian subspace, $\Gamma \subset V$ is a self-dual lattice equipped with a lifting to $\mathcal{H}(V)$. Recall that a lifting homomorphism $\Gamma \to \mathcal{H}(V)$ corresponds to a map $\alpha : \Gamma \to U(1)$ satisfying the equation (1.2.2). Since a choice of lifting is important for compatibility conditions, we will often write $(\Gamma, \alpha)$

instead of $\Gamma$. Thus, the condition of compatibility between $L$ and $(\Gamma, \alpha)$ is that $L \cap \Gamma$ is a lattice in $L$ and $\alpha|_{L\cap\Gamma} = 1$. In this situation we have a canonical Haar measure on $V/L$ (resp., $V/\Gamma$) such that the lattice $\Gamma/L\cap\Gamma$ has covolume 1 (resp., the volume of $V/\Gamma$ is equal to 1). Thus, we can rewrite our intertwining operators $R(L,\Gamma)$ and $R(\Gamma, L)$ in terms of the spaces $\mathcal{F}(L)$ and $\mathcal{F}(\Gamma)$ (which we represent now as subspaces of functions on $V$):

$$R(L,\Gamma)f(v) = \sum_{\gamma\in\Gamma/\Gamma\cap L} \alpha(\gamma)\exp(\pi i E(\gamma, v)) f(v+\gamma),$$

$$R(\Gamma, L)\phi(v) = \int_{L/\Gamma\cap L} \exp(\pi i E(l, v))\phi(v+l)dl,$$

where $f \in \mathcal{F}(L)$, $\phi \in \mathcal{F}(\Gamma)$, the measure on $L/\Gamma\cap L$ is normalized by the condition that the total volume is equal to 1.

Now let $L_1$ and $L_2$ be a pair of Lagrangian subspaces compatible with $(\Gamma, \alpha)$. Since the spaces $V/L_i$ are equipped with canonical Haar measures, the intertwining operator $R(L_1, L_2)$ can be considered as an operator from $\mathcal{F}(L_1)$ to $\mathcal{F}(L_2)$. More precisely, we get

$$R(L_1, L_2)\phi(v) = |\Gamma/(\Gamma\cap L_1 + \Gamma\cap L_2)|^{-\frac{1}{2}} \cdot \int_{L_2} \exp(\pi i E(l_2, v))\phi(v+l_2)dl_2$$

where $dl_2$ is the Haar measure on $L_2$ with respect to which the lattice $\Gamma \cap L_2$ has covolume 1. Note that we have $A(L_1,\Gamma, L_2) \simeq \Gamma \cap (L_1 + L_2)/(\Gamma \cap L_1 + \Gamma \cap L_2)$, so this group is finite. Also it is easy to see that

$$q_{L_1,\Gamma,L_2}(\gamma) = \alpha(\gamma)\exp(-\pi i Q_{L_1,L_2}(\gamma)), \tag{4.5.1}$$

where $Q_{L_1,L_2}$ is the quadratic form on $L_1 + L_2$ defined by $Q_{L_1,L_2}(l_1 + l_2) = E(l_1, l_2)$ where $l_1 \in L_1$, $l_2 \in L_2$. Now applying Theorem 4.5, we get

$$\begin{aligned} c(L_1, \Gamma, L_2) &= \gamma(q_{L_1,\Gamma,L_2}) \\ &= |A(L_1,\Gamma,L_2)|^{-1} \textstyle\sum_{\gamma\in\Gamma\cap(L_1+L_2)/(\Gamma\cap L_1+\Gamma\cap L_2)} \alpha(\gamma)\exp(-\pi i Q_{L_1,L_2}(\gamma)). \end{aligned}$$

**Remark.** In the case $L_1 \cap L_2 = 0$ the quadratic function $q_{L_1,\Gamma,L_2}$ can be interpreted geometrically as follows. From the data $(V, E, \Gamma, \alpha)$ as above we can construct the $U(1)$-torsor

$$\mathcal{T} = \Gamma\backslash\mathcal{H}(V) \to T$$

over the symplectic torus $T = V/\Gamma$ (in Exercise 1 of Chapter 2, we used the corresponding complex line bundle). Every real Lagrangian subspace $L \subset V$ compatible with $(\Gamma, \alpha)$ gives a Lagrangian subtorus $\overline{L} = L/(\Gamma\cap L)$ together with a section of $\mathcal{T}$ over $\overline{L}$ (induced by the lifting of $L$ to $\mathcal{H}(V)$). Now if we have two transversal Lagrangian subspaces $L_1$ and $L_2$ compatible with $(\Gamma, \alpha)$ then we can consider the finite subgroup

$$K = \overline{L_1} \cap \overline{L_2} = (L_1 + \Gamma) \cap (L_2 + \Gamma)/\Gamma \subset T.$$

We have two sections $\sigma_1$ and $\sigma_2$ of $\mathcal{T}$ over $K$ induced by the sections of $\mathcal{T}$ over $\overline{L_1}$ and $\overline{L_2}$. The ratio $\sigma_1/\sigma_2$ is a function on $K$ with values in $U(1)$. It is easy to check that under the natural isomorphism

$$K \simeq (L_1 + \Gamma) \cap L_2/(\Gamma \cap L_2) \simeq A(L_1, \Gamma, L_2)$$

the function $\sigma_1/\sigma_2$ gets identified with $q_{L_1,\Gamma,L_2}$

It is also instructive to consider intertwining operators for a pair of compatible self-dual lattices $\Gamma_1$ and $\Gamma_2$ in $V$. Here compatibility means that $\Gamma_1$ and $\Gamma_2$ are commensurable and that the lifting homomorphisms $\sigma : \Gamma_i \to \mathcal{H}(V)$, $i = 1, 2$, agree on $\Gamma_1 \cap \Gamma_2$. As before we can trivialize $\mathbb{R}_{>0}$-torsors $\operatorname{meas}(V/\Gamma_1)$ and $\operatorname{meas}(V/\Gamma_2)$ by choosing measures with total volume 1. Unraveling the definition, we can rewrite the canonical intertwining operator $R(\Gamma_1, \Gamma_2) : \mathcal{F}(\Gamma_1) \to \mathcal{F}(\Gamma_2)$ as follows:

$$R(\Gamma_1, \Gamma_2)\phi(h) = |\Gamma_2/\Gamma_1 \cap \Gamma_2|^{-\frac{1}{2}} \cdot \sum_{\gamma_2 \in \Gamma_2/\Gamma_1 \cap \Gamma_2} \phi(\sigma(\gamma_2)h).$$

Theorem 4.5 can be applied to compute constants $c(L, \Gamma_1, \Gamma_2)$ and $c(\Gamma_1, \Gamma_2, \Gamma_3)$, where $L$ is a Lagrangian subspace, $\Gamma_i$ are self-dual lattices, in terms of Gauss sums.

### 4.6. Application to Computation of Gauss Sums

Let us fix a self-dual lattice $\Gamma \subset V$ equipped with a lifting to $\mathcal{H}(V)$ induced by a map $\alpha : \Gamma \to U(1)$. For every pair of real Lagrangian subspaces $L_1, L_2$ compatible with $(\Gamma, \alpha)$ let us set $b(L_1, L_2) = c(L_1, \Gamma, L_2)$. Using Theorem 4.8 we get the following relation between constants $b(L_1, L_2)$ and the Maslov index.

**Proposition 4.9.** *Let $L_1$, $L_2$, and $L_3$ be a triple of Lagrangian subspaces compatible with $(\Gamma, \alpha)$. Then*

$$b(L_1, L_2)b(L_2, L_3)b(L_3, L_1) = \exp\left(\frac{\pi i}{4} m(L_1, L_2, L_3)\right).$$

*Proof.* Applying (4.2.3) to $(L_1, L_2, L_3, \Gamma)$ we get

$$c(L_1, L_2, \Gamma)c(L_2, L_3, \Gamma) = c(L_1, L_2, L_3)c(L_1, L_3, \Gamma).$$

It remains to use (4.2.2) and Theorem 4.8. □

From this we can recover the following classical result.

**Corollary 4.10.** *Let $d > 0$ be an even integer. Then*

$$\sum_{n \in \mathbb{Z}/d\mathbb{Z}} \exp\left(\pi i \frac{n^2}{d}\right) = \exp\left(\frac{\pi i}{4}\right) \sqrt{d}.$$

*Proof.* Let us consider the symplectic space $V = \mathbb{R}e_1 \oplus \mathbb{R}e_2$, where $E(e_1, e_2) = 1$. We can take $\Gamma = \mathbb{Z}e_1 \oplus \mathbb{Z}e_2, \alpha(me_1 + ne_2) = (-1)^{mn}$. Now let us apply Proposition 4.9 to $L_1 = \mathbb{R}e_1$, $L_2 = \mathbb{R}e_2$ and $L_3 = \mathbb{R}(e_1 + de_2)$ (we use the fact that $d$ is even in checking that $L_3$ is compatible with $(\Gamma, \alpha)$). We have $\Gamma = \Gamma \cap L_1 + \Gamma \cap L_2 = \Gamma \cap L_2 + \Gamma \cap L_3$, hence $b(L_1, L_2) = b(L_2, L_3) = 1$. On the other hand, since $m(L_1, L_2, L_3) = -1$, from Proposition 4.9 we get $b(L_1, L_3) = \exp(\pi i/4)$. Finally, we have

$$\Gamma/(\Gamma \cap L_1 + \Gamma \cap L_3) = \mathbb{Z}e_1 \oplus \mathbb{Z}e_2/(\mathbb{Z}e_1 \oplus d\mathbb{Z}e_2) \simeq \mathbb{Z}/d\mathbb{Z}$$

and $Q_{L_1, L_3}(ne_2) = -n^2/d$, so by formula (4.5.1) we get

$$b(L_1, L_3) = \gamma(q_{L_1, \Gamma, L_3}) = d^{-\frac{1}{2}} \cdot \sum_{n \in \mathbb{Z}/d\mathbb{Z}} \exp\left(\pi i \frac{n^2}{d}\right).$$ □

In Appendix B we will present a more general computation of Gauss sums using similar arguments.

## 4.7. More on Gauss Sums

In this section we will show that Gauss sums corresponding to quadratic functions on abelian groups are roots of unity.

**Lemma 4.11.** *Let $q : A \to U(1)$ be a quadratic function. Then for every $a \in A$ and $n \in \mathbb{Z}$ one has*

$$q(na) = q(a)^{\frac{n(n+1)}{2}} q(-a)^{\frac{n(n-1)}{2}}.$$

*In particular, if $q$ is a quadratic form then*

$$q(na) = q(a)^{n^2}.$$

The proof is an easy induction in $n$ using the identity $q((n+1)a) = q(na)q(a)\langle a, a\rangle^n$, where $\langle\cdot,\cdot\rangle$ is given by (4.3.1)

**Lemma 4.12.** *Let $q : A \to U(1)$ be a nondegenerate quadratic function. Then there exists a nondegenerate quadratic form $q_0 : A \to U(1)$ and an element $a_0 \in A$ such that $q(a) = q_0(a + a_0)/q_0(a_0)$ for all $a \in A$.*

*Proof.* Let us set $l(a) = q(a)/q(-a)$. Then $l : A \to U(1)$ is a homomorphism. Since the pairing $\langle\cdot,\cdot\rangle$ associated with $A$ is nondegenerate, there exists an element $\widetilde{a}$ such that $l(a) = \langle\widetilde{a}, a\rangle$. We claim that $\widetilde{a}$ belongs to the subgroup $2A \subset A$. Indeed, $2A$ is the orthogonal complement to $A_2 = \ker([2] : A \to A)$ with respect to $\langle\cdot,\cdot\rangle$. Now our claim follows from the fact that $l|_{A_2} \equiv 1$. It remains to choose $a_0 \in A$ such that $\widetilde{a} = 2a_0$ and to set $q_0(a) = q(a - a_0)/q(-a_0)$. □

Now we can prove our main result about the constants $\gamma(q)$.

**Theorem 4.13.** *Let $A$ be a finite abelian group of order $N$, $q$ a nondegenerate quadratic function on $A$. Then $\gamma(q)^{\mathrm{lcm}(2N,8)} = 1$, where* lcm *denotes the least common multiple. If $q$ is a nondegenerate quadratic form on $A$ then $\gamma(q)^8 = 1$. If in addition $|A|$ is odd then $\gamma(q)^4 = 1$.*

*Proof.* Using Lemma 4.12 and Exercise 6, we immediately reduce ourselves to the case when $q$ is a nondegenerate quadratic form. In this case we will employ the weak version of the so called "Zarhin trick" (cf. [136]). It is well known that there exists four integers $(a, b, c, d)$ such that $a^2+b^2+c^2+d^2 \equiv -1 \bmod (2N)$. Indeed, by the Chinese remainder theorem, it suffices to show that for every prime power $p^n$ there exists $(a, b, c, d)$ with $a^2+b^2+c^2+d^2 \equiv -1 \bmod (p^n)$. In the case when $p$ is odd, we can take $c = d = 0$, and use the fact that a residue $r \in \mathbb{Z}/p^n\mathbb{Z}$ is a square if and only if its reduction modulo $p$ is. On the other hand, in the case $p = 2$ we can take $b = 2$, $c = d = 1$ and use the fact that $-7$ is a square modulo $2^n$.

Now let $M$ be the $4 \times 4$-matrix with integer coefficients corresponding to the operation of multiplication by the quaternion $a + bi + cj + dk$. The crucial property of $M$ is the following equality:

$$ {}^t MM = (a^2 + b^2 + c^2 + d^2)\,\mathrm{Id}\,. \tag{4.7.1}$$

Let us consider the homomorphism $f : A^{\oplus A} \to A^{\oplus 4}$ given by $M$. Then

(4.7.1) together with Lemma 4.11 imply that $f$ is an automorphism and

$$f^*(q^{\oplus 4}) = \left(q^{a^2+b^2+c^2+d^2}\right)^{\oplus 4}.$$

By Exercise 6 we have $q^{2N} = 1$, so we can replace $a^2+b^2+c^2+d^2$ by $-1$. Hence, applying Proposition 4.6 (i) and (iii) we get

$$\gamma(q)^4 = \gamma(q^{\oplus 4}) = \gamma((q^{-1})^{\oplus 4}) = \gamma(q^{-1})^4 = \gamma(q)^{-4}.$$

If $N$ is odd, we can find a pair of integers $(a, b)$ such that $a^2+b^2 \equiv -1(N)$ and then apply a similar argument starting from the $2\times 2$-matrix of multiplication by the complex number $a+bi$ (note that for odd $N$ we have $q^N = 1$). □

**Corollary 4.14.** *Let* $(\Gamma, \alpha)$, $(L_1, L_2)$ *be as in Section 4.5. Assume in addition that* $\alpha^2 = 1$. *Then* $c(L_1, \Gamma, L_2)^8 = 1$.

*Proof.* Indeed, by Proposition (4.4), $q_{L_1,\Gamma,L_2}$ is a quadratic form. □

## Exercises

1. Let $(V, E)$ be a symplectic vector space, $L$ and $L'$ be a pair of Lagrangian subspaces in $V$ such that $L \cap L' = 0$. Let us identify $L'$ with the dual space to $L$ as follows:

$$L' \to L^* : l' \mapsto E(?, l').$$

We can identify the space $\mathcal{F}(L)$ with $L^2(L')$ by restricting a function $f \in \mathcal{F}(L)$ from $V$ to $L'$. Similarly we identify $\mathcal{F}(L')$ with $L^2(L)$. Then the intertwining operator $R(L', L)$ can be considered as acting from $L^2(L)$ to $L^2(L') = L^2(L^*)$.

(a) Show that $R(L', L)$ coincides with the Fourier transform

$$f(l) \mapsto \hat{f}(l^*) = \int_l f(l)\exp(2\pi i\langle l^*, l\rangle)dl,$$

(b) Let $\Gamma \subset L$ be a lattice, $\Gamma^\perp \subset L^*$ be a dual lattice. Prove the Poisson summation formula:

$$\widehat{\delta_\Gamma} = \delta_{\Gamma^\perp},$$

where $\delta_\Gamma = \sum_{\gamma\in\Gamma}\delta(\gamma)$ is the delta-function of $\Gamma$, $\delta_{\Gamma^\perp}$ is the delta-function of $\Gamma^\perp$. Here both parts should be considered as distributions on the Schwarz spaces of functions rapidly descreasing (with all derivatives) at infinity. [*Hint*: Use the fact that $\mathcal{F}_{-\infty}(L)^{\Gamma\oplus\Gamma^\perp}$ is 1-dimensional.]

2. Let $V = \mathbb{R}e_1 + \mathbb{R}e_2$ be the symplectic vector space with the basis $(e_1, e_2)$ such that $E(e_1, e_2) = 1$. Consider the following Lagrangian subspaces in $V : L_1 = \mathbb{R}e_1, L_2 = \mathbb{R}(e_1+e_2), L_3 = \mathbb{R}e_2$. Show that $m(L_1, L_2, L_3) = 1$. Now check Theorem 4.8 in this case by applying both sides to the function $f(xe_1 + ye_2) = \exp(-\pi ixy - \pi y^2) \in \mathcal{F}(L_1)$ and using the equality

$$\widehat{\exp(-\pi ax^2)} = \frac{1}{\sqrt{a}} \exp(-\pi a^{-1} y^2)$$

where $a \in \mathbb{C}$, $\mathrm{Re}(a) > 0$.

3. Let $L_1, L_2, L_3$ be a triple of Lagrangian subspaces in a symplectic vector space $V$.
   (a) Prove that the Maslov index is equal to the signature of the quadratic form $Q$ on $L_1 \oplus L_2 \oplus L_3$ given by $Q(x_1, x_2, x_3) = E(x_1, x_2) + E(x_2, x_3) + E(x_3, x_1)$. [*Hint*: Changing the variable $x_2$ to $-x_2$ we see that $m(L_1, L_2, L_3)$ is equal to the signature of the quadratic form $-E(x_1, x_2) - E(x_2, x_3) + E(x_3, x_1) = -E(x_1, x_2) + E(x_3, x_1 + x_2)$ on $L_1 \oplus L_2 \oplus L_3$. Set $I = 0 \oplus 0 \oplus L_3 \subset L_1 \oplus L_2 \oplus L_3$. Show that $I$ is isotropic and that $I^\perp = I \oplus \ker(d_3)$.]
   (b) Construct an isomorphism of $\ker Q$ with $L_1 \cap L_2 \oplus L_2 \cap L_3 \oplus L_3 \cap L_1$. [*Hint*: $\ker Q$ is the space of triples $(x_1, x_2, x_3)$, such that $x_i \in L_i$, $x_1 - x_2 \in L_3$, $x_3 - x_2 \in L_1$ and $x_1 - x_3 \in L_2$. The required isomorphism sends $(x_1, x_2, x_3)$ to $(y_1, y_2, y_3)$ ($y_1 \in L_2 \cap L_3$, etc.), where $y_1 = x_2 + x_3 - x_1$, $y_2 = x_3 + x_1 - x_2$, $y_3 = x_1 + x_2 - x_3$.]
   (c) Deduce from (b) that $m(L_1, L_2, L_3)$ is invariant under deformations that leave constant dimensions of pairwise intersections.
4. Let $L_1, L_2, L_3$ be a triple of compatible Lagrangian subgroups (not necessarily linear subspaces) in a symplectic vector space $V$. Show that $A(L_1, L_2, L_3) \simeq \mathbb{R}^l \times F$ for some finite group $F$. [*Hint*: By self-duality of $A(L_1, L_2, L_3)$, it suffices to prove that the group $\pi_0(A(L_1, L_2, L_3))$ is finite. Now one can use the complex (4.3.2) and the additive function $A \mapsto \mathrm{rk}\, \pi_0(A)$ defined on the category of locally compact abelian groups with finitely generated $\pi_0$. In addition, one has to use the following property: if $I \subset V$ is an isotropic subgroup then $\mathrm{rk}\, \pi_0(I) = \mathrm{rk}\, \pi_0(I^\perp)$.]
5. Let $\Gamma \subset V$ be a lattice in a symplectic vector space $(V, E)$, such that $E|_{\Gamma\times\Gamma}$ takes integer values, and let $T = V/\Gamma$ be the corresponding symplectic torus. Let $L_1, L_2 \subset V$ be a pair of transversal real Lagrangian subspaces, such that $L_i = \mathbb{R}(L_i \cap \Gamma)$, $i = 1, 2$. Show that the intersection of the corresponding Lagrangian subtori $\overline{L}_1 = L_1/\Gamma \cap L_1$ and $\overline{L}_2 =$

$L_2/\Gamma \cap L_2$ in $T$ is a finite subgroup $K \subset T$, isomorphic to $\Gamma/\Gamma \cap L_1 + \Gamma \cap L_2$. Let $\widehat{K}$ be the dual group. Construct a symmetric homomorphism $K \to \widehat{K}$, which in the case of a self-dual lattice $\Gamma$ coincides with the isomorphism induced by the non-degenerate quadratic form $q_{L_1,\Gamma,L_2}$ (see Section 4.5).

In the remaining exercises $q$ denotes a nondegenerate quadratic function on a finite abelian group $K$.

6. Assume that $K$ is annihilated by $N \in \mathbb{Z}$.
   (a) Show that $q^{2N} \equiv 1$.
   (b) Show that if $N$ is odd then $q^N \equiv 1$.
7. For every $k \in K$ let us define the function $kq$ on $K$ by the formula

$$(kq)(k') = q(k')\langle k, k'\rangle,$$

where $\langle\ ,\ \rangle$ is the symmetric pairing associated with $q$. Show that $kq$ is a non-degenerate quadratic function and that

$$\gamma(kq) = q(k)^{-1}\gamma(q).$$

8. Assume that $I \subset K$ is a subgroup such that $q|_I \equiv 1$. Let $I^\perp$ be the orthogonal complement to $I$ with respect to the pairing $\langle\ ,\ \rangle$. Then the restriction of $q$ to $I^\perp$ descends to a function $q_I$ on $I^\perp/I$. Show that $q_I$ is a non-degenerate quadratic function and that

$$\gamma(q_I) = \gamma(q).$$

## Appendix B. Gauss Sums Associated with Integral Quadratic Forms

In this appendix we are going to use intertwining operators for Heisenberg group representations to compute some Gauss sums associated with integral quadratic forms. The main difference from the particular case considered in Section 4.6 is that we will have to use nonstandard liftings of real Lagrangian subspaces to the Heisenberg group.

Let $M$ be a free $\mathbb{Z}$-module of finite rank, $s : M \times M \to \mathbb{Z}$ be a non-degenerate symmetric bilinear form of signature $\tau$. Let $M' \subset M_{\mathbb{R}} := M \otimes \mathbb{R}$ be the dual lattice to $M$ with respect to $s$, that is, $M' = \{x \in M_{\mathbb{R}} \mid s(x, M) \subset \mathbb{Z}\}$. Choose an element $w \in M'$ such that $s(x, x) \equiv s(x, w) \bmod 2$ for all $x \in M$ (such $w$ is uniquely determined modulo $2M'$). Now we can define a quadratic function $q : M'/M \to U(1)$ by the formula

$$q(x) = \exp(\pi i(s(x, x) - s(x, w))).$$

The corresponding bilinear pairing is $\langle x, y\rangle = \exp(2\pi i s(x, y))$, so $q$ is non-degenerate. Also, $q(-x) = q(x)$, so $q$ is a quadratic form. The following theorem was proven in a particular case by Van der Blij [20] (see Corollary 4.16). Turaev [127] found the general formula presented here and observed that Van der Blij's proof still works in this case.

**Theorem 4.15.** *With the above notation one has*

$$\gamma(q) = \exp\left(\frac{\pi i}{4}(\tau - s(w, w))\right).$$

*Proof.* Set $L = M_{\mathbb{R}}$, and let us consider the symplectic vector space $V = L \oplus L$, where the symplectic form is given by $E((l_1, l_2), (l_1', l_2')) = s(l_1, l_2') - s(l_1', l_2)$. Let us denote by $\Delta : L \to L \oplus L$ the diagonal map $l \mapsto (l, l)$. Note that its image is a Lagrangian subspace. Let us also consider the self-dual lattice $\Gamma := M \oplus M' \subset V$. The idea is to deduce the desired equality from (4.2.3) applied to a triple of Lagrangian subspaces $L_1 := L \oplus 0$, $L_2 := 0 \oplus L$, $\Delta(L)$, and to a self-dual lattice $\Gamma$. However, we have to be careful with choosing compatible liftings of these Lagrangian subgroups to subgroups in $\mathcal{H}(V)$. For $L_1$ and $L_2$ we take standard liftings. For $\Gamma$ we take the lifting associated with the isotropic decomposition $M \oplus M'$ (see Section 3.2). Finally, we define the lifting homomorphism of $\Delta(L)$ using $w$:

$$\Delta(l) \mapsto (\exp(\pi i s(l, w)), \Delta(l)).$$

To remember this special choice of a lifting homomorphism we will write $\Delta(L)_w$, while $\Delta(L)$ will mean the same Lagrangian subspace equipped with the standard lifting. It is easy to check that the subgroups $L_1$, $L_2$, $\Delta(L)_w$ and $\Gamma$ are pairwise compatible. Indeed, the only nonobvious check is compatibility of $\Gamma$ and $\Delta(L)_w$. The intersection $\Gamma \cap \Delta(L)_w$ consists of elements $\Delta(m)$, where $m \in M$. The lifting of $\Gamma$ maps such an element to $(\exp(\pi i s(m,m)), \Delta(m))$, so the compatibility follows from the defining property of $w$. It is easy to see that $A(L_1, L_2, \Gamma) = A(L_2, \Gamma, \Delta(L)_w) = 0$. Therefore, the equation (4.2.3) reduces in this case to

$$c(L_1, \Gamma, \Delta(L)_w) = c(L_1, L_2, \Delta(L)_w).$$

By Theorem 4.5, the left-hand side is equal to the Gauss sum of the quadratic function $q_{L_1,\Gamma,\Delta(L)_w}$. We find easily that

$$\begin{aligned} A(L_1, \Gamma, \Delta(L)) &= \Gamma/(\Gamma \cap L_1 + \Gamma \cap \Delta(L)) = M \oplus M'/(M \oplus M) \simeq M'/M, \\ q_{L_1,\Gamma,\Delta(L)_w}(x) &= (1, (x,0)) \cdot (1, (0,x)) \cdot (\exp(-\pi i s(x,w), (-x,-x)) \\ &= \exp(\pi i s(x,x) - \pi i s(x,w)) \in U(1). \end{aligned}$$

To compute $c(L_1, L_2, \Delta(L)_w)$ we compare it with $c(L_1, L_2, \Delta(L))$. Note that the quadratic form $q_{L_1,L_2,\Delta(L)}$ is just $s(x,x)$ on $L$. Hence, using Theorem 4.8 we get

$$c(L_1, L_2, \Delta(L)) = \exp\left(\frac{\pi i \tau}{4}\right).$$

Now we note that our Lagrangian subgroup $\Delta(L)_w$ in $\mathcal{H}(V)$ is conjugate to the subgroup $\Delta(L)$. In fact, we have two natural elements $h_1, h_2 \in \mathcal{H}(V)$ such that

$$\Delta(L)_w = h_1 \Delta(L) h_1^{-1} = h_2 \Delta(L) h_2^{-1},$$

namely, $h_1 = (1, (w/2, 0))$ and $h_2 = (1, (0, -w/2))$. We have the induced intertwining operators

$$R_i = R_i(\Delta(L), \Delta(L)_w) : \phi \mapsto \big(h \mapsto \phi(h_i^{-1}h)\big), i = 1, 2.$$

These operators differ by a scalar which we will presently determine. Since

$$h_2^{-1}h_1 = (\exp(-\pi i s(w/2, w/2)), (w/2, w/2)),$$

it follows that $\exp(\frac{\pi i}{4} s(w,w)) h_2^{-1} h_1$ belongs to $\Delta(L)$. Hence, for $\phi \in \mathcal{F}(\Delta(L))$ we have

$$\phi(h_2^{-1}h) = \phi\big((h_2^{-1}h_1)h_1^{-1}h\big) = \exp\left(-\frac{\pi i}{4} s(w,w)\right)\phi(h_1^{-1}h).$$

Thus, $R_2 = \exp(-\frac{\pi i}{4} s(w, w)) R_1$. On the other hand, since $h_1 \in L_1, h_2 \in L_2$, it follows that

$$R(L_i, \Delta(L)_w) = R_i \circ R(L_i, \Delta(L)),\ i = 1, 2.$$

Therefore,

$$\begin{aligned} c(L_1, L_2, \Delta(L)_w) &= \exp\left(-\frac{\pi i}{4} s(w, w)\right) c(L_1, L_2, \Delta(L)) \\ &= \exp\left(\frac{\pi i}{4}(\tau - s(w, w))\right). \end{aligned}$$

□

In the particular case when the discriminant of $s$ is odd, we get the following formula originally proven by Van der Blij in [20].

**Corollary 4.16.** *Assume that the discriminant $D$ of $s$ is odd and let $w \in M$ be such that $s(x, x) \equiv s(x, w) \bmod 2$ for $x \in M$. Then*

$$|D|^{-\frac{1}{2}} \cdot \sum_{x \in M'/M} \exp(4\pi i s(x, x)) = \exp\left(\frac{\pi i}{4}(\tau - s(w, w))\right).$$

*Proof.* Note that $|M'/M| = D$. The fact that $|M'/M|$ is odd implies easily that $w \in M$ with required property exists. By Theorem 4.15 we have

$$\sum_{x \in M'/M} \exp(\pi i (s(x, x) - s(x, w))) = |D|^{\frac{1}{2}} \exp\left(\frac{\pi i \tau}{4}\right).$$

Since 2 is invertible in $M'/M$, we can substitute $2x$ instead of $x$ in the LHS. □

# 5

# Theta Functions II: Functional Equation

Throughout this chapter we fix a symplectic space $(V, E)$. Recall that additional data used to define the theta series is a triple $((\Gamma, \alpha), J, L)$, where $\Gamma$ is a self-dual lattice in $V$ equipped with a map $\alpha : \Gamma \to U(1)$ satisfying (1.2.2), $J$ is a complex structure strictly compatible with $E$, $L$ is a Lagrangian subspace in $V$ compatible with $(\Gamma, \alpha)$. In this chapter we derive functional equations describing the change of the theta series when either $L$ or $\Gamma$ changes (in the latter case we consider the change of $\Gamma$ to a commensurable lattice). The idea is to consider the theta series $\theta^{\alpha}_{H,\Gamma,L}$ (multiplied by an exponential factor) as an element in the realization $\mathcal{F}(\Gamma)$ of the irreducible representation of $\mathcal{H}(V)$, associated with the lattice $\Gamma$ (see Chapter 2). Then this element can be characterized up to a scalar by its invariance with respect to some Lie subalgebra $P_J$ of Lie $\mathcal{H}(V)$ associated with the complex structure $J$. On the other hand, one can easily construct a $P_J$-invariant element $f_L$ in the realization $\mathcal{F}(L)$ of Schrödinger representation of $\mathcal{H}(V)$ associated with $L$. It turns out that these two $P_J$-invariant elements correspond to each other under the isomorphism between $\mathcal{F}(L)$ to $\mathcal{F}(\Gamma)$ constructed in Chapter 4. If $L' \subset V$ is another Lagrangian subspace compatible with $(\Gamma, \alpha)$, then one can easily find the proportionality coefficient between vectors $f_L$ and $f_{L'}$, where the spaces $\mathcal{F}(L)$ and $\mathcal{F}(L')$ are identified as in Chapter 4. Now the functional equation for theta series follows from the results of Chapter 4, where the constant $c(L, L', \Gamma)$ measuring the proportionality coefficient between intertwining operators $R(L', \Gamma) \circ R(L, L')$ and $R(L, \Gamma)$ is expressed as a Gauss sum (an 8th root of unity). Choosing a symplectic basis in $\Gamma$ and parametrizing complex structures on $V$ in a natural way we deduce the classical form of the functional equation for theta series. The functional equation describing the change of $\theta_{H,\Gamma,L}$ when $\Gamma$ changes to a commensurable lattice $\Gamma'$ is derived along similar lines. It uses the constant $c(L, \Gamma, \Gamma')$ that was introduced and computed in Chapter 4.

## 5.1. Theta Series and Intertwining Operators

Let $\mathcal{H}(V)$ be the Heisenberg group of $(V, E)$, $J$ a complex structure on $V$ strictly compatible with $E$. Recall that this means that the corresponding Hermitian form $H$ on $V$ with $\operatorname{Im} H = E$ is positive. Let $\Gamma \subset V$ be a self-dual lattice, $\alpha : \Gamma \to U(1)$ a map satisfying equation (1.2.2), and $L \subset V$ a Lagrangian subspace compatible with $(\Gamma, \alpha)$.

In this situation the theta series $\theta^{\alpha}_{H,\Gamma,L}$ is a nonzero generator of the space $\mathrm{Fock}_{-\infty}(V, J)^{\Gamma}$ (see Section 2.5). We are going to show how to obtain this element naturally using the intertwining operator $R(L, \Gamma) : \mathcal{F}(L) \to \mathcal{F}(\Gamma)$.

We start with the following observation: the function

$$(\lambda, v) \mapsto \lambda^{-1} \exp\left(-\frac{\pi}{2} H(v, v)\right) \theta^{\alpha}_{H,\Gamma,L}(v)$$

is a generator of the space of functions $\phi$ on $\mathcal{H}(V)$ that are invariant under the distribution $P^r_J$ and under the action of $\Gamma$ by right translations and satisfy $\phi(\lambda h) = \lambda^{-1}\phi(h)$. The map $\phi(h) \mapsto \phi(h^{-1})$ provides an isomorphism of this space with the space of functions $\phi$ on $\mathcal{H}(V)$ that are invariant under the action of $\Gamma$ by left translations and under the distribution $P^l_J$ (left-invariant distribution defined by the subalgebra $P_J \subset \mathrm{Lie}(\mathcal{H}(V))_{\mathbb{C}}$) and satisfy $\phi(\lambda h) = \lambda\phi(h)$. But the latter space is precisely the space of vectors in the $\mathcal{H}(V)$-representation $\mathcal{F}(\Gamma)$ annihilated by the action of the subalgebra $P_J$. We can replace $\mathcal{F}(\Gamma)$ by an isomorphic $\mathcal{H}(V)$-representation $\mathcal{F}(L)$, so it suffices to construct an element in $\mathcal{F}(L)$ annihilated by $P_J$. Let us consider the function

$$f_L(v) = \exp\left(-\frac{\pi}{2}(H - S_L)(v, v)\right),$$

where $S_L$ is a $\mathbb{C}$-linear symmetric form on $V$ extending $H|_{L\times L}$. Then it is easy to check that $f_L$ is an element of the space $\mathcal{F}(L)$ and that $f_L$ is annihilated by $P_J$ (the latter condition is equivalent to the fact that $\exp(\frac{\pi}{2} H) f_L$ is holomorphic). Thus, we can get an element in $\mathrm{Fock}^{\Gamma}_{-\infty}$ by applying to $f_L$ the above sequence of isomorphisms. In fact, one gets precisely the theta series: it is easy to check that

$$\exp\left(-\frac{\pi}{2} H(v, v)\right)\theta^{\alpha}_{H,\Gamma,L}(-v) = R(L, \Gamma) f_L(v). \tag{5.1.1}$$

## 5.2. Existence of Compatible Lagrangian Subspace

Let $\Gamma \subset V$ be a self-dual lattice, $\alpha : \Gamma \to U(1)$ be a map satisfying (1.2.2). Assume in addition that $\alpha^2 \equiv 1$.

The equation (1.2.2) implies that $\alpha|_{2\Gamma} = 1$, so $\alpha$ descends to a well-defined map $\overline{\alpha} : \Gamma/2\Gamma \to \{\pm 1\}$. Identifying $\Gamma/2\Gamma$ with $(\mathbb{Z}/2\mathbb{Z})^{2n}$

(where $n = \dim_{\mathbb{C}} V$) we can consider $\overline{\alpha}$ as a quadratic form $(\mathbb{Z}/2\mathbb{Z})^{2n} \to \{\pm 1\}$ whose associated pairing is equal to $\exp(\pi i E)$. Self-duality of $\Gamma$ implies that this pairing is nondegenerate. It is well known that every nondegenerate quadratic form $q : (\mathbb{Z}/2\mathbb{Z})^{2n} \to \{\pm 1\}$ by a change of variables can be brought either to the form $(-1)^{\sum_{i=1}^{n} x_i y_i}$ or to the form $(-1)^{x_1^2+y_1^2+\sum_{i=1}^{n} x_i y_i}$. The former ones are called *even* while the latter ones are called *odd*. A more conceptual way to distinguish the two types is to say that the form $q$ is even if and only if there exists a subgroup $I \subset (\mathbb{Z}/2\mathbb{Z})^{2n}$ such that $q|_I \equiv 1$ and $I$ is Lagrangian with respect to the symplectic form $e(x, y) = q(x+y)q(x)q(y)$ (see Exercise 1).

**Theorem 5.1.** *The quadratic form $\overline{\alpha}$ is even if and only if there exists a Lagrangian subspace $L \subset V$ compatible with $(\Gamma, \alpha)$ in the sense of Section 4.5.*

*Proof.* The "if" part follows immediately from the fact that $\alpha$ vanishes on the subgroup $\Gamma \cap L/2\Gamma \cap L \subset \Gamma/2\Gamma$ (which has order $2^n$). Conversely, assume that the form $\alpha : \Gamma/2\Gamma \to \mu_2$ is even. Then there exists a direct sum decomposition $\Gamma/2\Gamma = I_1 \oplus I_2$ such that $\alpha|_{I_1} = \alpha|_{I_2} = 1$. Now it is easy to show (exercise!) that such a decomposition can be lifted to an isotropic decomposition $\Gamma = \Gamma_1 \oplus \Gamma_2$. It remains to take $L = \mathbb{R}\Gamma_1$. □

## 5.3. Functional Equation

Let $(H, \Gamma, \alpha)$ be as in Section 5.1. Assume in addition that $\alpha^2 \equiv 1$ and the corresponding quadratic form $\overline{\alpha}$ is even. Recall that for every Lagrangian subspace $L \subset V$ compatible with $(\Gamma, \alpha)$ the theta series $\theta^{\alpha}_{H,\Gamma,L}$ generates the space $T(H, \Gamma, \alpha)$ (see Chapter 3). Thus, if we have another Lagrangian $L'$ compatible with $(\Gamma, \alpha)$ then we necessarily have

$$\theta^{\alpha}_{H,\Gamma,L'} = \lambda \cdot \theta^{\alpha}_{H,\Gamma,L}$$

for some constant $\lambda \in \mathbb{C}^*$. Our goal is to determine this constant (or at least its square).

For every pair $M_1, M_2$ of free $\mathbb{Z}$-modules of rank $n = \dim V$ in $V$ such that each $M_i$ generates $V$ over $\mathbb{C}$, we define $\det_{M_1}(M_2) \in \mathbb{C}^*/\{\pm 1\}$ as follows: choose arbitrary bases of $M_i$ and write the transition matrix from the basis in $M_1$ to that in $M_2$, then take its determinant. Up to a sign this number doesn't depend on a choice of bases in $M_i$. To get rid of the sign ambiguity one needs to choose an orientation in $M_1 \oplus M_2$ (then the choice of bases above should be compatible with this orientation).

**Theorem 5.2.** *One has*

$$\theta^{\alpha}_{H,\Gamma,L'} = \zeta \cdot \det_{\Gamma\cap L}(\Gamma \cap L')^{\frac{1}{2}} \theta^{\alpha}_{H,\Gamma,L}, \tag{5.3.1}$$

*where $\zeta^8 = 1$. Assume in addition that $L \cap L' = 0$. Then we have an orientation on $\Gamma \cap L \oplus \Gamma \cap L'$ induced by the symplectic form, hence the 4th root of unity $\zeta^2$ is well defined. In this situation one has*

$$\zeta^2 = i^n \cdot c(L', \Gamma, L)^2,$$

*where $c(L', \Gamma, L)$ is equal to the Gauss sum (4.5.2), $n = \dim_{\mathbb{C}} V$.*

*Proof of the theorem.* The proof is based on the formula

$$\exp\left(-\frac{\pi}{2}H\right)\theta^{\alpha}_{H,\Gamma,L} = R(L, \Gamma) f_L,$$

which follows from (5.1.1) since $\alpha^2 = 1$. We claim that

$$R(L, L') f_L = c \cdot f_{L'}, \tag{5.3.2}$$

where the measures on $V/L$ and $V/L'$ are normalized using lattices in these spaces and the constant $c \in \mathbb{C}^*$ differs from $\det_{\Gamma\cap L}(\Gamma \cap L')^{-\frac{1}{2}}$ by an 8th root of unity. Indeed, it suffices to prove this when $L \cap L' = 0$. In this case

$$R(L, L') f_L = d^{\frac{1}{2}} \cdot \int_{l' \in L'} f_L(x + l') \exp(\pi i E(l', x)) dl',$$

where $dl'$ is the measure on $L'$ with respect to which the covolume of $\Gamma \cap L'$ is equal to 1, $d = |\Gamma/(\Gamma \cap L + \Gamma \cap L')|$. Thus, the computation essentially reduces to the standard computation of the Fourier transform of the exponent of a quadratic function so we get (5.3.2) with

$$c = d^{\frac{1}{2}} \cdot \int_{L'} \exp\left(-\frac{\pi}{2}(H - S_L)(l', l')\right) dl' = \left(\frac{d}{\Delta}\right)^{\frac{1}{2}},$$

where

$$\Delta = \det\left(\frac{1}{2}(H - S_L)(e'_i, e'_j)\right)$$

for some basis $(e'_i)$ of $\Gamma \cap L'$. Let $(e_i)$ be a basis of $\Gamma \cap L$, such that $((e_i); (e'_j))$ is a positively oriented basis of $V$. Then we can write $e'_i = \sum_j a_{ij} e_j$ so that

$$\Delta = \det(a_{ij}) \cdot \det\left(\frac{1}{2}(H - S_L)(e_i, e'_j)\right).$$

Now we observe that

$$\frac{1}{2}(H - S_L)(l, v) = i \cdot E(l, v)$$

for any $l \in L$, $v \in V$. Therefore,

$$\det\left(\frac{1}{2}(H - S_L)(e_i, e'_j)\right) = i^n \det(E(e_i, e'_j)) = i^n |(\Gamma \cap L')^\vee/(\Gamma \cap L)|,$$

where the embedding $\Gamma \cap L \to (\Gamma \cap L')^\vee$ is induced by $E$. Since $\Gamma \cap L' \subset \Gamma$ is a Lagrangian sublattice, we have an isomorphism $(\Gamma \cap L')^\vee \simeq \Gamma/(\Gamma \cap L)$, so that

$$|(\Gamma \cap L')^\vee/(\Gamma \cap L)| = |\Gamma/(\Gamma \cap L + \Gamma \cap L')| = d.$$

It follows that

$$c = (i^n \det(a_{ij}))^{-\frac{1}{2}}$$

as we claimed.

Now we have

$$\begin{aligned}\exp\left(-\frac{\pi}{2}H(x,x)\right)\theta^{\alpha}_{H,\Gamma,L'}(x) &= R(L', \Gamma) f_{L'} = c^{-1} R(L', \Gamma) R(L, L') f_L \\ &= c^{-1} c(L, \Gamma, L')^{-1} R(L, \Gamma) f_L \\ &= c^{-1} c(L', \Gamma, L) \exp\left(-\frac{\pi}{2}H(x,x)\right)\theta^{\alpha}_{H,\Gamma,L}(x).\end{aligned}$$

It remains to note that $c(L', \Gamma, L)^8 = 1$ by Corollary 4.14. □

## 5.4. Group Action

Let $\Gamma$ be a self-dual lattice in a symplectic vector space $(V, E)$, $\alpha : \Gamma \to \pm 1$ be a map satisfying (1.2.2). Then we can consider the group $G(\Gamma, \alpha)$ of symplectic automorphisms of $\Gamma$ preserving $\alpha$.

**Theorem 5.3.** *The group $G(\Gamma, a)$ acts transitively on the set of Lagrangian subspaces $L \subset V$ that are compatible with $(\Gamma, \alpha)$.*

The proof is based on the following lemma.

**Lemma 5.4.** *Let $L$ be a Lagrangian subspace compatible with $(\Gamma, \alpha)$. Then there exists an $E$-isotropic decomposition $\Gamma = \Gamma_1 \oplus \Gamma_2$ such that $\Gamma_1 = \Gamma \cap L$ and $\alpha = \alpha_0(\Gamma_1, \Gamma_2)$.*

*Proof.* Let us set $\Gamma_1 = \Gamma \cap L$. First we claim that there exists an isotropic subgroup $\Gamma_2 \subset \Gamma$ such that $\Gamma = \Gamma_1 \oplus \Gamma_2$. Indeed, since the natural map $\Gamma^\vee \to \Gamma_1^\vee$ is surjective, from self-duality of $\Gamma$ we deduce that the natural map $\Gamma/\Gamma_1 \to (\Gamma \cap L)^\vee : x \mapsto E(x, \cdot)$ is an isomorphism. Let $\widetilde{\Gamma}_2 \subset \Gamma$ be some complement to $\Gamma_1$ in $\Gamma$, so that $\Gamma = \Gamma_1 \oplus \widetilde{\Gamma}_2$. Then the restriction of $E$ to $\Gamma_1 \times \widetilde{\Gamma}_2$ is a perfect pairing. Let us choose a bilinear form $B$ on $\widetilde{\Gamma}_2$ such that $E|_{\widetilde{\Gamma}_2 \times \widetilde{\Gamma}_2}$ is a skew-symmetrization of $B$, i.e., $E(x, y) = B(x, y) - B(y, x)$ for $x, y \in \widetilde{\Gamma}_2$. Let $f : \widetilde{\Gamma}_2 \to \Gamma_1$ be a homomorphism such that $E(f(x), y) = -B(x, y)$ for $x, y \in \widetilde{\Gamma}_2$. Then $\Gamma_2 = \{f(x) + x, x \in \widetilde{\Gamma}_2\}$ is an isotropic complement to $\Gamma_1$.

The map $\alpha$ differs from $\alpha_0(\Gamma_1, \Gamma_2)$ by a homomorphism $\Gamma \to \pm 1$ which is trivial on $\Gamma_1$. Now we claim that $\alpha = \alpha_0(\Gamma_1, \Gamma_2')$ for an appropriate isotropic sublattice $\Gamma_2'$ of the form $(f(\gamma_2) + \gamma_2, \gamma_2 \in \Gamma_2)$ where $f : \Gamma_2 \to \Gamma_1$ is a symmetric homomorphism (note that $\Gamma_1$ can be identified with $\Gamma_2^\vee$ via $E$). Indeed, we have

$$\alpha_0(\Gamma_1, \Gamma_2')(\gamma) = \alpha_0(\Gamma_1, \Gamma_2)(\gamma) \cdot \exp(\pi i E(f(\gamma_2), \gamma_2)),$$

where $\gamma = \gamma_1 + \gamma_2$ and $\gamma_i \in \Gamma_i$. It remains to notice that every homomorphism from $\Gamma_2$ to $\pm 1$ has form $\gamma_2 \mapsto \exp(\pi i E(f(\gamma_2), \gamma_2))$ for some symmetric homomorphism $f$. □

*Proof of Theorem 5.3.* By Lemma 5.4 it suffices to prove that $G(\Gamma, \alpha)$ acts transitively on the set of $E$-isotropic decompositions $\Gamma = \Gamma_1 \oplus \Gamma_2$ such that $\alpha = \alpha_0(\Gamma_1, \Gamma_2)$. Given two such decomposition $\Gamma = \Gamma_1 \oplus \Gamma_2 = \Gamma_1' \oplus \Gamma_2'$ we can choose a symplectic automorphism $g : \Gamma \to \Gamma$ such that $g(\Gamma_i) = \Gamma_i'$ for $i = 1, 2$. Since $\alpha$ is determined in terms of either of these two decompositions, the element $g$ will automatically belong to the subgroup $G(\Gamma, \alpha)$. □

Let us denote by $\mathcal{D}$ the space of complex structures $J$ on $V$ strictly compatible with $E$. Recall that strict compatibility means that $J$ is compatible with $E$ and $E(Jx, x) > 0$. Thus, $\mathcal{D}$ can be identified with an open subset in the Lagrangian Grassmannian of $V \otimes_{\mathbb{R}} \mathbb{C}$ (see Section 2.4). In particular, $\mathcal{D}$ has a natural complex structure. The group $G(\Gamma, \alpha)$ acts on $\mathcal{D}$ by holomorphic automorphisms: an element $g \in G(\Gamma, \alpha)$ sends $J \in \mathcal{D}$ to $gJg^{-1}$. We have the induced action of $G(\Gamma, \alpha)$ on the space $\mathcal{O}^*(\mathcal{D})$ of invertible holomorphic functions on $\mathcal{D}$: $g \cdot \phi(J) = \phi(g^{-1}Jg)$. For every $J \in \mathcal{D}$ let us denote by $H_J$ the $J$-Hermitian form on $V$ with $\operatorname{Im} H_J = E$. We want to look at $\theta$ as a function of $J$, so we fix $\Gamma$, $E$ and $\alpha$ once and for all, and rearrange our notation as follows: $\theta_L(v, J) := \theta^{\alpha}_{H_J,\Gamma,L}(v)$ where $L \subset V$ is a Lagrangian compatible with $(\Gamma, \alpha)$. Then for a pair of Lagrangians $(L, L')$ compatible with $(\Gamma, \alpha)$

and a complex structure $J \in \mathcal{D}$, we can consider the nonzero constant

$$c_{L,L'}(J) = \frac{\theta_{L'}(\cdot, J)}{\theta_L(\cdot, J)}.$$

Theorem 5.2 tells us that

$$c_{L,L'}(J) = \zeta \cdot \det{}^J_{\Gamma\cap L}(\Gamma \cap L')^{\frac{1}{2}},$$

where $\zeta^8 = 1$, and gives an expression for $\zeta^2$ in the case $L \cap L' = 0$ (the superscript $J$ means that the complex structure $J$ is used when computing the relative determinant). Now let us set $L' = gL$ for $g \in G(\Gamma, \alpha)$. Since $\theta_{gL}(v, J) = \theta_L(g^{-1}v, g^{-1}Jg)$, we have

$$\theta_L(g^{-1}v, g^{-1}Jg) = c_L(g)(J)\theta_L(v, J), \tag{5.4.1}$$

where $c_L(g)(J) := c_{L,gL}(J)$. This equation implies that $g \mapsto c_L(g)$ as a 1-cocycle of the group $G(\Gamma, \alpha)$ with coefficients in $\mathcal{O}^*(\mathcal{D})$. On the other hand, Theorem 5.3 implies that the cohomology class of $c_L(g)$ does not depend on $L$.

Let us choose an orientation on $L$. Then for every $g \in G(\Gamma, \alpha)$ we have the natural orientation on $gL$ induced by the isomorphism $g : L \to gL$. Let us define $\det^J_{(\Gamma\cap L)}(\Gamma \cap gL)$ using these orientations. Note that this number does not depend on a choice of orientation on $L$. Then we claim that

$$g \mapsto \left(J \mapsto \det{}^J_{\Gamma\cap L}(\Gamma \cap gL)\right)$$

is a 1-cocycle with values in $\mathcal{O}^*(\mathcal{D})$. For a complex number $z = x + iy$, a vector $v \in V$ and a complex structure $J \in \mathcal{D}$, let us denote $z *_J v := xv + yJv$. Then for any $g \in G(\Gamma, \alpha)$ we have $g(z *_J v) = z *_{gJg^{-1}} (gv)$. Using this observation we can prove our claim as follows. Let $(e_i)$ be a basis of $\Gamma \cap L$. Then $(ge_i)$ is a basis of $\Gamma \cap gL$. By the definition, $\det^J_{\Gamma\cap L}(\Gamma \cap gL) = \det A^g(J)$, where the complex matrix $A^g(J)$ is defined by $ge_i = \sum_j A^g(J)_{ij} *_J e_j$. Now for a pair of elements $g_1, g_2 \in G(\Gamma, \alpha)$ we have

$$\begin{aligned} g_1g_2e_i &= \sum_j g_1\left(A^{g_2}\left(g_1^{-1}Jg_1\right)_{ij} *_{g_1^{-1}Jg_1} e_j\right) = \sum_j A^{g_2}\left(g_1^{-1}Jg_1\right)_{ij} *_J (g_1e_j) \\ &= \sum_{j,k} \left(A^{g_2}\left(g_1^{-1}Jg_1\right)_{ij} A^{g_1}(J)_{jk}\right) *_J e_k. \end{aligned}$$

Hence,

$$A^{g_1g_2}(J) = A^{g_2}\left(g_1^{-1}Jg_1\right)A^{g_1}(J).$$

Passing to determinants we finish the proof of our claim.

Now the second part of Theorem 5.2 implies that for every $g \in G(\Gamma, \alpha)$ such that $gL$ is transversal to $L$, one has

$$c_L(g)(J)^2 = i^n c(gL, \Gamma, L)^2 \cdot \epsilon(g) \cdot \det{}^J_{\Gamma \cap L}(\Gamma \cap gL),$$

where $\epsilon(g) = \pm 1$ is the difference between the orientation of $V = L \oplus gL$ induced by some orientation of $L$ and the natural orientation of $V$ induced by the symplectic form ($\epsilon(g)$ doesn't depend on a choice of orientation on $L$). It follows that the map

$$g \mapsto i^n c(gL, \Gamma, L)^2 \epsilon(g)$$

extends to a character of $G(\Gamma, \alpha)$ of order 4. It is instructive to compare this conclusion with the result of Proposition 4.9. Let us set $L_1 = g_1 g_2 L$, $L_2 = g_1 L$, $L_3 = L$, where $g_1, g_2 \in G(\Gamma, \alpha)$ are such that $L_i$ are pairwise transversal. Then we derive that

$$\exp\left(-\frac{\pi i}{2} m(g_1 g_2 L, g_1 L, L)\right) = s(g_1 g_2) s(g_1)^{-1} s(g_2)^{-1}, \tag{5.4.2}$$

where

$$s(g) = i^n \epsilon(g)$$

for $g \in G(\Gamma, \alpha)$ such that $gL$ is transversal to $L$. Note that the lattice $\Gamma$ plays no role in this formula. In fact, it holds for arbitrary $g_1, g_2$ in the symplectic group, such that the Lagrangian subspaces $L_1, L_2, L_3$ are pairwise transversal (cf. [84], (1.7.8)).

### 5.5. Theta Series for Commensurable Lattices

In Section 5.3 we studied what happens with the theta series when we change a Lagrangian subspace $L$ (and fix all the other data). Now we are going to fix the data $(V, H, L)$ and change the lattice $\Gamma$ and its lifting $\alpha$. Namely, assume that $\Gamma'$ is another self-dual lattice (with respect to $E$) equipped with a lifting $\alpha'$ such that $\alpha'|_{\Gamma' \cap L} \equiv 1$. We assume also that $\Gamma$ and $\Gamma'$ are compatible, i.e., $\Gamma$ and $\Gamma'$ are commensurable and $\alpha|_{\Gamma \cap \Gamma'} = \alpha'|_{\Gamma \cap \Gamma'}$.

**Theorem 5.5.** *One has*

$$\theta^{\alpha'}_{H,\Gamma',L} = c(L, \Gamma', \Gamma) \cdot \frac{|\Gamma \cap L / \Gamma \cap \Gamma' \cap L|^{\frac{1}{2}}}{|\Gamma' \cap L / \Gamma \cap \Gamma' \cap L|^{\frac{1}{2}}} \cdot |\Gamma' / \Gamma \cap \Gamma'|^{-\frac{1}{2}}$$
$$\cdot \sum_{\gamma' \in \Gamma' / \Gamma \cap \Gamma'} U_{(\alpha'(\gamma'), \gamma')} \theta^{\alpha}_{H,\Gamma,L},$$

*where* $c(L, \Gamma', \Gamma)$ *is equal to the Gauss sum associated with the quadratic function* $q_{L,\Gamma',\Gamma}$ *on the finite group* $L \cap (\Gamma + \Gamma')/(L \cap \Gamma + L \cap \Gamma')$ *(see Section 4.3).*

*Proof.* This essentially follows from the equality (5.1.1) applied to $\Gamma$ and $\Gamma'$ and from the equality

$$R(L, \Gamma') = c(L, \Gamma', \Gamma)R(\Gamma, \Gamma') \circ R(L, \Gamma).$$

One only has to be careful about one point: the canonical measures on $V/L$ induced by the lattices $\Gamma/\Gamma \cap L$ and $\Gamma'/\Gamma' \cap L$ are different. This accounts for the constant $|\Gamma \cap L/\Gamma \cap \Gamma' \cap L|^{\frac{1}{2}}/|\Gamma' \cap L/\Gamma \cap \Gamma' \cap L|^{\frac{1}{2}}$. Also, we have to note that under the isomorphism $f \mapsto \exp(-\frac{\pi}{2}H(v, v))f(-v)$ between $\mathrm{Fock}(V, J)$ and $\mathcal{F}^+(J)$ the operators $U_h$ of the Fock representation correspond to the operators $\phi(h') \mapsto \phi(h^{-1}h')$. The latter operators are precisely the ones used in the definition of $R(\Gamma, \Gamma')$. □

**Corollary 5.6.** *Assume in addition that* $(\Gamma + \Gamma') \cap L = \Gamma \cap L + \Gamma' \cap L$. *Then*

$$\theta^{\alpha'}_{H,\Gamma',L} = |\Gamma' \cap L/\Gamma \cap \Gamma' \cap L|^{-1} \cdot \sum_{\gamma' \in \Gamma'/\Gamma \cap \Gamma'} U_{(\alpha'(\gamma'), \gamma')}\theta^{\alpha}_{H,\Gamma,L}$$

*Proof.* Indeed, according to Theorem 4.5, in this case $c(L, \Gamma', \Gamma) = 1$. On the other hand, since the lattice $\Gamma_0 = \Gamma \cap \Gamma' + (\Gamma + \Gamma') \cap L$ is self-dual, we have

$$|\Gamma'/\Gamma \cap \Gamma'| = |\Gamma_0/\Gamma \cap \Gamma'| = |(\Gamma \cap L + \Gamma' \cap L)/\Gamma \cap \Gamma' \cap L|.$$

Using the exact sequence

$$0 \to \Gamma \cap L/\Gamma \cap \Gamma' \cap L \to (\Gamma \cap L + \Gamma' \cap L)/\Gamma \cap \Gamma' \cap L$$
$$\to \Gamma' \cap L/\Gamma \cap \Gamma' \cap L \to 0$$

we get

$$\frac{|\Gamma \cap L/\Gamma \cap \Gamma' \cap L|^{\frac{1}{2}}}{|\Gamma'/\Gamma \cap \Gamma'|^{\frac{1}{2}}} = |\Gamma' \cap L/\Gamma \cap \Gamma' \cap L|^{-\frac{1}{2}},$$

which gives the required form of the constant factor. □

As in Section 5.4, we can rewrite the equation of Theorem 5.5 in terms of the action of the symplectic group on $\mathcal{D}$. Namely, let $\mathrm{Sp}(\Gamma \otimes \mathbb{Q})$ be the group of symplectic automorphisms of $\Gamma \otimes \mathbb{Q}$. Then for a self-dual lattice $\Gamma \subset V$ and for an element $g \in \mathrm{Sp}(\Gamma \otimes \mathbb{Q})$, the lattice $g\Gamma$ is again self-dual and commensurable with $\Gamma$. Thus, if we start with the data $((\Gamma,\alpha), J, L)$ as above and set $\Gamma'=g\Gamma$, $\alpha'=\alpha\circ g^{-1}$, then we can apply the equation above provided that $\alpha'|_{\Gamma'\cap L}\equiv 1$ and $\alpha'|_{\Gamma\cap\Gamma'}=\alpha|_{\Gamma\cap\Gamma'}$. The former condition is equivalent to the compatibility of $g^{-1}L$ with $(\Gamma,\alpha)$. The latter condition is equivalent to the equality $\alpha(\gamma)=\alpha(g(\gamma))$ for $\gamma\in g^{-1}\Gamma\cap\Gamma$. Denoting as before $\theta_L(v, J)=\theta^{\alpha}_{H_J,\Gamma,L}(v)$ (where $(\Gamma,\alpha)$ is fixed) and using the equality $\theta^{\alpha\circ g^{-1}}_{H_J,g\Gamma,L}(v)=\theta_{g^{-1}L}(g^{-1}v, g^{-1}Jg)$, we can rewrite the equation of Theorem 5.5 in the form

$$\theta_{g^{-1}L}(g^{-1}v, g^{-1}Jg) = c(L, g\Gamma, \Gamma)\cdot\frac{|\Gamma\cap L/\Gamma\cap g\Gamma\cap L|^{\frac{1}{2}}}{|g\Gamma\cap L/\Gamma\cap g\Gamma\cap L|^{\frac{1}{2}}}$$
$$\cdot|g\Gamma/\Gamma\cap g\Gamma|^{-\frac{1}{2}}\cdot\sum_{\gamma\in\Gamma/g^{-1}\Gamma\cap\Gamma}\left(U^J_{(\alpha(\gamma),g\gamma)}\theta_L\right)(v, J)$$

(where $U^J$ denote the operators of the Fock representation associated with $J$). On the other hand, since the Lagrangian $g^{-1}L$ is compatible with $(\Gamma,\alpha)$, we have

$$\theta_L(g^{-1}v, g^{-1}Jg) = c_{g^{-1}L,L}(g^{-1}Jg)\cdot\theta_{g^{-1}L}(g^{-1}v, g^{-1}Jg).$$

Combining this with the previous equation we get

$$\theta_L(g^{-1}v, g^{-1}Jg)=c_{g^{-1}L,L}(g^{-1}Jg)\cdot c(L, g\Gamma,\Gamma)\times\frac{|\Gamma\cap L/\Gamma\cap g\Gamma\cap L|^{\frac{1}{2}}}{|g\Gamma\cap L/\Gamma\cap g\Gamma\cap L|^{\frac{1}{2}}}$$
$$\cdot|g\Gamma/\Gamma\cap g\Gamma|^{-\frac{1}{2}}\cdot\sum_{\gamma\in\Gamma/g^{-1}\Gamma\cap\Gamma}\left(U^J_{(\alpha(\gamma),g\gamma)}\theta_L\right)(v, J).$$

Let us replace $g$ by $g^{-1}$. Then we obtain the functional equation

$$\begin{aligned}&\theta_L(gv, gJg^{-1})\\&\quad= c_{gL,L}(gJg^{-1})\cdot c(L, g^{-1}\Gamma,\Gamma)\times\frac{|\Gamma\cap L/\Gamma\cap g^{-1}\Gamma\cap L|^{\frac{1}{2}}}{|g^{-1}\Gamma\cap L/\Gamma\cap g^{-1}\Gamma\cap L|^{\frac{1}{2}}}\\&\qquad\cdot|\Gamma/\Gamma\cap g\Gamma|^{-\frac{1}{2}}\cdot\textstyle\sum_{\gamma\in\Gamma/g\Gamma\cap\Gamma}\left(U^J_{(\alpha(\gamma),g^{-1}\gamma)}\theta_L\right)(v, J),\end{aligned}\tag{5.5.1}$$

where an element $g\in\mathrm{Sp}(\Gamma\otimes\mathbb{Q})$ is such that $gL$ is compatible with $(\Gamma,\alpha)$ and $\alpha(g\gamma)=\alpha(\gamma)$ whenever both parts are defined. In the case when $g$ preserves the lattice $\Gamma$, this is the usual functional equation. On the other hand, if $g\in\mathrm{Sp}(\Gamma\otimes\mathbb{Q})$ is such that $gL=L$ (and $\alpha(g\gamma)=\alpha(\gamma)$ for $\gamma\in\Gamma\cap g^{-1}\Gamma$), then the equation simplifies as follows:

$$\theta_L(gv, gJg^{-1}) = c(L, g^{-1}\Gamma, \Gamma) \cdot \frac{|g\Gamma \cap L/g\Gamma \cap \Gamma \cap L|^{\frac{1}{2}}}{|\Gamma \cap L/g\Gamma \cap \Gamma \cap L|^{\frac{1}{2}}} \cdot |\Gamma/\Gamma \cap g\Gamma|^{-\frac{1}{2}}$$
$$\times \sum_{\gamma \in \Gamma/g\Gamma \cap \Gamma} \left(U^J_{(\alpha(\gamma), g^{-1}\gamma)}\theta_L\right)(v, J).$$

Another interesting special case is when $gJg^{-1} = J$. Then $g$ can be considered as an element of $\mathrm{End}(A) \otimes \mathbb{Q}$, where $A = V/\Gamma$ is an abelian variety corresponding to $J$, such that $g$ preserves the polarization on $A$. In this case, the equation (5.5.1) is an explicit form of the action of $g$ on theta series.

## 5.6. Classical Theta Series

In this section following the tradition, we denote the dimension of $V$ by $g$ (hopefully this will not lead to confusion with the notation for a group element).

Let us fix a symplectic vector space $(V, E)$ with a self-dual lattice $\Gamma$ and a map $\alpha : \Gamma \to \pm 1$ satisfying (1.2.2). Assume in addition that $\alpha$ is even. Then according to Theorem 5.1 and Lemma 5.4, we can choose an isotropic decomposition $\Gamma = \Gamma_1 \oplus \Gamma_2$ such that $\alpha = \alpha_0(\Gamma_1, \Gamma_2)$. We set $L_i = \mathbb{R}\Gamma_i$, $i = 1, 2$. Let us choose a basis $e_1, \dots, e_g$ of $\Gamma_2$. Let $f_1, \dots, f_g$ be the dual basis of $\Gamma_1$, so that $E(f_i, e_j) = \delta_{ij}$. Let $J$ be a complex structure on $V$ strictly compatible with $E$. Then $J$ is determined uniquely by the complex Lagrangian subspace $P_J \subset V \otimes_{\mathbb{R}} \mathbb{C}$, consisting of elements $v \otimes 1 + Jv \otimes i$, where $v \in V$. Since $P_J \cap L_i \otimes_{\mathbb{R}} \mathbb{C} = 0$ for $i = 1, 2$, $P_J$ is the graph of a symmetric homomorphism $\phi_J : L_1 \otimes_{\mathbb{R}} \mathbb{C} \to L_2 \otimes_{\mathbb{R}} \mathbb{C}$. This homomorphism can be described by a symmetric matrix $Z = (Z_{ij}) \in \mathrm{Mat}_g(\mathbb{C})$ such that $\phi(f_j \otimes 1) = -\sum_i e_i \otimes Z_{ij}$. We claim that $Z$ is also characterized by the equations

$$f_j = \sum_{i=1^g} Z_{ij} *_J e_i, \ j = 1, \dots, g. \tag{5.6.1}$$

Indeed, by the definition, $P_J$ is spanned by the elements $f_j \otimes 1 - \sum_i e_i \otimes Z_{ij}$. It remains to use a natural isomorphism of $\mathbb{C}$-vector spaces $V_J \widetilde{\to} V \otimes_{\mathbb{R}} \mathbb{C}/P_J$ where $V_J$ denotes $V$ considered as a complex vector space via $J$. It is easy to check that if we identify $V$ with $\mathbb{C}^g$ using $(e_i)$ as the complex basis of $V$, then the $J$-Hermitian form $H$ such that $\mathrm{Im}\, H = E$ is given by the matrix $(\mathrm{Im}\, Z)^{-1}$. Also, under this identification the lattice $\Gamma$ becomes $Z\mathbb{Z}^g \oplus \mathbb{Z}^g \subset \mathbb{C}^g$ with $\Gamma_1 = Z\mathbb{Z}^g, \Gamma_2 = \mathbb{Z}^g$. Therefore, the isomorphism class of data $(V, E, J, \Gamma, \alpha)$

can be recovered from the matrix $Z$ which should be an element of the Siegel upper half-space $\mathfrak{H}_g$, consisting of complex $g \times g$ matrices, such that $Z^t = Z$ and $\operatorname{Im} Z > 0$.

Conversely, for every element $Z \in \mathfrak{H}_g$ we have the lattice $\Gamma(Z) = Z\mathbb{Z}^g + \mathbb{Z}^g$ in $\mathbb{C}^g$ and the Hermitian form $H_Z$ on $\mathbb{C}^g$ given by the matrix $(\operatorname{Im} Z)^{-1}$ in the standard basis $(e_i)$, such that $f_1 = Ze_1, \ldots, f_g = Ze_g, e_1, \ldots, e_g$ is the symplectic basis of $\Gamma(Z)$ (with respect to the symplectic form $E_Z = \operatorname{Im} H_Z$). In particular, the decomposition $\Gamma(Z) = Z\mathbb{Z}^g \oplus \mathbb{Z}^g$ is isotropic, so we have the corresponding map $\alpha_0 : \Gamma(Z) \to \{\pm 1\}$. One also has the corresponding decomposition $\mathbb{C}^g = Z\mathbb{R}^g \oplus \mathbb{R}^g$ into Lagrangian summands. For $x \in \mathbb{C}^g$ we use the notation $x = Zx_1 + x_2$ where $x_1, x_2 \in \mathbb{R}^g$. Now for any $c \in \mathbb{C}^g$ one has

$$U_{(1,c)}\theta^{\alpha_0}_{H_Z,\Gamma(Z),\mathbb{R}^g}(x) = \exp\left(\frac{\pi}{2}S_Z(x,x) - \pi i (c_1)^t \cdot c_2\right)\theta\begin{bmatrix} c_1 \\ c_2 \end{bmatrix}(x, Z),$$

where $S_Z(x, y) = x^t(\operatorname{Im} Z)^{-1}y$, $\theta\left[\begin{smallmatrix} c_1 \\ c_2 \end{smallmatrix}\right](x, Z)$ is the classical theta series with characteristics:

$$\theta\begin{bmatrix} c_1 \\ c_2 \end{bmatrix}(x, Z) = \sum_{l \in \mathbb{Z}^g} \exp(\pi i (l + c_1)^t Z (l + c_1) + 2\pi i (x + c_2)(l + c_1))$$

for $x \in \mathbb{C}^g$. In particular,

$$\theta^{\alpha_0}_{H_Z,\Gamma(Z),\mathbb{R}^g}(x) = \exp\left(\frac{\pi}{2}S_Z(x,x)\right)\theta(x, Z),$$

where $\theta(x, Z) := \theta\left[\begin{smallmatrix} 0 \\ 0 \end{smallmatrix}\right](x, Z)$. Now let us fix the data $(V, E, \Gamma, \alpha)$ and the symplectic basis $(f_1, \ldots, f_g, e_1, \ldots, e_g)$ as above. The correspondence $J \mapsto Z$ defines an isomorphism $\mathcal{D} \simeq \mathfrak{H}_g$. In particular, the group $\operatorname{Sp}(\Gamma)$ of symplectic automorphisms of $\Gamma$ acts on $\mathfrak{H}_g$. To write the formula for this action we represent an element $T \in \operatorname{Sp}(\Gamma)$ by a $2g \times 2g$ matrix with respect to the basis $(f_1, \ldots, f_g, e_1, \ldots, e_g)$. We write this matrix in the block form: $T = \left(\begin{smallmatrix} A & B \\ C & D \end{smallmatrix}\right)$, where $A, B, C, D \in \operatorname{Mat}(g, \mathbb{Z})$. Let us set $J' = TJT^{-1}$ and let $Z'$ be the corresponding element of the Siegel space. Applying $T$ to (5.6.1) we get

$$T(f_j) = \sum_{i=1^g} Z_{ij} *_{J'} T(e_i).$$

Substituting the expressions for $T(e_i)$ and $T(f_j)$ in terms of $A, B, C, D$ and using (5.6.1) for the data $(J', Z')$ we get the matrix equation

$$Z'A + C = DZ + Z'BZ.$$

Hence, $T(Z) = Z' = (DZ - C)(-BZ + A)^{-1}$. Now we are going to interpret the functional equation (5.4.1) as the transformation formula for the classical theta series considered as a function of $Z \in \mathfrak{H}_g$ and $v \in \mathbb{C}^g$ under the action of the subgroup of $\mathrm{Sp}(\mathbb{Z}^{2g})$ preserving the quadratic form $\sum_{i=1}^{g} x_i y_i$ modulo 2 (we will denote this subgroup by $G_{\theta,g}$). Let us rewrite (5.4.1) as

$$\theta_L(v, J) = c_L(T)(J')\theta_L(Tv, J'),$$

where $T \in G_{\theta,g}$. Note that the identification of $V$ with $\mathbb{C}^g$ corresponding to the complex structure $J$ sends the Lagrangian subspace $L$ spanned by $e_1, \dots, e_g$ to $\mathbb{R}^g$. Moreover, the real coordinates $\left[\begin{smallmatrix} x_1 \\ x_2 \end{smallmatrix}\right]$ on $\mathbb{C}^g$ given by $x = Zx_1 + x_2$ are precisely the coordinates with respect to the basis $(f_1, \dots, f_g, e_1, \dots, e_g)$. Thus, we get

$$\exp\left(\frac{\pi}{2} S_Z(x, x)\right)\theta(x, Z) = c_L(T)(J')\exp\left(\frac{\pi}{2} S_{Z'}(x', x')\right)\theta(x', Z'), \quad (5.6.2)$$

where $Z' = T(Z)$, $x' = Z'(Ax_1 + Bx_2) + (Cx_1 + Dx_2)$. We are going to express everything in terms of $x'$ and $Z'$. First, we have $Z = (Z'B + D)^{-1}$ $(Z'A + C)$. Next, we have

$$x' = (Z'A + C)x_1 + (Z'B + D)x_2 = (Z'B + D)(Zx_1 + x_2) = (Z'B + D)x,$$

hence $x = (Z'B + D)^{-1}x'$. Also, $c_L(T)(J') = \zeta \cdot \det(Z'B + D)^{\frac{1}{2}}$, where $\zeta^8 = 1$. It remains to compute $S_{Z'}(x', x') - S_Z(x, x)$. This is done in the following lemma.

**Lemma 5.7.** *One has*

$$S_{Z'}(x', x') - S_Z(x, x) = 2i(x')^t \cdot B(Z'B + D)^{-1} \cdot x'.$$

*Proof.* First we claim that $H_{Z'}(x', x') = H_Z(x, x)$. Indeed, this follows from equalities

$$H_J(v, v) = E(Jv, v) = E((TJT^{-1})(Tv), Tv) = H_{TJT^{-1}}(Tv, Tv).$$

Next, we observe that

$$(S_Z - H_Z)(x, x) = (S_Z - H_Z)(x, Zx_1) = 2ix^t \cdot x_1,$$

where $x = Zx_1 + x_2$. Therefore,

$$S_{Z'}(x', x') - S_Z(x, x) = 2i \cdot [(x')^t \cdot x_1' - x^t \cdot x_1].$$

Note that $T^{-1} = \left(\begin{smallmatrix} D^t & -B^t \\ -C^t & A^t \end{smallmatrix}\right)$. Hence, $x_1 = D^t x'_1 - B^t x'_2$. On the other hand, $x = (Z'B + D)^{-1}x'$. Therefore,

$$\begin{aligned} x^t \cdot x_1 &= (x')^t (B^t Z' + D^t)^{-1}(D^t x'_1 - B^t x'_2) \\ &= (x')^t (B^t Z' + D^t)^{-1}((B^t Z' + D^t)x'_1 - B^t x') \\ &= (x')^t \cdot x'_1 - (x')^t (B^t Z' + D^t)^{-1} B^t x' \\ &= (x')^t \cdot x'_1 - (x')^t B(Z'B + D)^{-1} x' \end{aligned}$$

and the result follows. □

Combining the above calculations and replacing $(x', Z')$ by $(x, Z)$, we can rewrite the equation (5.6.2) in the form

$$\begin{aligned} &\theta((ZB + D)^{-1}x, (ZB + D)^{-1}(ZA + C)) \\ &\quad = \zeta \cdot \det(ZB + D)^{\frac{1}{2}} \exp(\pi i [x^t \cdot B(ZB + D)^{-1} \cdot x])\theta(x, Z), \end{aligned} \tag{5.6.3}$$

where $\zeta^8 = 1$. This is the classical form of the functional equation for theta series.

**Remark.** The above construction identifies the set of isomorphism classes of data $(V, H, \Gamma, \alpha)$ where $\alpha^2 \equiv 1$ and $\overline{\alpha}$ is even, with the quotient of $\mathfrak{H}_g$ by $G_{\theta,g}$ (see Exercise 3). This quotient should be considered as an orbifold (roughly speaking, this is an additional structure that remembers stabilizer subgroups of points in $\mathfrak{H}_g$ under the action of $G_{\theta,g}$). The functional equation (5.6.3) implies that $\theta(0, Z)$ defines a section of a line bundle $L$ on $\mathfrak{H}_g/G_{\theta,g}$. Moreover, the pull-back of this line bundle to $\mathfrak{H}_g$ is trivial and (5.6.3) gives an expression for the corresponding 1-cocycle of $G_{\theta,g}$ with values in $\mathcal{O}^*(\mathfrak{H}_g)$. One can deduce from this that $L^8$ is isomorphic to $\omega^4$, where $\omega$ is the determinant of the relative cotangent bundle of the universal abelian variety over $\mathfrak{H}_g/G_{\theta,g}$ (see [90]).

## Exercises

1. (a) Let $q : (\mathbb{Z}/2\mathbb{Z})^{2g} \to \mu_2$ be a nondegenerate quadratic form. Let $q^{-1}(1)$ be the set of vectors $v \in (\mathbb{Z}/2\mathbb{Z})^{2g}$ such that $q(v) = 1$. Show that

$$|q^{-1}(1)| = \begin{cases} 2^{g-1}(2^g + 1), & q \text{ is even}, \\ 2^{g-1}(2^g - 1), & q \text{ is odd}. \end{cases}$$

(b) Prove that $\gamma(q) = 1$ if $q$ is even and $\gamma(q) = -1$ if $q$ is odd. Thus, $\gamma(q)$ coincides with the Arf-invariant of $q$.
(c) Show that $q$ is even if and only if there exists a subgroup $I \subset (\mathbb{Z}/2\mathbb{Z})^{2g}$ of order $2^g$ such that $q|_I \equiv 1$.

2. (a) For a pair of Lagrangians $L, L' \subset V$ there is a canonical orientation of the space $L \oplus L'$ induced by the isomorphism

$$L/(L \cap L') \oplus L'/(L \cap L') \simeq (L + L')/(L \cap L')$$

and by the symplectic structure on the latter space. Thus, if $L$ and $L'$ are compatible with $(\Gamma, \alpha)$ then we can correctly define $\det_{\Gamma \cap L}(\Gamma \cap L')$. Show that

$$\zeta^2 = i^{n-\dim(L \cap L')} \cdot c(L', \Gamma, L)^2$$

where $\zeta$ is the root of unity in the functional equation (5.3.1).
(b) Deduce from (a) that (5.4.2) holds for arbitrary $g_1, g_2 \in G(\Gamma, \alpha)$, where

$$s(g) = i^{n-\dim(L \cap gL)} \cdot \epsilon(g),$$

(as before $\epsilon(g)$ is the difference between two natural orientations on $L \oplus gL$).

3. (a) Show that isomorphism classes of data $(V, H, \Gamma, \alpha)$, where $\alpha^2 \equiv 1$ and $\overline{\alpha}$ is even, are in bijection with the quotient of $\mathfrak{H}_g$ by the action of $\Gamma_{\theta,g}$.
(b) Write explicitly an isomorphism of two data

$$(\mathbb{C}^g, H_Z, \Gamma(Z), \alpha_{0,Z}) \simeq (\mathbb{C}^g, H_{T(Z)}, \Gamma(T(Z)), \alpha_{0,T(Z)})$$

for $T = \left(\begin{smallmatrix} A & B \\ C & D \end{smallmatrix}\right)$ in $\Gamma_{\theta,g}$.

4. Show that every isomorphism class of data $(V, H, \Gamma, \alpha)$, where $\alpha^2 \equiv 1$ and $\overline{\alpha}$ is *odd*, comes from some element $Z \in \mathfrak{H}_g$ as follows: $V = \mathbb{C}^g$, $H = (\operatorname{Im} Z)^{-1}$, $\Gamma = Z\mathbb{Z}^g + \mathbb{Z}^g$, $\alpha(Zn_1 + n_2) = (-1)^{n_1 \cdot n_2 + (n_1)_1 + (n_2)_2}$ where $n_1, n_2 \in \mathbb{Z}^g$.

5. (a) Let $\Gamma_\tau \subset \mathbb{C}$ be the lattice $\mathbb{Z} + \mathbb{Z}\tau$, where $\tau = \alpha + i\beta$ is an element of the upper half-plane. For $\omega = m + n\tau \in \Gamma_\tau$ let us denote $\chi(\omega) = (-1)^{mn}$. Prove that there exists a constant $c(\tau)$ such that

$$\sum_{\omega \in \Gamma_\tau} \chi(\omega) \exp\left(-\frac{\pi}{2\beta}(|\omega|^2 + 2\overline{\omega}u + u^2)\right) = c(\tau)\theta(u, \tau).$$

(b) Show that on the other hand,

$$\sum_{\omega\in\Gamma_\tau}\chi(\omega)\exp\left(-\frac{\pi}{2\beta}(|\omega|^2+2\overline{\omega}u+u^2)\right)=\theta\left(\left(\frac{\overline{\tau}i}{2\beta}u,\frac{i}{2\beta}u\right),Z\right),$$

where $Z$ is the following element in $\mathfrak{H}_2$:

$$Z=\frac{i}{2\beta}\begin{pmatrix}|\tau|^2 & \overline{\tau}\\ \overline{\tau} & 1\end{pmatrix}.$$

(c) Prove that the decomposition of $\mathbb{C}^2$ into the direct sum of the subspaces $(z_1=\tau z_2)$ and $(z_1=-\overline{\tau}z_2)$ is compatible with the lattice $\mathbb{Z}^2+Z\mathbb{Z}^2$. Then use the functional equation to deduce that

$$\sum_{\omega\in\Gamma_\tau}\chi(\omega)\exp\left(-\frac{\pi}{2\beta}|\omega|^2\right)=\zeta\cdot\sqrt{2\beta}\theta(0,\tau)\theta(0,-\overline{\tau}),$$

where $\zeta=\pm1$. Considering the limit $\beta\to\infty$ show that $\zeta=1$. Combining this with (a) we get the equality

$$\sum_{\omega\in\Gamma_\tau}\chi(\omega)\exp\left(-\frac{\pi}{2\beta}(|\omega|^2+2\overline{\omega}u+u^2)\right)=\sqrt{2\beta}\theta(0,-\overline{\tau})\theta(u,\tau).$$

(d) Prove the following identity (a particular case of Proposition 4.1 of [125]):

$$\sum_{\omega\in\Gamma_\tau}\chi(\omega)\exp\left(-\frac{\pi}{2\beta}(|\omega|^2+2\overline{\omega}u+2\omega\overline{v}+\overline{v}^2+2\overline{v}u+u^2)\right)$$
$$=\sqrt{2\beta}\theta(\overline{v},-\overline{\tau})\theta(u,\tau).$$

# 6

# Mirror Symmetry for Tori

In this chapter we describe a correspondence between symplectic tori and complex tori that is a particular case of mirror symmetry (see Section 6.1 for some discussion of the relevant parts of mirror symmetry). We also construct for such a dual pair $(T, T')$ (where $T$ is a symplectic torus, $T'$ is a complex torus) a correspondence between Lagrangian subspaces in $T$ and holomorphic vector bundles on $T'$. The natural context for these constructions is the following situation. Let $M$ be a symplectic manifold equipped with a fibration $p : M \to B$. We assume that every fiber of $p$ is a Lagrangian submanifold of $M$ and is isomorphic to a torus (we also usually assume that a Lagrangian section of $p$ is fixed). Then the total space of the dual torus fibration $p^\vee : M^\vee \to B$ has a natural complex structure. Also, there is a natural line bundle with connection $(\mathcal{P}, \nabla)$ on $M \times_B M^\vee$, such that its restriction to the fiber of the projection $M \times_B M^\vee \to M^\vee$ over a point $\xi \in M^\vee$ is the local system on the torus $p^{-1}(p^\vee(\xi))$ corresponding to $\xi$. To a Lagrangian submanifold $L \subset M$ that intersects all fibers of $p$ transversally, we associate a holomorphic vector bundle $\mathrm{Four}(L)$ on $M^\vee$ (called the *Fourier transform of* $L$) in the following way. Consider the restriction of $(\mathcal{P}, \nabla)$ to $L \times_B M^\vee$, and then take the push-forward of this line bundle to $M^\vee$. The $(0, 1)$-component of the connection on this vector bundle induced by $\nabla$ is flat, so it defines a holomorphic structure. The Dolbeault complex of $\mathrm{Four}(L)$ can be identified with the complex of rapidly decreasing differential forms on an (infinite) unramified covering of $L$ with a modified de Rham differential. The construction of $\mathrm{Four}(L)$ can be generalized to include an additional datum, a local system on $\mathcal{L}$. Also, the symplectic form on $M$ can have a complex part (which is a closed 2-form on $M$). The main example we are interested in is when $M$ is a torus $T$ equipped with a constant symplectic form, $p : T \to T/L_0$ is the fibration associated with a linear Lagrangian subtorus $L_0 \subset T$. Then the dual manifold $M^\vee$ is a complex torus. If we allow the symplectic form on $M = T$ to have a complex part, then we can obtain every complex torus in this way. In the next chapter we will show how to compute in this situation the cohomology of Four($L$) on $M^\vee$, where $L$ is a Lagrangian subtorus in $T$, transversal to $L_0$.

It turns out that the mirror symmetry between symplectic and complex tori is related to the usual duality of complex tori. Namely, we prove in Section 6.5 that dual complex tori are mirror dual to the same symplectic torus (with two different Lagrangian fibrations). As is explained below, this fact provides a link between Kontsevich's homological mirror conjecture and the Fourier–Mukai transform which will be considered in Chapter 11.

## 6.1. Categories behind Mirror Symmetry

This is a little digression providing a general context and motivation for what we are doing. The reader may skip it without any harm for understanding the other parts of this chapter.

Mirror symmetry is a many-facet correspondence between symplectic and complex geometry developed to understand formulas for the numbers of rational curves on some 3-dimensional Calabi–Yau manifolds (complex algebraic manifolds with trivial canonical bundle) discovered by physicists, see [134] and references therein. These formulas involve Hodge theory on *mirror dual* Calabi–Yau manifolds. At present we do not have a precise definition of the notion of a mirror dual pair of Calabi–Yau manifolds that would encompass all the known examples of such pairs (for a definition in the context of toric geometry, see [10]). Despite this fact, there has been a remarkable progress on understanding the mirror symmetry phenomenon. On the one hand, the problem of counting rational curves on Calabi–Yau manifolds has been embedded in the general context of the theory of Gromov–Witten invariants; see [72]. Then using the localization techniques, many of these invariants were computed confirming the predictions made by physicists; see [50, 71]. On the other hand, M. Kontsevich [70] presented the idea of homological mirror symmetry which states that there should be an equivalence of categories behind mirror duality, one category being the derived category of coherent sheaves on a Calabi–Yau manifold, another – the Fukaya category associated with a mirror dual manifold (which was invented by K. Fukaya [43]). Roughly speaking, objects of the Fukaya category are Lagrangian submanifolds equipped with some additional structure, while the morphisms are given by the Floer homology. However, the complete details of the definition of Fukaya category are very complicated (for example, it is not really a category, but an $A_\infty$-category). The most up-to-date version of this definition can be found in [45]. Another important idea, proposed by

A. Strominger, S.-T. Yau, and E. Zaslow [126], is that mirror-dual Calabi–Yau manifolds should be fibered over the same base in such a way that generic fibers are dual tori and each fiber of any of these two fibrations is a Lagrangian submanifold. For generic Calabi–Yau hypersurfaces in toric varieties, such fibrations, called *Lagrangian torus fibrations*, were constructed by W.-D. Ruan (see [120]). Their topology was extensively studied by M. Gross (see [53]).

In the case of abelian varieties there is a precise definition of mirror duality (see [51], [87]) which agrees with the Strominger–Yau–Zaslow approach via dual torus fibrations. In this case a version of Kontsevich's homological mirror conjecture was proven by M. Kontsevich and Y. Soibelman [73]. The correspondence between Lagrangian submanifolds in the symplectic torus and holomorphic vector bundles on the mirror dual complex torus considered below is one of the building blocks of their proof.

Kontsevich's homological mirror conjecture implies that Calabi–Yau manifolds that have the same mirror, should have equivalent derived categories of coherent sheaves. This was indeed checked in some cases. The most spectacular result in this direction is the proof by T. Bridgeland of the fact that birational Calabi–Yau threefolds have equivalent derived categories; see [27]. Historically, the first example of an equivalence between derived categories of coherent sheaves on different varieties was the Fourier–Mukai transform which is an equivalence between derived categories for dual abelian varieties (it will be studied in Chapter 11). This equivalence fits nicely in the above context, since as we prove in Section 6.5, complex dual tori have the same mirror.

It will be more convenient for us to leave the category of algebraic manifolds and to consider below more general mirror dual pairs consisting of a complex torus and a symplectic torus. Furthermore, a large part of the theory is carried out for an arbitrary Lagrangian torus fibration with smooth fibers. However, the reader will not lose much by assuming that our manifolds are tori themselves, since in the compact case this is the only interesting example when such fibrations exist.

## 6.2. Mirror Dual to a Lagrangian Torus Fibration

Let $(M, \omega)$ be a symplectic manifold, $p : M \to B$ be a smooth map such that all fibers $p^{-1}(b)$ are Lagrangian tori (thus, the restriction of $\omega$ to every fiber is zero). We also always assume (except for Section 6.4) that there exists a $C^{\infty}$-section $\sigma : B \to M$ of $p$, such that the submanifold $\sigma(B) \subset M$ is Lagrangian. Examples of smooth Lagrangian torus fibrations without such a section are presented in Exercise 1.

**Proposition 6.1.** *In the described situation, there is a natural symplectic isomorphism $M \simeq T^*B/\Gamma$, where $\Gamma \subset T^*B$ is a relative lattice in the cotangent bundle to $B$, locally generated by closed 1-forms. Under this isomorphism the section $\sigma(B)$ corresponds to the zero section in $T^*B$.*

*Proof.* Since $T_x p^{-1}(b)$ is a Lagrangian subspace in $T_x M$ for $x \in p^{-1}(b)$, we have an isomorphism

$$T_b B = T_x M / T_x p^{-1}(b) \tilde{\to} T_x^* p^{-1}(b) : v \mapsto \omega(\cdot, v).$$

This gives an action of $T_b^* B$ on $p^{-1}(b)$. Since we have a marked point $\sigma(b) \in p^{-1}(b)$, we can identify $p^{-1}(b)$ with the quotient of $T_b^* B$ by some lattice in such a way that the action of $T_b^* B$ becomes the usual action by translations (and $\sigma(b)$ corresponds to zero in $T_b^* B$). This construction globalizes to an isomorphism $M \simeq T^*B/\Gamma$ for some relative lattice $\Gamma \subset T^*B$. Locally we can choose a basis of $T^*B$ consisting of closed forms. Since the corresponding vector fields on $M$ are hamiltonian, the constructed morphism $T^*B \to M$ is compatible with symplectic structures. Therefore, translations by local sections of the lattice $\Gamma$ preserve the canonical symplectic form on $T^*B$. Equivalently, all local sections of $\Gamma$ are closed 1-forms. □

The isomorphism of the above proposition induces an identification of the tangent sheaf[5] $T_B$ on $B$ with $R^1 p_* \mathbb{R}$, such that the lattice $\Gamma$ is dual to the lattice $\Gamma^\vee = R^1 p_* \mathbb{Z} \subset R^1 p_* \mathbb{R} \simeq T_B$. There is a natural flat connection on $R^1 p_* \mathbb{R}$ with respect to which the lattice $\Gamma^\vee$ is horizontal. It induces a flat connection on $T_B$. The fact that local sections of $\Gamma$ are closed 1-forms implies that this connection on $T_B$ is *symmetric* (= torsion-free), i.e., it admits locally a horizontal basis of commuting vector fields. Therefore, the induced flat connection on $T_M$ is also symmetric.

From now on we assume that in addition $M$ is equipped with a covariantly constant 2-form $\xi$ and we set $\Omega = \omega + i\xi$ ($\Omega$ is automatically closed). We will refer to $\Omega$ as *complexified symplectic form* on $M$.

Now we consider the family of dual tori

$$M^\vee = TB/\Gamma^\vee \xrightarrow{p^\vee} B.$$

The connection on $T_B$ gives an isomorphism $T_y M^\vee = T_b B \oplus T_b B$ for any $y \in (p^\vee)^{-1}(b)$, such that the map $d_y p^\vee : T_y M^\vee \to T_b B$ coincides with the

[5] We reserve the notation $TB$ for the total space of the tangent bundle to $B$, while $T_B$ denotes the tangent sheaf.

projection to the first summand. Now we define an isomorphism

$$T_bB \oplus T_bB \xrightarrow{\sim} \mathrm{Hom}_{\mathbb{R}}(T_x p^{-1}(b), \mathbb{C}) : (v_1, v_2) \mapsto \Omega(\cdot, v_1) + i\omega(\cdot, v_2).$$

Using this isomorphism we obtain a complex structure $J$ on the space $T_bB \oplus T_bB$. Here is an explicit formula:

$$J(v_1, v_2) = (-v_2 - Av_1, v_1 + Av_2 + A^2v_1),$$

where $A : T_bB \to T_bB$ is the operator defined by the condition $\omega(\cdot, Av) = \xi(\cdot, v)$. Equivalently, this complex structure can be described by specifying the antiholomorphic subspace $P_J \subset T_yM^\vee{}_{\mathbb{C}}$. If we identify $T_yM^\vee$ with $T_bB \oplus T^*_{\sigma(b)}p^{-1}(b)$, then $P_J = \{(v, i\Omega(\cdot, v)), v \in T_bB_{\mathbb{C}}\}$.

**Proposition 6.2.** *The almost complex structure on $M^\vee$ defined above is integrable.*

*Proof.* This follows essentially from the fact that the connection on $T_B$ is symmetric and $\Omega$ is horizontal. The details are left to the reader. □

**Definition.** $M^\vee$ considered as a complex manifold, is called *mirror dual* to $(M, \Omega)$.

**Remarks.** 1. The summands of the decomposition $T_xM = T_{p(x)}B \oplus T_xp^{-1}(p(x))$ define integrable distributions on $M$, so we can decompose our 2-form $\Omega$ into types with respect to this decomposition: $\Omega = \Omega_{bb} + \Omega_{bf} + \Omega_{ff}$ (the subscript $b$ stands for "base", while $f$ stands for "fiber"). The definition of the complex structure on $M^\vee$ depends only on $\Omega_{bf}$, so we can assume from the start that $\Omega_{bb} = \Omega_{ff} = 0$.

2. It is easy to see that the construction of $M^\vee$ and the complex structure on it do not depend on a choice of a Lagrangian section $\sigma : B \to M$. This section will play an important role below in the construction of the real analogue of the Poincaré bundle.

## 6.3. Fourier Transform

There is a natural line bundle with connection $(\mathcal{P}, \nabla)$ on $M \times_B M^\vee$, which comes from the interpretation of fibers of $p^\vee$ as moduli spaces of unitary rank-1 local systems on the fibers of $p$. Here is an explicit construction. Using connections on $T^*B$ and $TB$ we can choose local isomorphisms $M \simeq X \times B$, $M^\vee \simeq X^\vee \times B$, where $X = V/\Gamma$ is a torus, $X^\vee$ is a dual torus, in such a way that $\sigma$ corresponds to $0 \times B \subset X \times B$. By the definition, the pull-back of $\mathcal{P}$ to $V \times X^\vee \times B$ is trivial, with the $\Gamma$-equivariant structure given by

$$\gamma(f)(v, x^\vee, b) = \exp(-2\pi i \langle x^\vee, \gamma \rangle) f(v - \gamma, x^\vee, b),$$

where $\gamma \in \Gamma$, and with the connection $d + 2\pi i \langle dx^\vee, v \rangle$ (where $(v, x^\vee, b)$ are coordinates on $V \times X^\vee \times B$). The claim is that these local data glue into a global one (for this a choice of a section $\sigma$ is important). Note that $\nabla$ is not flat although its restrictions to fibers of $p$ and $p^\vee$ are flat (and have unitary monodromies).

**Remark.** Assume that the base $B$ is a point and that the torus $X = V/\Gamma$ has a complex structure. Then the dual torus $X^\vee$ can be identified with the complex dual of $X$ (see Chapter 1). Furthermore, in this case the $\overline{\partial}$-component of $\nabla$ is integrable, so $\mathcal{P}$ can be considered as a holomorphic line bundle. It is easy to see that in fact $\mathcal{P}$ coincides with the Poincaré line bundle defined in Chapter 1.

Let $L \subset M$ be a submanifold such that $p|_L : L \to B$ is an unramified covering, and let $(\mathcal{L}, \nabla_{\mathcal{L}})$ be a complex vector bundle with connection on $L$. We define the Fourier transform of $(L, \mathcal{L})$ by the formula

$$\mathrm{Four}(L, \mathcal{L}) := (p_{M^\vee})_*((i \times \mathrm{id})^* \mathcal{P} \otimes p_L^* \mathcal{L}), \tag{6.3.1}$$

where $p_{M^\vee} : L \times_B M^\vee \to M^\vee$, $(i \times id) : L \times_B M^\vee \to M \times_B M^\vee$, and $p_L : L \times_B M^\vee \to L$ are the natural maps. The map $p_{M^\vee}$ is an unramified covering, so $\mathrm{Four}(L, \mathcal{L})$ is a bundle with connection on $M^\vee$.

**Theorem 6.3.** *The $\overline{\partial}$-component of the connection on* $\mathrm{Four}(L, \mathcal{L})$ *is flat (so* $\mathrm{Four}(L, \mathcal{L})$ *can be considered as a holomorphic vector bundle on $M^\vee$) if and only if* $\mathrm{curv}\, \nabla_{\mathcal{L}} = 2\pi \Omega|_L$.

*Proof.* Let us denote $L \times_B M = N$. Since the map $p_{M^\vee} : N \to M^\vee$ is an unramified covering, we have the induced complex structure on $N$. Now the $(0, 1)$-component of the connection on $\mathrm{Four}(L, \mathcal{L})$ is flat if and only if the same is true for the connection $\widetilde{\nabla} = (i \times \mathrm{id})^* \nabla \otimes \mathrm{id} + \mathrm{id} \otimes p_L^* \nabla_{\mathcal{L}}$ on the bundle $(i \times \mathrm{id})^* \mathcal{P} \otimes (p_L)^* \mathcal{L}$ over $N$. The subspace of antiholomorphic tangent vectors in $T_y M^\vee_{\mathbb{C}} \simeq T_b B_{\mathbb{C}} \oplus T^*_{\sigma(b)} p^{-1}(b)_{\mathbb{C}}$ (where $p^\vee(y) = b$) consists of vectors of the form $(v, i\Omega(\cdot, v))$, where $v \in T_b B_{\mathbb{C}}$. Similarly, for $(l, y) \in N$ (where $l \in L$, $y \in M^\vee$) we can identify $T_{(l,y)} N$ with $T_l L \oplus T^*_{\sigma(b)} p^{-1}(b)$, so that antiholomorphic tangent vectors in $T_{(l,y)} N_{\mathbb{C}}$ correspond to vectors of the form $(v, i\Omega(\cdot, dp(v)))$, where $v \in T_l L_{\mathbb{C}}$. Note that $\frac{\mathrm{curv}(\nabla)}{2\pi i}$ is the skew-symmetric bilinear form on $T(M \times_B M^\vee)$ obtained as the pull-back under the map

$$T(M \times_B M^\vee) \to Tp^{-1}(b) \oplus T^* p^{-1}(b)$$

of the natural symplectic form on the latter space. Therefore, the restriction of $\mathrm{curv}(\nabla)/2\pi i$ to the space of antiholomorphic tangent vectors (identified with $T_l L_{\mathbb{C}}$) is equal to $i\Omega|_L$. Hence, the restriction of $\mathrm{curv}(\widetilde{\nabla})$ to this space is equal to $-2\pi\Omega|_L + \mathrm{curv}(\nabla_{\mathcal{L}})$. □

In particular, if $\Omega|_L = 0$ and the connection $\nabla_{\mathcal{L}}$ is flat then $\mathrm{Four}(L, \mathcal{L})$ is a holomorphic vector bundle on $M^\vee$.

## 6.4. Twisted Case

As we have already remarked, Lagrangian torus fibrations do not always admit a Lagrangian section. In general, such a section exists only locally. Let us choose a covering $(U_i)_{i\in I}$ of $B$ and Lagrangians sections $\sigma_i : U_i \to M$ of $p$ over $U_i$. Although the construction of the mirror dual $M^\vee$ does not depend on a choice of Lagrangian section, the Poincare bundle $\mathcal{P}$ does depend on such a choice (since we trivialize $\mathcal{P}$ along this section). Let $\mathcal{P}_i$ be the Poincare bundle over $p^{-1}(U_i) \times_{U_i} (p^\vee)^{-1}(U_i)$, then for any open subset $V \subset U_i$ we denote by $\mathrm{Four}_i^V$ the corresponding Fourier transform over $V$. Now let $U_{ij} = U_i \cap U_j$. Then we can define a holomorphic line bundle $P_{ij}$ on $(p^\vee)^{-1}(U_{ij})$ as follows:

$$P_{ij} = \mathrm{Four}_i^{U_{ij}}(\sigma_j(U_{ij})).$$

One can check that there are canonical isomorphisms $\alpha_{ijk} : P_{ij} \otimes P_{jk} \widetilde{\to} P_{ik}$ over $(p^\vee)^{-1}(U_i \cap U_j \cap U_k)$, which are compatible over quadruple intersections. Using these data we can define the following category $\mathcal{C}$ of twisted coherent sheaves on $M^\vee$. An object of $\mathcal{C}$ is a collection of coherent sheaves $Q_i$ over $(p^\vee)^{-1}(U_i)$ and a collection of isomorphisms $Q_i \otimes P_{ij} \simeq Q_j$ over $(p^\vee)^{-1}(U_{ij})$ that are compatible with isomorphisms $\alpha_{ijk}$ over $U_i \cap U_j \cap U_k$. Now if $L \subset M$ is a lagrangian submanifold, étale over $B$, then its local Fourier transforms $\mathrm{Four}_i(L|_{p^{-1}(U_i)})$ glue into a locally free object of $\mathcal{C}$.

**Remark.** The reader familiar with the language of *gerbes* may notice that our data define a gerbe on $M^\vee$ with the band $\mathcal{O}^*$ (see [49]). It represents certain cohomology class $e \in H^2(M^\vee, \mathcal{O}^*)$. The category $\mathcal{C}$ is the standard twist of the category of coherent sheaves by an $\mathcal{O}^*$-gerbe. Its objects can be alternatively described using a sufficiently fine covering $(V_i)$ as collections of sheaves $Q_i$ on $V_i$ equipped with isomorphisms on intersections $V_i \cap V_j$, which are not compatible on triple intersections, but differ by a Cech 2-cocycle representing $e$.

## 6.5. Mirror Duality between Symplectic and Complex Tori

In this section we specialize to the case when $M = V/\Gamma$ is a real torus, $\Omega = \omega + i\xi$ is a complexified (constant) symplectic form on $V$. Let $L \subset V$ be an $\Omega$-Lagrangian subspace, i.e., a linear subspace of dimension $\dim V/2$ such that $\Omega|_L \equiv 0$. We assume also that $L$ is generated over $\mathbb{R}$ by $\Gamma \cap L$. In this situation we can consider the Lagrangian torus fibration $p : V/\Gamma \to V/(L+\Gamma)$. It does not necessarily admits a Lagrangian section. However, as we observed in Section 6.4, the construction of the mirror dual complex manifold $M^\vee$ can still be carried out. Furthermore, in this case $M^\vee$ turns out to be a complex torus that we are going to describe explicitly. First, we note that the natural map

$$\alpha : V \oplus V^* \to \mathrm{Hom}_{\mathbb{R}}(V, \mathbb{C}) : (v, v^*) \mapsto (x \mapsto \Omega(x, v) + iv^*(x))$$

is an isomoprhism of real vector spaces. Indeed, if $\alpha(v, v^*) = 0$ then $\mathrm{Re}\,\alpha(v, v^*)(x) = \omega(x, v) = 0$ for any $x \in V$, hence $v = 0$. Then $\alpha(v, v^*) = iv^* = 0$. Let $L^\perp \subset V^*$ be the orthogonal complement to $L$. Then $\alpha$ maps the subspace $L \oplus L^\perp \subset V \oplus V^*$ to the subspace $\mathrm{Hom}_{\mathbb{R}}(V/L, \mathbb{C}) \subset \mathrm{Hom}_{\mathbb{R}}(V, \mathbb{C})$. Passing to quotient-spaces we get an isomorphism

$$\alpha_L : V/L \oplus L^* \to \mathrm{Hom}_{\mathbb{R}}(L, \mathbb{C}), \tag{6.5.1}$$

which induces a complex structure $J_\Omega$ on $W = V/L \oplus L^*$. Also, we have the lattice $\Lambda = \Gamma/\Gamma \cap L \oplus (\Gamma \cap L)^\perp$ in $W$. Now it is easy to check that the complex torus $W/\Lambda$ is mirror dual to $(V/\Gamma, \Omega)$.

Let us denote $V^0 = L \oplus V/L$. Then there is a complexified symplectic structure $\Omega^0$ on $V^0$ such that $L$ and $V/L$ are Lagrangian subspaces and $\Omega^0(l, v \bmod L) = \Omega(l, v)$. Also we have the lattice $\Gamma^0 = \Gamma \cap L \oplus \Gamma/\Gamma \cap L$ in $V^0$. It is clear that the mirror of $(V^0/\Gamma^0, \Omega^0, L)$ is the same complex torus $W/\Lambda$.

**Theorem 6.4.** *The above construction induces a bijective correspondence between the following two kinds of data.*

Data I. *A real torus $V/\Gamma$ of dimension $2n$, a complexified symplectic form $\Omega$ on $V$, a decomposition $V = L \oplus L'$, where $L$ and $L'$ are Lagrangian subspaces for $\Omega$, such that $\Gamma = \Gamma \cap L \oplus \Gamma \cap L'$.*

Data II. *A complex torus $W/\Lambda$ of complex dimension n, a decomposition $\Lambda = \Lambda_1 \oplus i\Lambda_2$ into sublattices of rank n, such that $\mathbb{C}\Lambda_2 = W$.*

*Proof.* In the above construction we have $W = V/L \oplus L^*$, $\Lambda_1 = \Gamma/\Gamma \cap L \subset V/L$, $\Lambda_2 = (\Gamma \cap L)^{\perp} \subset L^*$. Since $\alpha_L$ maps $\mathbb{R}\Lambda_2$ onto $iL^* \subset \mathrm{Hom}_{\mathbb{R}}(L, \mathbb{C})$ it follows that $\Lambda_2$ generates $W$ over $\mathbb{C}$.

To go from Data II to Data I we set $L = (\mathbb{R}\Lambda_2)^*$, $L' = \mathbb{R}\Lambda_1$, $\Gamma_L = (\Lambda_2)^{\perp} \subset L$, $\Gamma_{L'} = \Lambda_1 \subset L'$. Thus, $\Gamma = \Gamma_L \oplus \Gamma_{L'}$ is a lattice in $V = L \oplus L'$. Let

$$\nu : \mathbb{R}\Lambda_1 \to W \simeq \mathbb{R}\Lambda_2 \otimes_{\mathbb{R}} \mathbb{C}$$

be the natural embedding. Then we define $\Omega$ by requiring that $L$ and $L'$ are Lagrangian and by setting $\Omega(l, l') = \langle \nu(l'), l \rangle$ for $l \in L$, $l' \in L'$. □

There is a natural duality functor on the category of symplectic tori: a symplectic structure on a real torus $V/\Gamma$ induces a symplectic structure on the dual torus $V^*/\Gamma^{\perp}$. One has to be more careful when extending this duality to complexified symplectic structures. Let $\Omega$ be a complexified symplectic structure on a torus $V/\Gamma$. We say that $\Omega$ is *nondegenerate* if the corresponding symplectic form on $V \otimes_{\mathbb{R}} \mathbb{C}$ is nondegenerate. In this case there is a natural symplectic form on the dual complex space $V^* \otimes_{\mathbb{R}} \mathbb{C}$, which induces a complexified symplectic structure $\Omega^*$ on the dual torus $V^*/\Gamma^{\perp}$.

**Proposition 6.5.** *1. Let $(V/\Gamma, \Omega, L, L')$ be data I. Then the mirror dual complex torus to $(V/\Gamma, \Omega, L')$ is complex dual to the mirror of $(V/\Gamma, \Omega, L)$.*

*2. Let $(W/\Lambda, \Lambda_1, \Lambda_2)$ be data II such that $\mathbb{C}\Lambda_1 = W$. Then the symplectic torus corresponding to $(W/\Lambda, \Lambda_1, \Lambda_2)$ is symplectic dual to the symplectic torus corresponding to $(W/\Lambda, i\Lambda_2, i\Lambda_1)$.*

*Proof.* 1. Recall that the dual complex torus to a complex torus $W/\Lambda$ is $\mathrm{Hom}_{\mathbb{C}-anti}(W, \mathbb{C})/\Lambda^{\perp}$ where $\Lambda^{\perp} = \{\phi : \mathrm{Im}(\phi(\Lambda)) \subset \mathbb{Z}\}$. By the definition, the mirror dual torus to $V/\Gamma$ associated with $L$ is $L' \oplus L^*/(\Gamma \cap L' \oplus (\Gamma \cap L)^{\perp})$, where the complex structure is induced by the isomorphism (6.5.1). Now the real vector spaces $L' \oplus L^*$ and $L \oplus (L')^*$ are naturally dual to each in a way compatible with lattices, so it remains to check that this isomorphism is an imaginary part of a Hermitian pairing between $\mathrm{Hom}(L, \mathbb{C})$ and $\mathrm{Hom}(L', \mathbb{C})$. Let us choose a basis $(e_i)$ in $L$ and let $(e_i')$ be the corresponding basis in $L'$ defined by the condition $\omega(e_i', e_j) = \delta_{ij}$. Then we define the required

Hermitian pairing by the formula

$$\langle \phi, \phi' \rangle = \sum_i \phi(e_i)\overline{\phi'(e_i')},$$

where $\phi \in \mathrm{Hom}(L, \mathbb{C})$, $\phi' \in \mathrm{Hom}(L', \mathbb{C})$. We claim that the imaginary part of this pairing is a canonical real pairing between $L' \oplus L^*$ and $L \oplus (L')^*$. We will only check that $\mathrm{Im}\langle L', L \rangle = 0$ and leave the remaining part of the proof to the reader. Indeed, for any $i$ we have

$$\mathrm{Im}\langle \Omega(\cdot, e_i'), \Omega(\cdot, e_i) \rangle$$
$$= \mathrm{Im} \sum_k \Omega(e_k, e_i')\overline{\Omega(e_k, e_i)} = \mathrm{Im}\, \Omega(e_i, e_i')\overline{\Omega(e_i', e_i)} = 0,$$

whereas for $i \neq j$ we have

$$\mathrm{Im}\langle \Omega(\cdot, e_i'), \Omega(\cdot, e_j) \rangle$$
$$= \mathrm{Im} \sum_k \Omega(e_k, e_i')\overline{\Omega(e_k, e_j)} = \mathrm{Im}(\Omega(e_i, e_i')\overline{\Omega(e_i', e_j)} + \Omega(e_j, e_i')\overline{\Omega(e_j', e_j)})$$
$$= \xi(e_i', e_j) + \xi(e_j, e_i') = 0.$$

2. The proof is straightforward. □

## 6.6. Fourier Coefficients

Let $p : M \to B$, $p^\vee : M^\vee \to B$ be as in Section 6.2, $i : L \to M$ be a submanifold of dimension $\dim B$ transversal to all fibers, $(\mathcal{L}, \nabla_{\mathcal{L}})$ be a complex vector bundle with connection on $L$ such that $\mathrm{curv}(\nabla_{\mathcal{L}}) = 2\pi\Omega|_L$. Then by Theorem 6.3 we have the corresponding holomorphic vector bundle $\mathrm{Four}(L, \mathcal{L})$ on $M^\vee$. We want to interpret sections of $\mathrm{Four}(L, \mathcal{L})$ in terms of $(L, \mathcal{L})$. First of all, by the definition, a global section of $\mathrm{Four}(L, \mathcal{L})$ is the same as a global section $s$ of $(i \times \mathrm{id})^*\mathcal{P} \times p_L^*\mathcal{L}$ on $L \times_B M^\vee$. For every point $l \in L$ let $s_l$ be the restriction of $s$ to the fiber of the projection $L \times_B M^\vee \to L$ over $l$. Then $s_l$ is a section of $\mathcal{L}_l \otimes \mathcal{P}|_{l \times X^\vee}$, where $X^\vee$ is the dual torus to $X = p^{-1}(p(l))$. Now the idea is that $\mathcal{P}|_{l \times X^\vee}$ is a trivial bundle on a torus $X^\vee$ so we can consider Fourier coefficients of a section $s_l$. However, we have to be careful: to get a canonical trivialization of $\mathcal{P}|_{l \times X^\vee}$ we have to choose a lifting of $l$ to the universal covering $u : V \to X$. Then the Fourier coefficients are numbered by the lattice $\Gamma \subset V$. The invariant way to say this is that the Fourier coefficients of $s_l$ constitute a function on $\pi^{-1}(l)$ with values in $\mathcal{L}_l$. Let $\widetilde{L}$ be the preimage of $L$ under the fiberwise universal covering $u : T^*B \to M$. Then the Fourier coefficients of $s$ give us a $C^\infty$-section $FC(s)$ of $u^*\mathcal{L}$. Since the Fourier coefficients of a $C^\infty$-function are rapidly decreasing we obtain that $FC(s)$ is rapidly decreasing along each fiber of $u$. However, we only used the fact that $s$ is smooth along fibers of $p^\vee$. Differentiating along the base $B$ (note that we can differentiate sections of $\mathcal{L}$ along vector fields on $B$ since $L \to B$ is unramified) we get that all higher derivatives of $FC(s)$ are rapidly decreasing along fibers of $u$. Thus, the map $s \mapsto FC(s)$ gives an isomorphism of $C^\infty(M^\vee, \text{Four}(L, \mathcal{L}))$ with the subspace $\mathcal{S}(T^*B, u^*\mathcal{L}) \subset C^\infty(T^*B, u^*\mathcal{L})$ consisting of sections $f$ such that for every differential operator $D$ on the base the section $Df$ rapidly descreases along every fiber of $u$. Similarly, we can identify smooth $k$-forms with values in $\text{Four}(L, \mathcal{L})$ with the similar space $\mathcal{S}(T^*B, \Omega^k \otimes u^*\mathcal{L})$. Now the direct computation shows that the operator $\overline{\partial}$ on $C^\infty(M^\vee, \text{Four}(L, \mathcal{L}))$ corresponds to the differential on $\mathcal{S}(T^*B, \Omega^\bullet \otimes u^*\mathcal{L})$ given by the following connection $\widetilde{\nabla}$ on $u^*\mathcal{L}$. First, let us define a complex 1-form $\eta^\Omega$ on $T^*B$ with the property that $d\eta^\Omega$ is the pull-back of $\Omega$. Namely, for a point $x \in T_b^*B$ we have a natural linear form $\nu_x$ on $T_bB$ defined as follows: identify $T_b^*B$ with $T_x p^{-1}(b)$ using $\omega$ and set $\nu_x(\delta b) = \Omega(x, \delta b)$ for $\delta b \in T_bB$. Now $\eta_x^\Omega$ is a composition of the projection $T_x(T^*B) \to T_bB$ with $\nu_x$. Note that $\operatorname{Re} \eta^\Omega$ is the canonical 1-form on $T^*B$. Now we have

$$\widetilde{\nabla} = u^*\nabla_{\mathcal{L}} - 2\pi\eta^\Omega|_{\widetilde{L}}.$$

Note that $\widetilde{\nabla}$ is flat since $\operatorname{curv} \nabla_{\mathcal{L}} = 2\pi\Omega|_L$. Thus, we arrive to the following result.

**Theorem 6.6.** *The Dolbeault complex of* $\text{Four}(L, \mathcal{L})$ *is isomorphic to* $\mathcal{S}(T^*B, \Omega^\bullet \otimes u^*\mathcal{L})$ *with differential induced by the connection* $\widetilde{\nabla}$.

In the case when $M$ is compact one can consider an analogue of Hodge theory for our complex $(\mathcal{S}(T^*B, \Omega^\bullet \otimes u^*\mathcal{L}), \widetilde{\nabla})$ (henceforward, we will call elements of this complex "rapidly decreasing" sections). The idea is that since $\operatorname{Re} \eta^\Omega$ grows linearly at infinity, the corresponding Laplace operator will look like $-\Delta + F$, where $F$ is some potential with quadratic growth at infinity. Such an operator behaves similarly to the Laplace operator on a compact manifold. In particular, one can replace the spaces $\mathcal{S}(T^*B, \Omega^\bullet \otimes u^*\mathcal{L})$ with the corresponding $L^2$-spaces. In the next chapter we will apply this result to compute cohomology of holomorphic line bundles on complex tori.

## Exercises

1. Let $V$ be an $\mathbb{R}$-vector space with the basis $e_1, \ldots, e_n, f_1, \ldots, f_n$, $\omega$ be a symplectic form on $V$ such that $\omega(e_i, e_j) = \omega(f_i, f_j) = 0$ and $\omega(f_i, e_j) = x_{ij}$ for some non-degenerate $n \times n$ matrix $X = (x_{ij})$. Let

$L \subset V$ be the $\mathbb{R}$-linear (Lagrangian) subspace generated by $e_1, \ldots, e_n$, $\Gamma \subset V$ be a lattice spanned by the vectors $(e_i, i = 1, \ldots, n)$ and $(f_j + \sum_i a_{ij} e_i, j = 1, \ldots, n)$ for some $n \times n$ matrix $A = (a_{ij})$. Prove that the projection $V/\Gamma \to V/(L + \Gamma)$ admits a Lagrangian $C^\infty$-section if and only if $A$ is a sum of a $\mathbb{Z}$-valued matrix with the matrix of the form $X^{-1}D$, where $D$ is a symmetric $n \times n$ matrix. Show also that when such a section exists it can be chosen to be linear. [*Hint*: Such a section should be induced by some $C^\infty$-map $\phi : V/L \to L$ such that its differential at every point is symmetric (with respect to the duality between $V/L$ and $L$) and such that $\phi$ is a sum of $\Gamma/\Gamma \cap L$-periodic map and a linear map sending $f_j$ to $\sum_i (a_{ij} + n_{ij}) e_i$ with some integer $n_{ij}$'s. Now one has to use the fact that a nonzero constant 2-form on a torus has a nontrivial de Rham cohomology class.]

2. Let $M = \mathbb{R}^2/\mathbb{Z}^2$ be the 2-dimensional torus equipped with the complexified symplectic form $\Omega = -2\pi i \tau dx \wedge dy$, where $\tau$ is an element of the upper half-plane. Consider the Lagrangian circles fibration $M \to \mathbb{R}/\mathbb{Z}$ given by the projection $(x, y) \mapsto x$.
   (a) Show that the dual complex torus $M^\vee$ is isomorphic to the elliptic curve $\mathbb{C}/(\mathbb{Z} + \tau\mathbb{Z})$.
   (b) Let $\widetilde{L} \subset \mathbb{R}^2$ be a line of rational slope $d/r$ where $d$ and $r$ are relatively prime integers, $r > 0$, $L \subset M$ be the image of $\widetilde{L}$ under the natural projection. Prove that $\mathrm{Four}(L)$ ($L$ is equipped with the trivial local system) is a holomorphic vector bundle of rank $r$ and degree $d$ on $M^\vee$. [*Hint*: Reduce to the case $r = 1$ and then compute $c_1(\mathrm{Four}(L))$.]

# 7

# Cohomology of a Line Bundle on a Complex Torus: Mirror Symmetry Approach

In this chapter we compute cohomology of a holomorphic line bundle on a complex torus $W/\Lambda$ with nondegenerate first Chern class $E$ (recall that we can identify $H^2(W/\Lambda, \mathbb{Z})$ with the group of skew-symmetric bilinear forms $\Lambda \times \Lambda \to \mathbb{Z}$). The main result is that this cohomology is always concentrated in one degree, which is equal to the number of negative eigenvalues of $H$, the Hermitian form on $W$ such that $E = \operatorname{Im} H$. Furthermore, in the case when $E$ is unimodular, the cohomology is 1-dimensional. First, we consider the case when our complex torus is mirror dual to some symplectic torus $T$ (with real symplectic form) and the holomorphic line bundle is the Fourier transform of some Lagrangian subtorus of $T$ (as defined in the previous chapter). In this case, we can identify the Dolbeualt complex of our line bundle with the complex of rapidly decreasing differential forms on $\mathbb{R}^g$ equipped with the differential $d + dQ$, where $Q$ is a nondegenerate quadratic form having the same number of negative squares as $H$. This immediately implies the result. In order to generalize this computation to an arbitrary holomorphic line bundle on $W/\Lambda$, one can try to present $W/\Lambda$ as the mirror dual to a complexified symplectic torus and the line bundle as the Fourier transform of some Lagrangian subtorus. However, we proceed in a more direct way (inspired by this idea): we choose an $E$-Lagrangian subtorus $U/\Lambda \cap U \subset W/\Lambda$ and make the partial Fourier transform of sections of the Dolbeault complex of our line bundle along the translations of $U/\Lambda \cap U$. In this manner we again reduce the problem to the calculation of cohomology of the complex of rapidly decreasing forms on $\mathbb{R}^g$ with the differential $d + dQ$ as in the above particular case.

## 7.1. Lagrangian Subtori and Holomorphic Line Bundles

Let us consider a real torus $V/\Gamma$ equipped with a (real) symplectic form $\omega$ on $V$ and an isotropic decomposition $V = L \oplus L'$ such that $\Gamma = \Gamma \cap L \oplus \Gamma \cap L'$.

Then we have a Lagrangian torus fibration $p : V/\Gamma \to V/(L+\Gamma) \simeq L'/\Gamma \cap L'$ and the natural section $\sigma : L'/\Gamma \cap L' \to V/\Gamma$. The mirror-dual complex torus is $W/\Lambda$, where $W = L' \oplus L^*$, $\Lambda = \Lambda_1 \oplus \Lambda_2$ with $\Lambda_1 = \Gamma \cap L'$, $\Lambda_2 = (\Gamma \cap L)^{\perp}$. Assume that we have a homomorphism of lattices $f : \Gamma \cap L' \to \Gamma \cap L$. It extends to an $\mathbb{R}$-linear map $f : L' \to L$ and we denote $L_f = \{f(l') + l', l' \in L'\} \subset V$. Note that $\omega$ induces an isomorphism $\nu : L' \to L^* : l' \mapsto \omega(\cdot, l')$ and $L_f$ is Lagrangian if and only if $f$ considered as a map from $L^*$ to $L$ is symmetric. Moreover, in this case $\overline{L_f} = L_f/\Gamma \cap L_f$ is a Lagrangian torus in $V/\Gamma$ intersecting each fiber of $p$ in one point. It follows that the corresponding Fourier transform Four($\overline{L_f}$) (where we equip $\overline{L_f}$ with a trivial rank-1 local system) is a holomorphic line bundle. Let us denote by $\pi : L'/\Gamma \cap L' \times L^* \to W/\Lambda$ the natural projection. By definition $\pi^*$ Four($\overline{L_f}$) is the trivial line bundle on $L'/\Gamma \cap L' \times L^*$ with the connection $d + 2\pi i \langle df(l'), l^* \rangle$ and the following action of $(\Gamma \cap L)^{\perp}$:

$$\gamma^* \phi(l', l^*) = \exp(-2\pi i \langle f(l'), \gamma^* \rangle) \phi(l', l^* - \gamma^*),$$

where $\gamma^* \in (\Gamma \cap L)^{\perp}$, $l^* \in L^*$, $l' \in L'$. In order to represent Four($\overline{L_f}$) in a more standard form we have to find a global nowhere vanishing holomorphic section of $\pi^*$ Four($\overline{L_f}$).

**Lemma 7.1.** *Let us define the quadratic form on* $L^*$ *by setting* $Q(l^*) = \langle l^*, f(\nu_\omega^{-1}(l^*)) \rangle$. *Then* $s(l', l^*) = \exp(-\pi Q(l^*))$ *is a global holomorphic section of* $\pi^*$ Four($\overline{L_f}$).

*Proof.* Recall that the complex structure on $L' \oplus L^*$ is induced by the isomorphism

$$L' \oplus L^* \to L' \otimes_{\mathbb{R}} \mathbb{C} : (l', l^*) \mapsto l' + i\nu^{-1}(l^*).$$

In other words, if we choose coordinates $(l'_i)$ and $(l^*_i)$ on $L'$ and $L^*$ which correspond to each other via $\nu$ then $z_i = l'_i + il^*_i$ are complex coordinates on $L' \oplus L^*$. If the map $f$ is given by some symmetric matrix $(f_{ij})$ then the connection on $\pi^*$ Four($\overline{L_f}$) has the form

$$\nabla = d + 2\pi i \sum_{ij} l^*_i f_{ij} dl'_j = d + \frac{\pi}{2} \sum_{ij} f_{ij}(z_i - \overline{z_i})(dz_j + d\overline{z_j}).$$

Hence, its (0, 1)-component is

$$\nabla^{0,1} = \overline{\partial} + \frac{\pi}{2} \sum_{ij} f_{ij}(z_i - \overline{z_i}) d\overline{z_j}.$$

On the other hand, we have

$$s = \exp\left(\frac{\pi}{4}\sum_{ij} f_{ij}(z_i - \overline{z_i})(z_j - \overline{z_j})\right).$$

Therefore,

$$\begin{aligned}\overline{\partial} s &= -\frac{\pi}{4}\sum_{ij} f_{ij}[(z_j - \overline{z_j})d\overline{z_i} + (z_i - \overline{z_i})d\overline{z_j}]s \\ &= -\frac{\pi}{2}\sum_{ij} f_{ij}(z_i - \overline{z_i})d\overline{z_j}s,\end{aligned}$$

or equivalently, $\nabla^{0,1}(s) = 0$. □

Multiplying a trivialization of $\pi^* \operatorname{Four}(\overline{L_f})$ by $\exp(\pi Q(l^*))$ we obtain the standard description of $\operatorname{Four}(\overline{L_f})$ as the descent of the trivial line bundle with a holomorphic group action (see Section 1.2). Namely, the action of the group $(\Gamma \cap L)^{\perp}$ in the new trivialization takes form

$$\gamma^*\phi(v) = \exp(-\pi Q(\gamma^*) - 2\pi i\langle \gamma^*, f(v)\rangle)\phi(v - i\nu^{-1}(\gamma^*)),$$

where $v = l' + i\nu^{-1}(l^*)$, $f$ extends to $L' \otimes_{\mathbb{R}} \mathbb{C}$ by $\mathbb{C}$-linearity.

It follows easily that the skew-symmetric form $E$ on $L' \oplus L^*$ representing $c_1(\operatorname{Four}(\overline{L_f}))$ is given by

$$E((l_1', l_1^*), (l_2', l_2^*)) = \langle l_2^*, f(l_1')\rangle - \langle l_1^*, f(l_2')\rangle.$$

The corresponding Hermitian form $H$ is recovered from its restriction to $L'$ which is given by

$$H(l_1', l_2') = E(il_1', l_2') = E(\nu(l_1'), l_2') = -\langle \nu(l_1'), f(l_2')\rangle.$$

## 7.2. Computation of Cohomology

Let us assume now that $f : \Gamma \cap L' \to \Gamma \cap L$ is an isomorphism. Then the form $E|_{\Lambda}$ is unimodular. On the other hand, in this case the preimage of $\overline{L_f}$ under the covering $u : V/\Gamma \cap L' \to V/\Gamma$ has one component isomorphic to $L_f$, so the cohomology of $\operatorname{Four}(L_f)$ can be computed as the cohomology of the "rapidly decreasing" de Rham complex corresponding to the connection $d - 2\pi\eta|_{L_f}$, where $\eta$ is the canonical 1-form on $L \oplus L'$ considered as a cotangent bundle to $L'$ (see Theorem 6.6). Thus, $\eta|_{L_f} = \langle d\nu(l'), f(l)\rangle$, so if we identify $L_f$ with $L'$ then we are reduced to considering the connection $d + \pi dH|_{L'}$. Note that $H|_{L'}$ is a nondegenerate quadratic form. We claim that the cohomology of the complex of "rapidly decreasing" forms on $L'$ with the differential $d + \pi dH|_{L'}$ is 1-dimensional and has degree equal to the number of negative squares in $H|_{L'}$. Let us give a proof of this statement in the case $\dim L' = 1$ (in general the proof is similar). We have to calculate the cohomology of the differential

$$\widetilde{d} : f \mapsto (f' + \epsilon x f)dx$$

on "rapidly decreasing" forms on $\mathbb{R}$, where $\epsilon = \pm 1$. We claim that $\exp(-\frac{x^2}{2})$ generates the kernel of $\widetilde{d}$ if $\epsilon = 1$ while $\exp(-\frac{x^2}{2})dx$ generates the cokernel of $\widetilde{d}$ if $\epsilon = -1$. To prove this one has to use an analogue of Hodge theory in this case. Namely, consider the conjugate operator $\widetilde{d}^*$ (on $L^2$-spaces). It is easy to see that

$$\widetilde{d}^*(f dx) = -f' + \epsilon x f.$$

Thus, the corresponding deformed Laplace operator $\Delta = \widetilde{d}\widetilde{d}^* + \widetilde{d}^*\widetilde{d}$ is given by

$$\Delta(f) = -f'' + (x^2 - \epsilon)f,$$
$$\Delta(f dx) = (-f'' + (x^2 + \epsilon)f)dx.$$

The spectrum of the operator $f \mapsto -f'' + x^2 f$ on $L^2(\mathbb{R})$ is well-known. Namely, its eigenfunctions are $H_n(x)\exp(-\frac{x^2}{2})$ with eigenvalues $2n+1$, $n = 0, 1, 2, \ldots$, where

$$H_n(x) = (-1)^n \exp(x^2)\frac{d^n}{dx^n}(\exp(-x^2))$$

are Hermite polynomials. It follows that for $\epsilon = 1$ the operator $\Delta$ has 1-dimensional kernel on $\Omega^0$ and no kernel on $\Omega^1$, while for $\epsilon = -1$ it's the other way round.

### 7.3. General Case

The above argument works for line bundles with unimodular Chern classes on complex tori that are mirror dual to tori with real symplectic form. It is possible to modify it in order to deal with complexified symplectic forms. Below we give a more direct calculation of the cohomology of line bundles on complex tori which uses the same main idea. The answer turns out to be the same as in the particular case considered above.

**Theorem 7.2.** *Let $T = W/\Lambda$ be a complex torus, $H$ be a nondegenerate Hermitian form on $W$ such that $E = \operatorname{Im} H$ takes integer values on $\Lambda$ and $E|_\Lambda$ is unimodular. Let $L$ be a holomorphic line bundle on $T$ with the first Chern class $E$. Let $m$ be the number of negative eigenvalues of $H$. Then $H^i(T, L) = 0$ for $i \neq m$ while $H^m(T, L)$ is 1-dimensional.*

*Proof.* Recall that according to the Appell-Humbert theorem the line bundle $L$ corresponds to the trivial line bundle on $W$ with a holomorphic action of $\Lambda$ so that sections of $L$ correspond to functions $f$ on $W$ satisfying equation

$$f(x+\lambda) = \alpha(\lambda)\exp\left(\frac{\pi}{2}H(\lambda,\lambda) + \pi H(x,\lambda)\right) f(x)$$

for $x \in W$, $\lambda \in \Lambda$, where $\alpha : \Lambda \to \mathbb{C}_1^*$ is some quadratic map such that $\alpha(\lambda_1+\lambda_2) = \alpha(\lambda_1)\alpha(\lambda_2)\exp(\pi i E(\lambda_1,\lambda_2))$. Furthermore, translating $L$ if necessary we can assume that there exists an $E$-Lagrangian subspace $U \subset W$ such that $U = \mathbb{R}(\Lambda \cap U)$ and $\alpha|_{\Lambda\cap U} = 1$ (this follows, e.g., from Theorem 5.1). Let us define a symmetric $\mathbb{C}$-linear form $S$ on $W$ by the condition $S|_{U\times U} = H|_{U\times U}$. Then for any section $f$ of $L$ the function $\widetilde{f}(x) = \exp(-\frac{\pi}{2}S(x,x))f(x)$ is $\Lambda \cap U$-periodic. Indeed, for $\lambda \in \Lambda \cap U$ we have

$$\widetilde{f}(x+\lambda)$$
$$= \exp\left(\frac{\pi}{2}H(\lambda,\lambda) + \pi H(x,\lambda) - \frac{\pi}{2}S(\lambda,\lambda) - \pi S(x,\lambda)\right)\widetilde{f}(x) = \widetilde{f}(x)$$

since $H|_{W\times U} = S|_{W\times U}$. Let us denote $U' = iU$, so that we have an isotropic decomposition $W = U \oplus U'$. Then we can write

$$\widetilde{f}(x) = \sum_{\lambda^* \in (\Lambda\cap U)^*} \phi_{\lambda^*}(x_2)\exp(2\pi i\langle\lambda^*, x_1\rangle)$$

where $x = x_1 + x_2$, $x_1 \in U$, $x_2 \in U'$. The quasi-periodicity equation on $f$ is equivalent to the following equation on $\widetilde{f}$:

$$\widetilde{f}(x+\lambda) = \overline{\alpha}(\lambda)\exp(\pi H(\lambda_2,\lambda_2) + 2\pi H(x,\lambda_2))\widetilde{f}(x),$$

where $\lambda = \lambda_1 + \lambda_2$, $\lambda_1 \in U$, $\lambda_2 \in U'$,

$$\overline{\alpha}(\lambda) = \alpha(\lambda)\cdot\exp(\pi i E(\lambda_1,\lambda_2)).$$

It is easy to check that $\overline{\alpha}$ descends to a well-defined quadratic function on $\Lambda/\Lambda \cap U$ (i.e., it depends only on $\lambda_2$). Notice that we have a natural lattice in $U'$ via the inclusion $\Lambda/\Lambda \cap U \hookrightarrow W/U \simeq U'$. Now the quasi-periodicity equation above is equivalent to the following condition on the Fourier coefficients $\phi_{\lambda^*}$:

$$\phi_{\lambda^*}(x_2+\lambda_2)\exp(2\pi i\langle\lambda^*,\lambda_1\rangle) = \overline{\alpha}(\lambda_2)\exp(\pi H(\lambda_2,\lambda_2) + 2\pi H(x_2,\lambda_2))\phi_{\lambda^*-\nu(\lambda_2)}(x_2)$$

for any $x_2 \in U'$, $\lambda_2 \in \Lambda/\Lambda\cap U \subset U'$, where $\nu : \Lambda/\Lambda\cap U \to (\Lambda\cap U)^*$ is a homomorphism induced by $E$, namely, $\langle\nu(\lambda),\lambda'\rangle = E(\lambda',\lambda)$, $\lambda_1 \in U/\Lambda\cap U$ is defined by the condition $\lambda_1+\lambda_2 \in \Lambda$. Since $\nu$ is surjective it follows that all the functions $\phi_{\lambda^*}$ can be expressed via $\phi_0$. On the other hand, from the above equation we deduce that $\phi_0(x_2)\exp(-\pi H(x_2,x_2))$ is "rapidly decreasing."

The above Fourier decomposition can be applied also to $(0,p)$-forms with coefficients in $L$. More precisely, the restriction map

$$\mathrm{Hom}_{\mathbb{C}-anti}(W,\mathbb{C}) \to \mathrm{Hom}_{\mathbb{R}}(U',\mathbb{C})$$

is an isomorphism, so we can consider $(0,p)$-forms on $W$ as functions on $U$ with values in complex $p$-forms on $U'$. Thus, we can associate to a $(0,p)$-form $\omega$ with values in $L$ a sequence $\phi_{\lambda^*}$ of the Fourier coefficients which are complex $p$-forms on $U'$. It is easy to see that the Fourier coefficients of $\overline{\partial}\omega$ are $\frac{1}{2}d\phi_{\lambda^*} - \pi\langle\lambda^*, ix_2\rangle$ (it is important here that we have chosen $U' = iU$). Thus, the operator $\overline{\partial}$ corresponds to the operator $\frac{1}{2}d$ on $\phi_0$. Now we write $\phi_0 = \psi\exp(\pi H(x_2,x_2))$, where $\psi$ is "rapidly decreasing" and we reduce the problem to calculating cohomology of "rapidly decreasing" forms with differential $d + \pi dH$ considered earlier. □

**Corollary 7.3.** *Let $L = L(H,\alpha)$ be a line bundle on a complex torus $T = W/\Lambda$ such that the Hermitian $H$ is nondegenerate. Then $H^i(T,L) = 0$ for $i \neq m$, where $m$ is the number of negative eigenvalues of $H$, while $H^m(T,L)$ is the Schrödinger representation of the Heisenberg group $G(E,\Lambda,\alpha^{-1})$. In particular, the dimension of $H^m(T,L)$ is equal to $\sqrt{\det(E(e_i,e_j))}$ where $E = \mathrm{Im}\,H$, $(e_i)$ is a basis of $\Lambda$ compatible with the orientation of $W$.*

*Proof.* Let $\Lambda^\perp \subset W$ be the lattice of $x \in W$ such that $E(x,\Lambda) \subset \mathbb{Z}$. Then $\Lambda^\perp/\Lambda$ is the finite abelian group of order $d^2 = \det(E(e_i,e_j))$ and the Heisenberg group $G(E,\Lambda,\alpha^{-1})$ of $\Lambda^\perp/\Lambda$ by $U(1)$. Let us choose a Lagrangian subgroup $I \subset \Lambda^\perp/\Lambda$ and its lifting to $G(E,\Lambda,\alpha^{-1})$. Then we have the corresponding self-dual lattice $\Lambda' = I + \Lambda \subset \Lambda^\perp \subset W$ and its lifting to $\mathcal{H}(W)$, extending the lifting of $\Lambda$ given by $\alpha^{-1}$. In other words, we obtain a map $\alpha' : \Lambda' \to U(1)$ satisfying (1.2.2) and such that $\alpha'|_\Lambda = \alpha$. It follows that our line bundle $L = L(H,\alpha)$ is a pull-back of the line bundle

$L' = L(H, \alpha')$ on $W/\Lambda'$ under the finite unramified covering of complex tori $g : W/\Lambda \to W/\Lambda'$. Note that $c_1(L') = E|_{\Lambda'}$ is unimodular, so by Theorem 7.2 $H^i(W/\Lambda', L') = 0$ for $i \neq m$ and $H^m(W/\Lambda', L')$ is 1-dimensional. But $H^i(W/\Lambda', L') \simeq H^i(W/\Lambda, L)^I$, where the action of $I$ on cohomology groups of $L$ is induced by the action of the Heisenberg group $G(E, \Lambda, \alpha^{-1})$ on it. Hence, $H^i(W/\Lambda, L) = 0$ for $i \neq m$, while $H^m(W/\Lambda, L)$ is the Schrödinger representation. □

## Exercises

1. (a) In the context of the proof of Theorem 7.2 prove the following formula:

$$\phi_{\nu(\lambda_2)}(x_2) = \widetilde{\alpha}(\lambda_2)\exp(-\pi H(\lambda_2, \lambda_2) + 2\pi H(x_2, \lambda_2))\phi_0(x_2 - \lambda_2),$$

where

$$\widetilde{\alpha}(\lambda_2) = \alpha(\lambda)\exp(-\pi i E(\lambda_1, \lambda_2))$$

is a well-defined quadratic function on $\Lambda/\Lambda \cap U$.

(b) Consider the following hermitian metric on the space of sections of $L$:

$$\langle f, g\rangle = \int_{W/\Lambda} f(x)\overline{g(x)}\exp(-\pi H(x, x))dx,$$

where the Haar measure $dx$ is normalized by the condition that the covolume of $\Lambda$ is equal to 1. Let $(\phi_{\lambda^*})$ and $(\psi_{\lambda^*})$ be the Fourier coefficients of $f$ and $g$ respectively. Prove that

$$\langle f, g\rangle = \int_{U'/(\Lambda/\Lambda\cap U)} \phi_0(x_2)\overline{\psi_0(x_2)}\exp(-2\pi H(x_2, x_2))dx_2,$$

where the Haar measure $dx_2$ is normalized by the condition that the covolume of $\Lambda/\Lambda \cap U$ is equal to 1.

(c) Choosing a metric on $U$ find a harmonic representative for the unique (up to a scalar) cohomology class of $L$.

2. Assume that the Hermitian form $H$ is positive definite. Check that the global section of the line bundle $L$ in the proof of Theorem 7.2 corresponding to $\phi_0 \equiv 1$ is the theta function $\theta^{\alpha}_{H,\Lambda,U}$.
3. Using a similar approach compute the cohomology of the structure sheaf on a complex torus.

# Part II

## Algebraic Theory

# 8

# Abelian Varieties and Theorem of the Cube

Starting from this chapter we always work over an algebraically closed field $k$. For schemes $S_1$ and $S_2$ over $k$ we denote $S_1 \times S_2 := S_1 \times_k S_2$. A *variety* is a reduced and irreducible scheme of finite type over $k$.

The algebraic definition of an abelian variety is very simple: it is a commutative group object in the category of complete varieties over $k$. The condition of completeness makes the geometry of abelian varieties much more rigid than that of arbitrary commutative group varieties over $k$. For example, any morphism between abelian varieties is a composition of a homomorphism with a translation. Another manifestation of this rigidity is the fact that line bundles on abelian varieties behave similarly to quadratic functions on abelian groups. Namely, for every line bundle $L$ on an abelian variety $A$ the line $\mathcal{B}_{x,y} = L|_e \otimes L|_{x+y} \otimes L^{-1}|_x \otimes L^{-1}|_y$ depends "bilinearly" on $(x, y) \in A \times A$, where $e \in A$ is the neutral element (this means that there is a canonical isomorphism $\mathcal{B}_{x+x',y} \simeq \mathcal{B}_{x,y} \otimes \mathcal{B}_{x',y}$ etc.) This is a consequence of the *theorem of the cube* stating (in its simplest version) that a line bundle on the product $X \times Y \times Z$ of three complete varieties is uniquely determined up to an isomorphism by its restrictions to $\{x\} \times Y \times Z$, $X \times \{y\} \times Z$ and $X \times Y \times \{z\}$, where $x \in X$, $y \in Y$ and $z \in Z$ are arbitrary points. It follows that the map $\phi_L : A(k) \to \mathrm{Pic}(A) : x \mapsto t_x^* L \otimes L^{-1}$ associated with a line bundle $L$ on $A$, is a homomorphism. We denote by $\mathrm{Pic}^0(A)$ the subgroup of $\mathrm{Pic}(A)$ consisting of $L$ with $\phi_L \equiv 0$. Line bundles in $\mathrm{Pic}^0(A)$ behave similarly to characters of abelian group. For example, for every nontrivial $L \in \mathrm{Pic}^0(A)$ one has $H^*(A, L) = 0$ (this is an analogue of the orthogonality of characters). The theorem of the cube implies that for every line bundle $L$ the image of $\phi_L$ belongs to $\mathrm{Pic}^0(A)$. In the next chapter we will show that there is an abelian variety $\hat{A}$ (called *dual to* $A$), such that $\mathrm{Pic}^0(A) = \hat{A}(k)$ and the map $\phi_L$ is induced by a homomorphism of abelian varieties $A \to \hat{A}$.

We use the results of our study of line bundles to prove that every abelian variety is projective. First, we prove the following criterion of ampleness for a line bundle $L$ on an abelian variety $A$: if $\phi_L$ has finite kernel and

$H^0(A, L) \neq 0$ then $L$ is ample. Then we check that these conditions are satisfied for $L = \mathcal{O}_A(D)$, where $D$ is an effective divisor in $A$ such that $A \setminus D$ is affine.

## 8.1. Group Schemes and Abelian Varieties

A *group scheme* $G$ over $k$ is a group object in the category of schemes over $k$. This means that we have a group law morphism $m : G \times G \to G$, a morphism $[-1]_G : G \to G$ of passing to inverse, and a neutral element $e_G \in G(k)$, satisfying the usual axioms. One can also easily formulate what it means that a group scheme is commutative. By a homomorphism of group schemes we mean a morphism $f : G \to H$ which sends $e_G$ to $e_H$, respects the group laws and passing to inverse on $G$ and $H$. A *subgroup scheme* in a group scheme $G$ is a subscheme $H \subset G$ equipped with a group scheme structure such that the natural embedding is a homomorphism. For every homomorphism of group schemes $f : G \to H$ the kernel $\ker(f) = f^{-1}(e_H)$ is a subgroup scheme in $G$. A homomorphism $f : G \to H$ is called an *epimorphism* if $f$ is surjective and the natural morphism of sheaves $\mathcal{O}_H \to f_*\mathcal{O}_G$ is injective (in other words, this is an epimorphism in the category of group schemes – see [54], Section 1.1 for a general concept of epimorphism in a category). If $f : G \to H$ is an epimorphism then for every group scheme $G'$ the subset $\mathrm{Hom}(H, G') \subset \mathrm{Hom}(G, G')$ consists of homomorphisms $\phi : G \to G'$, such that $\ker(f)$ is a subscheme of $\ker(\phi)$.

An *abelian variety* $A$ over $k$ is a group scheme which is a complete variety over $k$. As we will see below in this case $A$ is automatically a *commutative* group scheme. The group law on abelian varieties will be usually denoted additively. Sometimes, we will denote the neutral element in $A$ simply by $e$ or by 0. For every point $x \in A(k)$ we denote by $t_x : A \to A$ the morphism of translation by $x$, so that $t_x(y) = x + y$. For an integer $n$ we have the corresponding endomorphism $[n]_A : A \to A : x \mapsto nx$.

Other important examples of commutative group schemes are the *multiplicative group* $\mathbb{G}_m$ and the *additive group* $\mathbb{G}_a$. By definition $\mathbb{G}_m = \mathbb{A}^1_k \setminus \{0\}$ with the group law $(x, y) \mapsto x \cdot y$ while $\mathbb{G}_a = \mathbb{A}^1_k$ with the group law $(x, y) \mapsto x + y$.

When considering subgroup schemes of abelian varieties we will encounter commutative group schemes that are not necessarily connected. For such a group scheme $G$ the connected component of zero is denoted by $G^0$. If $G$ is Noetherian (e.g., if $G$ is a subgroup scheme of an abelian variety) then $G/G^0$ is finite.

## 8.2. Rigidity Lemma

Here it is.

**Lemma 8.1.** *Let $X$ be a complete variety, $Y$ and $Z$ be aribitrary varieties. Assume that a morphism $f : X \times Y \to Z$ contracts $X \times \{y\}$ for some $y \in Y$ to a point in $Z$. Then $f$ is a composition of the projection $p_2 : X \times Y \to Y$ and a morphism $Y \to Z$.*

*Proof.* Let us consider the subset $S \subset Y$ consisting of $y \in Y$ such that $f(X \times \{y\})$ is a point. Clearly, $S$ is closed. We claim that $S$ is also open. Indeed, assume that $y \in S$. Let $U_Z \subset Z$ be an affine open containing $f(X \times \{y\})$. It suffices to prove that there exists an open neighborhood $U_Y$ of $y$ in $Y$ such that $f(X \times U_Y) \subset U_Z$. Indeed, then for every function $\phi$ on $U_Z$ the function $f^*\phi$ on $X \times U_Y$ is a pull-back of some function on $U_Y$ due to completeness of $X$. Thus, $f(x, y)$ does not depend on $x$ for $y \in U_Y$. To find the neighborhood $U_Y$ as above let us denote by $T \subset X \times Y$ the complement to $f^{-1}(U_Z)$. Then $T$ is closed and does not intersect $X \times \{y\}$, hence $p_2(T) \subset Y$ is closed and we can take $U_Y = Y \setminus p_2(T)$. Thus, $S \subset Y$ is open and closed. By the assumption of the lemma, $S$ is non-empty, hence $S = Y$. Therefore, for all $x \in X$, $y \in Y$ we have $f(x, y) = f(x_0, y)$ where $x_0 \in X$ is a fixed point. □

**Proposition 8.2.** *(i) A morphism between abelian varieties $f : A \to B$ such that $f(e_A) = f(e_B)$, is a homomorphism.*

*(ii) The group law on an abelian variety is commutative.*

*Proof.* (i) This follows from the rigidity lemma applied to the morphism $A \times A \to B : (x, y) \mapsto f(x + y) - f(x) - f(y)$.

(ii) Apply (i) to the inversion morphism $[-1]_A : A \to A$. □

## 8.3. Theorem of the Cube

We present this theorem here in its simplest form (for a more general statement of this kind see [95]).

**Theorem 8.3.** *Let $X$ and $Y$ be complete varieties, $Z$ be arbitrary variety, $x_0$, $y_0$, $z_0$ be $k$-points of $X$, $Y$ and $Z$, respectively. Assume that a line bundle $L$ on $X \times Y \times Z$ has trivial restrictions to $\{x_0\} \times Y \times Z$, $X \times \{y_0\} \times Z$ and $X \times Y \times \{z_0\}$. Then $L$ is trivial.*

*Proof.* The proof consists of two steps.

**Step 1.** *Reduction to the case when $X$ is a smooth curve.* For every point $x \in X$ there exists an irreducible curve $C \subset X$ passing through $x$ and $x_0$. Assuming that the theorem is true for line bundles on $\widetilde{C} \times Y \times Z$ where $\widetilde{C} \to C$ is the normalization of $C$, we deduce that the pull-back of $L$ to $\widetilde{C} \times Y \times Z$ is trivial. In particular, $L|_{\{x\}\times Y\times Z}$ is trivial. Thus, we have showed that for every $(x, z) \in X \times Z$ the restriction $L|_{\{x\}\times Y\times\{z\}}$ is trivial. Since $Y$ is complete this implies that $L$ is a pull-back of a line bundle $L'$ on $X \times Z$. But then $L' \simeq L|_{X\times\{y_0\}\times Z}$, so it is trivial.

**Step 2.** *Proof in the case when $X$ is a curve.* Now we assume that $X$ is a smooth complete curve. The idea is to consider $L$ as a family of line bundles $L_{y,z}$ on $X$ parametrized by $Y \times Z$. These bundles are trivialized at the point $x_0 \in X$. On the other hand, for $y = y_0$ or $z = z_0$ the bundle $L_{y,z}$ is trivial. In particular, all line bundles $L_{y,z}$ have degree zero so our family gives a morphism $f : Y \times Z \to J$, where $J$ is the Jacobian of $X$ (the moduli space of line bundles of degree zero on $X$; see Chapter 16). Since $f(\{y_0\} \times Z)$ is a point, the rigidity lemma implies that $f(y, z) = f(y, z_0) = 0$, so $f$ maps everything to the zero point in $J$, hence, $L$ is trivial (here we use the trivialization at $x_0 \in X$). □

## 8.4. Line Bundles on Abelian Varieties

The general theorem of the cube implies the following "quadratic behavior" of line bundles on abelian varieties. From now on we denote by $A^n$ the cartesian product $A \times A \times \cdots A$ ($n$ times).

**Theorem 8.4.** *Let $L$ be a line bundle on abelian variety $A$ and let $p_i : A^3 \to A$, $i = 1, 2, 3$ be projections. Then the line bundle*

$$(p_1 + p_2 + p_3)^*L \otimes (p_1 + p_2)^*L^{-1} \otimes (p_2 + p_3)^*L^{-1} \otimes (p_1 + p_3)^*L^{-1} \otimes p_1^*L \otimes p_2^*L \otimes p_3^*L$$

*on $A^3$ is trivial.*

*Proof.* Apply Theorem 8.3 to the line bundle in question taking $e_A$ as a marked point in $A$. □

**Remarks.** 1. The isomorphism of the above theorem depends on a choice of trivialization of the fiber $L|_0$, where $0 \in A$ is the neutral element. In other words, there is a canonical isomorphism

$$(p_1 + p_2 + p_3)^* L \otimes (p_1 + p_2)^* L^{-1} \otimes (p_2 + p_3)^* L^{-1} \otimes (p_1 + p_3)^* L^{-1}$$
$$\otimes\, p_1^* L \otimes p_2^* L \otimes p_3^* L \simeq L|_0 \otimes_k \mathcal{O}_A \qquad (8.4.1)$$

2. One can restate Theorem 8.4 as some kind of bilinearity of the line bundle

$$\Lambda(L) = m^* L \otimes p_1^* L^{-1} \otimes p_2^* L^{-1} \qquad (8.4.2)$$

on $A \times A$, see (9.3.3). The general context for the study of such bilinearity conditions is provided by the notion of biextension that will be considered in Chapter 10.

**Corollary 8.5.** *For every line bundle $L$ on abelian variety $A$ and $n \in \mathbb{Z}$ there is an isomorphism*

$$[n]_A^* L \simeq L^{\frac{n^2+n}{2}} \otimes [-1]_A^* L^{\frac{n^2-n}{2}},$$

*where $[n]_A : A \to A$ is the morphism of multiplication by $n$.*

The proof is similar to that of Lemma 4.11.

**Corollary 8.6.** *Let $L$ be a line bundle on an abelian variety $A$. Let us define the map*

$$\phi_L : A(k) \to \mathrm{Pic}(A) : x \mapsto t_x^* L \otimes L^{-1},$$

*where $t_x : A \to A$ is the translation by $x$. Then $\phi_L$ is a homomorphism. Furthermore, for every point $y \in A$ we have $\phi_{t_y^* L} = \phi_L$.*

We denote by $\mathrm{Pic}^0(A) \subset \mathrm{Pic}(A)$ the subgroup consisting of such $L$ that $\phi_L = 0$. By definition $L \in \mathrm{Pic}^0(A)$ if and only if $L$ is *homogeneous*, i.e., $t_x^* L \simeq L$ for all $x \in A(k)$. As we will see later, there exists an abelian variety $\hat{A}$ such that $\mathrm{Pic}^0(A) = \hat{A}(k)$ ($\hat{A}$ is called the *dual abelian variety* to $A$). One checks immediately that for every $L \in \mathrm{Pic}^0(A)$ the line bundle $\Lambda(L)$ on $A^2$ defined by (8.4.2) is trivial. Note also that the map $L \mapsto \phi_L$ is a homomorphism, that is,

$$\phi_{L \otimes M} = \phi_L + \phi_M.$$

It follows that $\phi_{t_x^* L \otimes L^{-1}} = 0$, i.e., the image of $\phi_L$ is always contained in $\mathrm{Pic}^0(A)$. The following proposition shows that the invariant $\phi_L$ is stable under deformations of a line bundle $L$.

**Proposition 8.7.** *Let $L$ be a line bundle on $A \times S$ where $S$ is an arbitrary variety. Assume that $L|_{A\times\{s_0\}} \in \mathrm{Pic}^0(A)$ for some point $s_0 \in S$. Then $L|_{A\times\{s\}} \in \mathrm{Pic}^0(A)$ for every point $s \in S$.*

*Proof.* Consider the line bundle $M = (p_1 + p_2)^*L \otimes p_1^*L^{-1} \otimes p_2^*L^{-1}$ on $A \times A \times S$. By assumption, $M|_{A\times A\times\{s_0\}}$ is trivial. On the other hand, the restrictions $M|_{\{e_A\}\times A\times S}$ and $M|_{A\times\{e_A\}\times S}$ are also trivial. Thus, by the theorem of the cube we conclude that $M$ is trivial. Now triviality of $M|_{A\times A\times\{s\}}$ for a point $s \in S$ means that $L|_{A\times\{s\}}$ lies in $\mathrm{Pic}^0(A)$. □

The triviality of the line bundle (8.4.2) on $A \times A$ for $L \in \mathrm{Pic}^0(A)$ suggests that we can consider a line bundle in $\mathrm{Pic}^0(A)$ as a categorification of the character of an abelian group. The following theorem can be considered as an analogue of the orthogonality relation for characters.

**Theorem 8.8.** *For a nontrivial line bundle $L \in \mathrm{Pic}^0(A)$ one has $H^*(A, L) = 0$.*

*Proof.* For $L \in \mathrm{Pic}^0(A)$ we have an isomorphism

$$H^*(A, L) \otimes H^*(A, L) \simeq H^*(A^2, p_1^*L \otimes p_2^*L) \simeq H^*(A^2, m^*L)$$
$$\simeq H^*(A, \mathcal{O}) \otimes H^*(A, L).$$

If $H^0(A, L) = 0$ this immediately implies that $H^*(A, L) = 0$. Now assume that $H^0(A, L) \neq 0$. Since we have an isomorphism

$$H^0(A, L) \otimes H^0(A, L) \simeq H^0(A, \mathcal{O}) \otimes H^0(A, L)$$

the dimension of $H^0(A, L)$ should be equal to 1. On the other hand, since $L$ is homogeneous, it is generated by global sections, hence, it is trivial. □

## 8.5. Ampleness

For a line bundle $L$ on an abelian variety we denote by $K(L) \subset A(k)$ the kernel of $\phi_L$. Lemma 8.9 below implies that $K(L)$ is a Zariski closed subgroup of $A$. In Chapter 9 we will refine this definition and construct a subgroup scheme supported on $K(L)$ but for the rest of this chapter we will equip $K(L)$ with the reduced scheme structure. By definition, $K(L) = A$ if and only if $L \in \mathrm{Pic}^0(A)$.

**Lemma 8.9.** *Let $X$ be a complete variety, $T$ be an arbitrary variety, $L$ be a line bundle on $X \times T$. Then the subset $T_1 \subset T$ consisting of $t \in T$ such that the restriction $L|_{X\times\{t\}}$ is trivial, is closed.*

*Proof.* $T_1$ is the intersection of the closed subset consisting of $t$ such that $H^0(L|_{X\times\{t\}}) \neq 0$ with the closed subset consisting of $t$ such that $H^0(L^{-1}|_{X\times\{t\}}) \neq 0$. □

If $f : B \to A$ is a homomorphism of abelian varieties then the homomorphism $\phi_{f^*L}$ is equal to the composition

$$B \xrightarrow{f} A \xrightarrow{\phi_L} \mathrm{Pic}^0(A) \xrightarrow{f^*} \mathrm{Pic}^0(B).$$

In particular, we have the inclusion $f^{-1}(K(L)) \subset K(f^*L)$. This simple remark is used in the proof of the following result.

**Proposition 8.10.** *Let $L$ be a line bundle on an abelian variety $A$ such that $H^0(A, L) \neq 0$. Then the restriction $L|_{K(L)}$ is trivial.*

*Proof.* Let $B \subset A$ be the connected component of $K(L)$ containing zero. Then $B$ is an abelian subvariety of $A$. For every $x \in A$ we have the inclusion $B = K(t_x^*L) \cap B \subset K(t_x^*L|_B)$. Hence, $t_x^*L|_B \in \mathrm{Pic}^0(B)$ for all $x \in A$. Let $s$ be a nonzero section of $L$. For generic $x \in A$ the section $t_x^*s$ of $t_x^*L$ does not vanish on $B$. Therefore, $t_x^*L|_B$ (being homogeneous) is trivial for such $x$. Now Lemma 8.9 implies that $t_x^*L|_B$ is trivial for all $x \in A$. □

**Theorem 8.11.** *Let $L$ be a line bundle on an abelian variety $A$. Then $L$ is ample if and only if there exists $n > 0$ such that $H^0(A, L^n) \neq 0$ and $K(L^n)$ is finite.*

*Proof.* Assume that $L$ is ample. Then $L^n$ is very ample for $n \gg 0$, hence, $H^0(A, L^n) \neq 0$. Now Proposition 8.10 implies that $K(L^n)$ is finite.

Conversely, assume that $L = \mathcal{O}_A(D)$ where $D$ is an effective divisor, such that $K(L)$ is finite. We claim that the linear system $|2D|$ is base-point free and does not contract any curves. Indeed, since for all $x \in A$ the divisor $t_x^*D + t_{-x}^*D$ is rationally equivalent to $2D$ there are no base points. Now assume that the morphism corresponding to $|2D|$ contracts an irreducible curve $C \subset A$. Let $f : \widetilde{C} \to A$ be the (finite) morphism from the normalization of $C$ to $A$. Then for generic $x \in A$ the pull-backs of the divisors $t_x^*D$ and $t_{-x}^*D$ by $f$ are effective divisors on $\widetilde{C}$. Since their sum should be rationally equivalent to zero, we obtain that for generic $x$ the divisor $t_x^*D$ does not intersect $C$. Changing $D$ by $t_x^*D$ if necessary we can assume that $D \cap C = \emptyset$. Let $D = \sum n_i D_i$ where $D_i$ are irreducible. Since the line bundle $f^*\mathcal{O}_A(D_i)$ is trivial, this implies that for every $x \in A$ the line bundle $f^*\mathcal{O}_A(t_x^*D_i)$ has degree zero. Therefore, for every $x \in A$ the divisor $t_x^* D_i$ either contains $C$ or does not intersect it. We claim that this implies that $D_i$ is invariant under translations by $x_1 - x_2$, where $x_1, x_2 \in C$. Indeed, let $y$ be a point $D_i$. Then for every $x_2 \in C$ the intersection $C \cap t_{x_2-y}^* D_i$ contains $x_2$, hence, it contains the entire curve $C$, Therefore, $x_1 - x_2 + y \in D_i$ for every $y \in D_i$ and every $x_1, x_2 \in C$, i.e., $t_{x_1-x_2}^* D_i = D_i$. Hence, we obtain that $D$ is invariant under translations by $x_1 - x_2$ for all $x_1, x_2 \in C$. This contradicts to finiteness of $K(L)$. □

**Theorem 8.12.** *Every abelian variety is projective.*

*Proof.* We want to find an effective divisor $D \in A$ such that the complement $U = A \setminus D$ is affine. This can be done as follows. First, choose an affine open $U' \subset A$ and an effective divisor $D'$ such that $A = U' \cup D'$. Then take a nonzero $f \in \mathcal{O}(U')$ that vanishes on $D' \cap U'$ and set $D = D' \cup \overline{Z_f}$, where $Z_f \subset U'$ is the zero divisor of $f$, $\overline{Z_f} \subset A$ is its closure in $A$. We claim that $L = \mathcal{O}_A(D)$ is ample. Indeed, by the previous theorem it suffices to check that $K(L)$ is finite. Translating $D$ if necessary we can assume that $U$ contains zero. Let $B \subset K(L)$ be the connected component of zero. By Proposition 8.10, the restriction of $\mathcal{O}(D)$ to $B$ is trivial. Since $B$ is not contained in $D$ this implies that $B$ does not intersect $D$. So the complete variety $B$ is contained in the affine variety $U$. Therefore, $B$ is a point. □

**Corollary 8.13.** *For every $n > 0$ the subgroup $A_n = \ker([n]_A : A \to A)$ is finite.*

*Proof.* Let $L$ be an ample line bundle on $A$. Then $[n]_A^* L$ is also ample by Corollary 8.5. Hence, $A_n \subset K([n]_A^* L)$ is finite. □

## 8.6. Complex Tori Case

Most of the results of this chapter can be transferred to the category of complex tori (not necessarily algebraic).

For example, the analogue of Theorem 8.4 holds for holomorphic line bundles on a complex torus $V/\Gamma$. Indeed, this follows immediately from Appell-Humbert's description of line bundles by pairs $(H, \alpha)$ (where $H$ is a Hermitian form on $V$, $\alpha$ is a quadratic function $\Gamma \to U(1)$ compatible with the symplectic form $E = \operatorname{Im} H$) and from Exercise 1 of Chapter 1.

The definitions and properties of $\phi_L$ and $K(L)$ also work perfectly in the context of complex tori. The homomorphism $\phi_L$ for $L = L(H, \alpha)$ can be computed using Exercise 2 of Chapter 1. Namely, one has

$$\phi_L(v) = L(0, \nu_v),$$

where $\nu_v(\gamma) = \exp(2\pi i E(v, \gamma))$, $\gamma \in \Gamma$, $v \in V/\Gamma$. This implies that $\phi_L = 0$ if and only if $L$ has form $L(0, \alpha)$, i.e., when it is topologically trivial. Thus, for a complex torus $T$, the group $\mathrm{Pic}^0(T)$ of holomorphic line bundles with $\phi_L = 0$ can be identified with the dual complex torus $T^\vee$. We also derive that the subgroup $K(L) \subset V/\Gamma$ is equal to $\Gamma^\perp/\Gamma$ where $\Gamma^\perp = \{v \in V : E(v, \Gamma) \subset \mathbb{Z}\}$. In particular, $K(L(H, \alpha))$ is finite if and only if the symplectic form $E$ (or equivalently, the Hermitian form $H$) is nondegenerate, $K(L(H, \alpha))$ is trivial if and only if $E$ is unimodular on $\Gamma$. In view of Corollary 7.3, the analogue of Theorem 8.11 for complex tori is equivalent to the Lefschetz theorem stating that a line bundle $L(H, \alpha)$ is ample if and only if $H$ is positive-definite (see Section 3.4). Finally, the proof of Theorem 8.8 works literally in the context of complex tori.

## Exercises

1. Let $f : A \to B$ be a homomorphism of abelian varieties. Consider its Stein decomposition $A \xrightarrow{\widetilde{f}} \widetilde{B} \xrightarrow{p} B$ where the morphism $p$ is finite, while the morphism $\widetilde{f}$ has the property $\widetilde{f}_*\mathcal{O}_A \simeq \mathcal{O}_{\widetilde{B}}$. Prove that $\widetilde{B}$ has a structure of an abelian variety such that $\widetilde{f}$ is a homomorphism and that $\ker(\widetilde{f})$ is an abelian subvariety of $A$.
2. Let $A$ and $B$ be abelian varieties. Show that the $\mathbb{Z}$-module $\mathrm{Hom}(A, B)$ has no torsion.
3. Let $C$ be an elliptic curve and let $L$ be a line bundle on $C$.
   (a) Show that $\phi_L$ depends only on the degree of $L$.
   (b) Assume that $\deg L = 1$. Show that for all $p \in C$ the line bundle $\phi_L(p)$ is isomorphic to $\mathcal{O}_C(e - p)$, where $e \in C$ is a neutral element.
   (c) Let $d = \deg(L)$. Show that $K(L)$ coincides with $E_d = \{x \in E : dx = e\}$.
4. Let $L$ be a line bundle on an abelian variety $A$.
   (a) Show that one has the following isomorphism of line bundles on $A^4$:

$$(p_1 + p_2 + p_3 + p_4)^*L \otimes (p_1 + p_4)^*L \otimes (p_2 + p_4)^*L \otimes (p_3 + p_4)^*L$$
$$\simeq (p_1 + p_2 + p_4)^*L \otimes (p_1 + p_3 + p_4)^*L \otimes (p_2 + p_3 + p_4)^*L \otimes p_4^*L.$$

As we will see in Chapter 12 this isomorphism is at the core of Riemann's quartic theta relation.

(b) Assume that $L$ is equipped with an isomorphism $[-1]_A^*L \simeq L$ (in this case $L$ is called *symmetric*), Show that there is an isomorphism on $A^2$

$$(p_1 + p_2)^*L \otimes (p_1 - p_2)^*L \simeq L^2 \boxtimes L^2.$$

5. Let $C$ be an elliptic curve. For every point $x \in C$ we denote by $\Delta_x \subset C \times C$ the graph of the translation by $x$, i.e., the set of points of the form $(y, x + y)$ where $y \in C$. In particular, $\Delta_0 = \Delta$ is the usual diagonal.
   (a) Express the line bundle $\mathcal{O}_{C\times C}(\Delta_x - \Delta)$ in the form $L \boxtimes M$ where $L$ and $M$ are some line bundles on $C$.
   (b) Show that for all $x \in C$ and all line bundles $L$ on $C$ the line bundle $L \boxtimes t_x^*L(-\Delta)$ on $C \times C$ can be obtained from $L \boxtimes L(-\Delta)$ by translation.
6. Check that for $L = L(H, \alpha)$ the isomorphism (8.4.1) is compatible with canonical hermitian metrics on both sides.
7. Show that for a line bundle $L$ on an abelian variety $A$ and an integer $n \neq 0$ one has $[n]_A^*L \simeq L^{n^2} \otimes M$ for some $M \in \mathrm{Pic}^0(A)$.

# 9

# Dual Abelian Variety

In this chapter we sketch a construction of the dual abelian variety $\hat{A}$ to a given abelian variety $A$, parametrizing line bundles in $\mathrm{Pic}^0(A)$. The idea of the construction is to use the map $\phi_L : A(k) \to \mathrm{Pic}^0(A)$ associated with an ample line bundle $L$. We show that this map is surjective and that its kernel is the set of $k$-points of a closed subgroup scheme $K(L) \subset A$. Furthermore, since $L$ is ample, $K(L)$ is finite. The variety $\hat{A}$ is obtained by taking the quotient of $A$ by the action of $K(L)$: $\hat{A} = A/K(L)$. By construction, we have $\hat{A}(k) = \mathrm{Pic}^0(A)$. Moreover, using the descent theory we can define a line bundle $\mathcal{P}$ on $A \times \hat{A}$ (called the *Poincaré bundle*), such that the restriction of $\mathcal{P}$ to $A \times \{\xi\}$ is the line bundle in $\mathrm{Pic}^0(A)$ corresponding to $\xi \in \hat{A}(k)$. In fact, the variety $\hat{A}$ represents the functor that associates to a $k$-scheme $S$ the set of families of line bundles on $A$ parametrized by $S$ (trivialized along $\{0\} \times S \subset A \times S$), such that each line bundle of this family belongs to $\mathrm{Pic}^0(A)$, and $\mathcal{P}$ is the corresponding universal family. However, we postpone the proof of this theorem until Chapter 11, where we will introduce the Fourier–Mukai transform (which is a functor between derived categories $D^b(A)$ and $D^b(\hat{A})$ defined using $\mathcal{P}$). We will see that this theorem is closely related to the involutivity of the Fourier–Mukai transform.

In 1-dimensional case we show that $\hat{A} \simeq A$, that is, every elliptic curve is self-dual. In Section 9.5, as an application of duality, we show how to construct the quotient of an abelian variety by an abelian subvariety. Also, we prove Poincaré's complete reducibility theorem asserting that every abelian subvariety $B \subset A$ admits a complementary abelian subvariety $C \subset A$, such that $A = B + C$ and $B \cap C$ is finite. Finally, in the case $k = \mathbb{C}$ we identify the duality of abelian varieties with the duality of complex tori considered in Section 1.4.

## 9.1. Quotient by the Action of a Finite Group

Let $X$ be a variety equipped with an action of a finite group $G$. We want to define the quotient variety $X/G$ together with a projection $p : X \to X/G$.

The universal property characterizing the pair $(p, X/G)$ should be the following: for every $G$-equivariant morphism of varieties $f : X \to Y$ where $G$ acts trivially on $Y$, there is a unique morphism $\overline{f} : X/G \to Y$ such that $f = \overline{f} \circ p$. The uniqueness of a quotient with this property is clear. The existence is easy to establish in the affine case. Namely, if $X = \operatorname{Spec}(A)$, then we set $X/G = \operatorname{Spec}(A^G)$ where $A^G \subset A$ is the ring of invariants of $G$ in $A$. Let $B \subset A^G$ be the subalgebra generated over $k$ by elementary symmetric functions of all the collections $(g(a))_{g \in G}$ where $a \in A$. Then $B$ is finitely generated and $A$ is integral over $B$. In particular, $A$ is finite $B$-module. Hence, $A^G$ is a finite $B$-module. It follows that $A^G$ is finitely generated and $A$ is a finite $A^G$-module. It is clear that the universal property holds for morphisms into affine varieties. It remains to notice that the construction is compatible with localization of the following kind. Let $f \subset A^G$ be a nonzero element. Then we have $(A_f)^G = (A^G)_f$. Using this localization one can easily prove the universal property in general. Note that our argument implies that the quotient morphism $p : X \to X/G$ is finite and surjective. We claim that the fibers of $p$ are precisely $G$-orbits in $X$. Indeed, assume that two points $x, x' \in X$ have distinct orbits: $Gx \neq Gx'$. Then by the Chinese remainder theorem there exists a function $a \in A$ such that $a|_{Gx} \equiv 1$ while $a|_{Gx'} \equiv 0$. Now the function $\prod_{g \in G} g(a)$ is a $G$-invariant function separating $Gx$ from $Gx'$.

To define $X/G$ in the case when $X$ is not affine, the natural idea is to cover $X$ by $G$-invariant open affine subsets $X_i$ and then glue together the quotients $X_i/G$. This is possible under the additional assumption that $X$ is quasi-projective. Indeed, in this case every $G$-orbit $Gx$ (being finite) is contained in some open affine subset $U \subset X$. Then $\cap_{g \in G} gU$ is a $G$-invariant open affine neighborhood of $x$. Thus, $X$ admits a covering by $G$-invariant open affine subsets $X_i$. So we can define $X/G$ by gluing the open pieces $X_i/G$. The remaining problem is to show that the obtained scheme $X/G$ is separated, i.e., that the diagonal $\Delta_{X/G}$ is closed in $X/G \times X/G$. It suffices to prove that for every $i, j$ the set $\Delta_{X/G} \cap (X_i/G \times X_j/G)$ is closed in $X_i/G \times X_j/G$. But this set is the image of the closed subset $\Delta_{X/G} \cap X_i \times X_j \subset X_i \times X_j$ under the quotient morphism $X_i \times X_j \to X_i/G \times X_j/G$. It remains to use the fact that the quotient morphism is finite and therefore closed.

## 9.2. Quotient by the Action of a Finite Group Scheme

By a *finite group scheme* we mean a group scheme $G$ over $k$ such that the canonical morphism $G \to \operatorname{Spec}(k)$ is finite. The degree of this morphism is called the order of $G$. In characteristic $p$ there exist nonreduced finite group schemes, e.g., the kernel of the homomorphism $[p]_{\mathbb{G}_m} : \mathbb{G}_m \to \mathbb{G}_m$.

Furthermore, nonreduced finite group schemes appear naturally as subgroup schemes in abelian varieties. We want to be able to define a quotient of an abelian variety by such a subgroup scheme. More generally, one can define a quotient of a scheme $X$ by the *action* of a finite group scheme $G$ assuming that every orbit is contained in an open affine subset (e.g., when $X$ is quasi-projective). Here, by an action of a group scheme $G$ on a scheme $X$ we mean a morphism $a : G \times X \to X$ compatible with the group law and with the neutral element in an obvious way.

In the case when $X$ is affine, following the same approach as in Section 9.1 we have to define the ring of $G$-invariant functions on $X$. The correct definition is the following: consider the action morphism $a : G \times X \to X$, let also $p_2 : G \times X \to X$ be the projection. Then a function $f \in \mathcal{O}(X)$ is called $G$-invariant if $a^* f = p_2^* f$ in $\mathcal{O}(G \times X)$. $G$-invariant functions form a subring $\mathcal{O}(X)^G \subset \mathcal{O}(X)$ and we set $X/G = \mathrm{Spec}(\mathcal{O}(X)^G)$. One can check that this construction is compatible with $G$-invariant localization. In the nonaffine case we glue $X/G$ from open pieces corresponding to $G$-invariant open affine covering of $X$. The quotient morphism $p : X \to X/G$ is finite and surjective. As a topological space $X/G$ is the quotient of the topological space underlying $X$ by the action of the finite group underlying $G$, while the structure sheaf on $X/G$ coincides with the subsheaf of $G$-invariants in $p_*(\mathcal{O}_X)$. Furthermore, in the case when the action of $G$ on $X$ is free, i.e., the natural morphism $(a, p_2) : G \times X \to X \times X$ is a closed embedding, the image of this morphism coincides with the fibered product $X \times_{X/G} X$ and the quotient morphism $p : X \to X/G$ is flat of degree equal to the order of $G$. We skip the proofs of these facts (see [95], 12).

The general construction of the quotient can be applied to construct the quotient of an abelian variety $A$ by a finite subgroup scheme $K \subset A$. Namely, the restriction of the group law morphism $m : A \times A \to A$ gives a free action of $K$ on $A$. Furthermore, since $A$ is projective, there exists a quotient $A/K$ which is a complete variety. Since $A/K \times A/K$ can be identified with the quotient of $A \times A$ by $K \times K$, we can easily define the group law on $A/K$ such that the natural projection $p : A \to A/K$ is a homomorphism. Thus, $A/K$ acquires the structure of abelian variety. Furthermore, in this case $K$ coincides with $\ker(p)$.

## 9.3. Construction of the Dual Abelian Variety

The idea is that in order to introduce a structure of variety on $\mathrm{Pic}^0(A)$ we can use the map $\phi_L : A(k) \to \mathrm{Pic}^0(A)$ for an ample line bundle $L$. Our first claim is the following.

**Theorem 9.1.** *Let $L$ be a line bundle on $A$ such that $K(L)$ is finite. Then $\phi_L$ is surjective.*

*Proof.* Consider the line bundle

$$\Lambda(L) = m^*L \otimes p_1^*L^{-1} \otimes p_2^*L^{-1} \tag{9.3.1}$$

on $A \times A$. For every $M \in \mathrm{Pic}^0(A)$ we can compute cohomology of $\Lambda(L) \otimes p_1^*M$ in two ways using Leray spectral sequences associated with $p_1$ and $p_2$. We have:

$$\Lambda(L) \otimes p_1^*M|_{x\times A} \simeq t_x^*L \otimes L^{-1}$$

$$\Lambda(L) \otimes p_1^*M|_{A\times x} \simeq t_x^*L \otimes L^{-1} \otimes M.$$

Assume that $M$ does not lie in the image of $\phi_L$. Then $H^*(A, \Lambda(L)\otimes p_1^*M|_{A\times x})$ for every $x \in A$, hence, $H^*(A \times A, \Lambda(L) \otimes p_1^*M) = 0$. On the other hand, since $H^*(A, \Lambda(L) \otimes p_1^*M|_{x\times A}) \neq 0$ for a finite number of $x$ we deduce that $Rp_{1*}(\Lambda(L) \otimes p_1^*M) \simeq Rp_{1*}(\Lambda(L)) \otimes M$ is supported at a finite number of points. Hence, the Leray spectral sequence for $p_1$ degenerates and we should have $Rp_{1*}(\Lambda(L)) = 0$ which is obviously a contradiction (it has a nonzero restriction to the neutral element $0 \in A$ since $\Lambda(L)|_{\{0\}\times A}$ is trivial). □

Line bundles $L_1$ and $L_2$ on a variety $X$ are called *algebraically equivalent* if there exists a variety $S$, two points $s_1, s_2 \in S$ and a line bundle $L_S$ on $X \times S$ such that $L_S|_{X\times\{s_i\}} \simeq L_i$ for $i = 1, 2$.

**Corollary 9.2.** *Two line bundles $L$ and $M$ on an abelian variety $A$ are algebraically equivalent if and only if $\phi_L = \phi_M$.*

*Proof.* The "only if" part follows from Proposition 8.7 of Chapter 8. To prove the "if" part we notice that the equality $\phi_L = \phi_M$ means that $L \otimes M^{-1} \in \mathrm{Pic}^0(A)$. Thus, it suffices to prove that every element of $\mathrm{Pic}^0(A)$ is algebraically equivalent to the trivial bundle. But this follows from Theorem 9.1. □

Next we are going to introduce the structure of a subscheme on $K(L)$ for any line bundle $L$ on $A$. This is done using the following general construction.

**Proposition 9.3.** *Let $X$ be a complete variety, $Y$ be an arbitrary scheme, $L$ be a line bundle on $X \times Y$. Then there exists a closed subscheme $Y_1 \subset Y$ such that for every scheme $Z$ a morphism $f : Z \to Y$ factors through $Y_1$ if and only if there exists a line bundle $K$ on $Z$ and an isomorphism $p_2^*K \simeq (\mathrm{id} \times f)^*L$.*

*Proof.* Let $S \subset Y$ be the subset consisting of points $y \in Y$ such that $L|_{X\times y}$ is trivial. According to Lemma 8.9, $S$ is closed, so our task is to equip $S$ with the subscheme structure and then to prove the universal property. The idea is that the line bundle $L_1$ on $X \times Z$ has form $p_2^*K$ for some line bundle $K$ on $Z$ if and only if $p_{2*}(L_1)$ is locally free of rank 1 and the natural morphism $p_2^*p_{2*}(L_1) \to L_1$ is an isomorphism. The problem is local in $Y$, so we can replace $Y$ by a sufficiently small affine open neighborhood of a point $y \in Y$. Clearly, it suffices to consider the case when $y \in S$. In this case the morphism $p_2^*p_{2*}L \to L$ is an isomorphism over $X \times y$. Since $X$ is complete, shrinking $Y$ to a smaller affine neighborhood of $y$ we can assume that the morphism $p_2^*p_{2*}L \to L$ is an isomorphism over the entire $X \times Y$. Let $Y = \operatorname{Spec} A$. Applying the proper base change (see Appendix C) to the projection $p_2 : X \times Y \to Y$ and $L$ we see that there exists a finite complex $P_0 \xrightarrow{d_1} P_1 \to \cdots \to P_n$ of finitely generated projective $A$-modules and an isomorphism of functors

$$H^i(X \times \operatorname{Spec}(B), L \otimes_A B) \simeq H^i(P_\bullet \otimes_A B)$$

on the category of $A$-algebras $B$. Let us consider the $A$-module $M := \operatorname{coker}(d_1^\vee : P_1^\vee \to P_0^\vee)$ (where for a projective $A$-module $P$ we denote $P^\vee = \operatorname{Hom}_A(P, A)$). Then for any $A$-algebra $B$ we have

$$\begin{aligned}\operatorname{Hom}_A(M, B) &\simeq \operatorname{Hom}_B(M \otimes_A B, B) \simeq \ker(P_0 \otimes_A B \to P_1 \otimes_A B)\\ &\simeq H^0(X \times \operatorname{Spec}(B), L \otimes_A B). \qquad (9.3.2)\end{aligned}$$

Let $\mathfrak{m} \subset A$ be the maximal ideal corresponding to $y$. Then we have $\dim \operatorname{Hom}_A(M, A/\mathfrak{m}) = \dim H^0(X \times y, L|_{X\times y}) = 1$ since $L|_{X\times y}$ is trivial. Hence, $M/\mathfrak{m}M$ is 1-dimensional. Shrinking $Y$ if necessary we can assume that $M$ is a cyclic $A$-module, i.e., $M = A/I$ for some ideal $I \subset \mathfrak{m}$. Now we define $Y_1 \subset Y$ to be the subscheme corresponding to the ideal $I$. Using the isomorphism (9.3.2) one can easily check that the universal property is satisfied. □

We define the subscheme structure on $K(L) \subset A$ by applying the above proposition to the line bundle $\Lambda(L)$ on $A \times A$ given by (9.3.1). Clearly, the underlying closed subset is precisely the subgroup we defined earlier. The proof of the fact that $K(L) \subset A$ is a subgroup scheme can be derived from the universal property in Proposition 9.3 and from the following isomorphism of line bundles on $A \times A \times A$:

$$(p_1, p_2 + p_3)^*\Lambda(L) \simeq (p_1, p_2)^*\Lambda(L) \otimes (p_1, p_3)^*\Lambda(L), \qquad (9.3.3)$$

which in turn follows easily from the theorem of the cube.

By definition, the restriction of $\Lambda(L)$ to $A \times K(L)$ is of the form $p_2^*N$ for some line bundle $N$ on $K(L)$. Restricting an isomorphism $\Lambda(L)|_{A\times K(L)} \simeq p_2^*N$ to $0 \times K(L)$ we see that, in fact, $N \simeq \Lambda(L)|_{0\times K(L)}$ is trivial (since $\Lambda(L)|_{0\times A}$ is trivial). Thus, the line bundle $\Lambda(L)|_{A\times K(L)}$ is trivial. Moreover, we have a canonical (up to a nonzero constant factor) trivialization of $\Lambda(L)|_{A\times K(L)}$ associated with a trivialization of $L_0$, the stalk of $L$ at $0 \in A$. Namely, a trivialization of $L_0$ induces a trivialization of $\Lambda(L)|_{0\times A}$, hence, a trivialization of $\Lambda(L)|_{0\times K(L)}$. Since $\mathcal{O}^*(A \times K(L)) = \mathcal{O}^*(K(L))$, the latter trivialization extends uniquely to a trivialization of $\Lambda(L)|_{A\times K(L)}$.

We define the dual abelian variety to $A$ as the quotient-scheme $\hat{A} = A/K(L)$ where $L$ is an ample line bundle on $L$. We are going to construct a certain line bundle $\mathcal{P}$ on $A \times \hat{A}$ (the universal family) as follows. Let us denote by $p : A \to A/K(L)$ the natural projection. The idea is to start with the bundle $\Lambda(L)$ on $A \times A$ and then equip it with descent data for the flat morphism $\mathrm{id} \times p : A \times A \to A \times A/K(L)$ (See Appendix C.) These data would define uniquely a line bundle $\mathcal{P}$ on $A \times \hat{A}$ such that $(\mathrm{id} \times p)^*\mathcal{P} \simeq \Lambda(L)$. Note that since $K(L)$ acts on $A$ freely, we have a natural identification of $A \times_{A/K(L)} A$ with $A \times K(L)$, so that the two projections to $A$ are given by $p_1 + p_2 : A \times K(L) \to A$ and by $p_1 : A \times K(L) \to A$. Thus, the descent data should be given by an isomorphism

$$a : (p_1, p_2 + p_3)^*\Lambda(L)|_{A\times A\times K(L)} \widetilde{\to} (p_1, p_2)^*\Lambda(L)|_{A\times A\times K(L)} \quad (9.3.4)$$

satisfying the natural cocycle condition on $A\times A\times K(L)\times K(L)$. The input for our descent data will be a trivialization of $L_0$. Starting with such a trivialization we get a canonical trivialization of $\Lambda(L)|_{A\times K(L)}$ as explained above. On the other hand, we get a canonical isomorphism (9.3.3) which restricts to a given trivialization of $L_0$ over $0\times 0\times 0$. Now we can define an isomorphism (9.3.4) as the composition of this isomorphism with the trivialization of $\Lambda(L)|_{A\times K(L)}$. Note that the restriction of $a$ to $0 \times A \times K(L)$ coincides with the natural isomorphism induced by the trivialization of $\Lambda(L)|_{0\times K(L)}$. Using this property one can easily check that the cocycle condition for the descent data is satisfied.

**Theorem 9.4.** *The variety $\hat{A}$ represents the functor*

$$S \mapsto \{L \in \mathrm{Pic}(A \times S) \mid L|_{A\times\{s\}} \in \mathrm{Pic}^0(A) \textit{ for all } s \in S,\ L|_{0\times S} \simeq \mathcal{O}_S\}$$

*so that $\mathcal{P} \in \mathrm{Pic}(A \times \hat{A})$ corresponds to the identity morphism $\hat{A} \to \hat{A}$.*

We postpone the proof of this theorem until Chapter 11. The universal family $\mathcal{P}$ on $A \times \hat{A}$ is called the *Poincaré bundle*. We always normalize it by requiring that the bundles $\mathcal{P}|_{0\times\hat{A}}$ and $\mathcal{P}|_{A\times 0}$ are trivial. For a point $x \in \hat{A}$ we denote by $\mathcal{P}_x$ the corresponding line bundle in $\mathrm{Pic}^0(A)$ (the restriction of $\mathcal{P}$ to $A \times \{x\}$). Considering $\mathcal{P}$ as a family of line bundles on $\hat{A}$ parametrized by $A$ we get a morphism $\mathrm{can}_A : A \to \hat{\hat{A}}$. It is easy to see that $\mathrm{can}_A$ is surjective (exercise!). Later we will prove that $\mathrm{can}_A$ is an isomorphism (see Section 10.4).

From the above construction and from Theorem 9.4 (that shows that $\hat{A}$ depends only on $A$) we immediately get that for every line bundle $L$ such that $K(L)$ finite, the map $\phi_L : A \to \hat{A}$ is a homomorphism of abelian varieties and $K(L)$ is its scheme-theoretic kernel.

Now assume that $L$ is an arbitrary line bundle on $A$ (so that $K(L)$ is not necessarily finite). Then we can consider $\Lambda(L)$ as a family of line bundles on $A$ parametrized by $A$, so by Theorem 9.4 it corresponds to a morphism $\phi_L : A \to \hat{A}$ which was previously defined only on the level of $k$-points. By definition we have the following isomorphism of line bundles on $A \times A$:

$$(\mathrm{id} \times \phi_L)^*\mathcal{P} \simeq \Lambda(L). \tag{9.3.5}$$

Let $f : A \to B$ be a homomorphism of abelian varieties. We can consider the line bundle $(\mathrm{id}_{\hat{B}} \times f)^*\mathcal{P}_B$, where $\mathcal{P}_B$ is the (normalized) Poincaré bundle on $\hat{B} \times B$, as a family of line bundles on $A$ parametrized by $\hat{B}$. Furthermore, these line bundles are trivialized at $0 \in A$ and over $0 \in \hat{B}$ we have the trivial bundle. Hence, by Theorem 9.4 this family defines a morphism $\hat{f} : \hat{B} \to \hat{A}$ such that

$$(\mathrm{id}_{\hat{B}} \times f)^*\mathcal{P}_B \simeq (\hat{f} \times \mathrm{id}_A)^*\mathcal{P}_A,$$

where $\mathcal{P}_A$ is the Poincaré bundle on $\hat{A} \times A$.

**Definition.** The morphism $\hat{f}$ constructed above is called the *dual* morphism to $f$.

On the level of points we have $\hat{f}(\xi) = f^*\xi$ for $\xi \in \mathrm{Pic}^0(B) = \hat{B}(k)$.

**Proposition 9.5.** *In the above situation for every line bundle $L$ on $B$ one has $\phi_{f^*L} = \hat{f} \circ \phi_L \circ f$.*

The proof is left to the reader.

**Definition.** A morphism $\phi : A \to \hat{A}$ is called *symmetric* if $\phi = \hat{\phi} \circ \mathrm{can}_A$ where $\mathrm{can}_A : A \to \widehat{\hat{A}}$ is the canonical identification.

It is easy to see that for every line bundle $L$ on $A$ the morphism $\phi_L : A \to \hat{A}$ is symmetric (this follows essentially from the symmetry of the line bundle $\Lambda(L)$ with respect to switching two factors of the product $A \times A$). Moreover, we will see in Theorem 13.7 that every symmetric morphism $A \to \hat{A}$ has form $\phi_L$ for some line bundle $L$ on $A$.

**Definition.** A *polarization* of $A$ is a symmetric isomorphism $\phi : A \to \hat{A}$ such that $\phi = \phi_L$ for some ample line bundle $L$ on $A$ (however, $L$ is not part of the data). A *principal polarization* of $A$ is a polarization $\phi : A \to \hat{A}$ which is an isomorphism.

Note that an ample line bundle $L$ defines a principal polarization if and only if $K(L) = 0$. In the case $k = \mathbb{C}$ a polarization on abelian variety $V/\Gamma$ is the same as a choice of a positive-definite Hermitian form $H$ on $V$ such that $\operatorname{Im} H$ takes integer values on $\Gamma$. Thus, the above definition of polarization agrees with the one given in Section 3.4.

## 9.4. Case of Elliptic Curve

Let $E$ be an elliptic curve. It is easy to see that every line bundle of degree zero on $E$ has form $\mathcal{O}_E(p - e)$ for unique point $p \in E$ (here $e \in E$ is the neutral element). An easy generalization of this fact is that $E$ together with the family $\mathcal{P} = \mathcal{O}_{E\times E}(\Delta - p_1^{-1}(e) - p_2^{-1}(e))$ represents the functor of families of line bundles of degree zero on $E$ considered in Theorem 9.4. Indeed, let $L$ be any such family parametrized by a variety $S$. In other words, $L$ is a line bundle on $S \times E$, trivialized along $S \times e$, such that $\deg(L|_{s\times E}) = 0$ for all $s \in S$. Let us denote by $p_E$, $p_S$ the projections from $S \times E$ to $E$ and $S$ respectively. Consider the line bundle $L(p_E^{-1}(e))$. This line bundle has degree 1 on every fiber $s \times E$. Hence, $M = Rp_{S*}(L(p_E^{-1}(e)))$ is a line bundle concentrated in degree zero. Thus, the line bundle $L' = L(p_E^{-1}(e)) \otimes p_S^* M^{-1}$ has a canonical section $s : \mathcal{O} \to L'$ which does not vanish on every fiber $s \times E$. Let $D \subset S \times E$ be the divisor of zeroes of $s$. Then the scheme-theoretic intersection of $D$ with every fiber $s \times E$ is a point in $E$. Hence, the projection $p_S|_D : D \to S$ is an isomorphism. Thus, $D$ is the graph of a morphism $S \to E$. It is easy to check that our family is induced by the universal one via this morphism.

Let us compare this construction with the one given in Section 9.3. We can start with the line bundle $L = \mathcal{O}_E(e)$. Then by definition $\hat{E} = E/K(L)$ with the universal family induced by $\Lambda(L)$. We claim that $K(L)$ is the trivial subgroup of $E$ (taking into account the scheme structure). Indeed, let $S$ be a scheme equipped with a line bundle $K$ and a morphism $f : S \to E$ such that $p_2^*K \simeq (\mathrm{id} \times f)^*\Lambda(L)$ on $E \times S$. This means that

$$\mathcal{O}_{E\times S}((p_1 + fp_2)^{-1}(e)) \simeq p_2^*K'\big(p_1^{-1}(e)\big)$$

for some $K' \in \mathrm{Pic}(S)$. Considering the push-forward of this isomorphism to $S$ we see that $K'$ is trivial, so we have an isomorphism

$$M := \mathcal{O}_{E\times S}((p_1 + fp_2)^{-1}(e)) \simeq \mathcal{O}_{E\times S}\big(p_1^{-1}(e)\big).$$

Thus, we get two sections $s_1, s_2$ of $M$ vanishing on divisors $(p_1 + fp_2)^{-1}(e)$ and $p_1^{-1}(e)$, resp. The corresponding sections $\overline{s}_1, \overline{s}_2$ of $p_{2*}(M) \simeq \mathcal{O}_S$ do not vanish anywhere on $S$. It follows that $s_1 = \phi s_2$ for some invertible function $\phi$ on $S$. But this implies that zero loci of $s_1$ and $s_2$ coincide (as subschemes of $E \times S$), therefore $f$ factors through $e \in E$.

By definition, we have an isomorphism

$$\Lambda(\mathcal{O}_E(e)) \simeq \mathcal{O}_{E\times E}\big(m^{-1}(e) - p_1^{-1}(e) - p_2^{-1}(e)\big)$$

on $E \times E$. Hence, $\Lambda(\mathcal{O}_E(e))$ differs from $\mathcal{O}_{E\times E}(\Delta - p_1^{-1}(e) - p_2^{-1}(e))$ by the automorphism $\mathrm{id} \times [-1]_E : E^2 \to E^2$.

## 9.5. Quotient by an Abelian Subvariety

Let $A$ be an abelian variety, $B \subset A$ an abelian subvariety. We claim that there exists a surjective homomorphism of abelian varieties $p : A \to C$ such that $B = \ker(p)$. In this situation we will say that there is an exact sequence of abelian varieties

$$0 \to B \to A \to C \to 0.$$

For the proof we can obviously assume that $\dim B < \dim A$. Let $i : B \hookrightarrow A$ be the embedding. Consider the dual morphism $\hat{i} : \hat{A} \to \hat{B}$. We claim that $\hat{i}$ is surjective. Indeed, let us choose an ample line bundle $L$ on $A$. Then we have the following commutative diagram:

$$\begin{array}{ccc} B & \xrightarrow{\phi_{L|_B}} & \hat{B} \\ {\scriptstyle i}\downarrow & & \uparrow{\scriptstyle \hat{i}} \\ A & \xrightarrow{\phi_L} & \hat{A} \end{array} \qquad (9.5.1)$$

in which horisontal arrows are surjective by Theorem 9.1. This immediately implies our claim. Now let $C'$ be the abelian variety dual to the connected component of zero of $\ker \hat{i}$. Composing the dual morphism to the embedding $(\ker \hat{i})^0 \hookrightarrow \hat{A}$ with the canonical map $A \to \hat{\hat{A}}$ we get a morphism $p' : A \to C'$. The same argument as above shows that $p'$ is surjective. Also, by construction we have $B \subset \ker(p')$. Let $A \xrightarrow{p} C \to C'$ be the Stein factorization of $p'$, so that $p$ has connected fibers while $C \to C'$ is finite. Then according to Exercise 1 of Chapter 8, $C$ has a structure of an abelian variety, such that $p$ is a homomorphism and its kernel is an abelian subvariety of $A$. Now we have an inclusion $B \subset \ker(p)$ of abelian subvarieties of the same dimension. Hence, $B = \ker(p)$.

**Proposition 9.6.** *Let $i : B \hookrightarrow A$ be an embedding of abelian varieties. Then there exists a homomorphism $p : A \to B$ such that $p \circ i = [n]_B$ for some integer $n > 0$.*

*Proof.* The diagram (9.5.1) provides us with a homomorphism $p' : A \to \hat{B}$ such that $p' \circ i = \phi_M$ for some ample line bundle $M$ on $B$. Since the kernel $K(M)$ of $\phi_M$ is finite, it is annihilated by some integer $n > 0$. In other words, $K(M) \subset B_n$ for some $n > 0$. This means that there is a morphism $\pi : \hat{B} \to B$ such that $\pi \circ \phi_M = [n]_B$. Now we set $p = \pi \circ p'$. □

The following statement is called Poincaré's complete reducibility theorem.

**Corollary 9.7.** *Let $B$ be an abelian subvariety in an abelian variety $A$. Then there exists an abelian subvariety $C \subset A$ such that $B + C = A$ and the intersection $B \cap C$ is finite.*

*Proof.* Take $C$ to be the connected component of 0 in $\ker(p)$ where $p$ is constructed in the above proposition. □

**Remark.** Note that the category $\mathcal{AV}$ of abelian varieties is additive, i.e., there is a natural structure of an abelian group on every set $\mathrm{Hom}(A, B)$, where $A$, $B$ are abelian varieties. Using the notion of an exact sequence introduced above we can consider $\mathcal{AV}$ as an exact category (i.e., a full subcategory of some abelian category). In particular, it makes sense to consider the $K$-theory of abelian varieties (over a given ground field $k$). For example, the corresponding group $K_0(\mathcal{AV})$ is generated by classes $[A]$ of simple abelian varieties, i.e., varieties with no proper nontrivial abelian subvarieties. It is an interesting question whether the equality $[A] = [B]$ in $K_0$ for simple abelian varieties $A$ and $B$ implies that they are isomorphic.

## 9.6. Comparison with the Transcendental Picture

Recall that in Section 1.4 we have defined a dual complex torus $T^\vee$ to a complex torus $T$ and the Poincaré bundle $\mathcal{P}^{hol}$ on $T \times T^\vee$ such that the restrictions $\mathcal{P}^{hol}_{T\times\{x\}}$ run through all holomorphic line bundles on $T$ which are topologically trivial. It is easy to see that in the case when $T$ is a complex abelian variety, $T^\vee$ is isomorphic to the dual abelian variety $\hat{T}$. One way to see this is to use Theorem 9.4: since $T^\vee$ is the base of the family of line bundles on $T$ given by $\mathcal{P}^{hol}$ and since by the GAGA principle these objects are algebraic (see [124]), we have a morphism $T^\vee \to \hat{T}$. Since this map is a bijection, it is an isomorphism. A more direct way is to use the construction of the dual abelian variety we gave earlier. Let $L = L(H, \alpha)$ be a line bundle on $T = V/\Gamma$, such that the Hermitian form $H$ is nondegenerate. Then we have $\hat{T} = V/\Gamma^\perp$, where $\Gamma^\perp = \{x \in V : E(x, \Gamma) \subset \mathbb{Z}\}$ (as usual, $E = \operatorname{Im} H$). Now the Hermitian form $H$ induces a complex-linear map $\phi_H : V \to \overline{V}^\vee$, which descends to a holomorphic isomorphism $\hat{T} \widetilde{\to} T^\vee$. Note that under this identification, for every line bundle $L' = L(H', \alpha')$ the symmetric morphism $\phi_{L'} : T \to \hat{T} \simeq T^\vee$ corresponds to the map induced by the Hermitian form $H'$.

Now let us specialize to the case of elliptic curve $T = \mathbb{C}/\Gamma$ where $\Gamma = \Gamma_\tau = \mathbb{Z} + \mathbb{Z}\tau$, $\tau$ is an element of the upper half-plane. We have $T^\vee = \mathbb{C}/\Gamma^\vee$ where $\Gamma^\vee = \{z \in \mathbb{C} : \operatorname{Im}(z\overline{\Gamma}) \subset \mathbb{Z}\}$. The perfect pairing between $\Gamma^\vee$ and $\Gamma$ is given by

$$\langle \gamma^\vee, \gamma \rangle = \operatorname{Im}(\gamma^\vee \cdot \overline{\gamma}).$$

It is easy to see that $\Gamma^\vee = \frac{1}{\operatorname{Im}(\tau)}\Gamma$. Hence, we have a natural isomorphism $\phi : T \to T^\vee : z \mapsto \frac{z}{\operatorname{Im}(\tau)}$. The pull-back of the Poincaré line bundle $\mathcal{P}$ under $\operatorname{id} \times \phi$ is the line bundle $L(H^{(2)}, \alpha^{(2)})$ on $T \times T = \mathbb{C} \times \mathbb{C}/\Gamma \times \Gamma$ where

$$H^{(2)}((z_1, z_2), (z_1', z_2')) = \frac{z_1\overline{z_2'} + z_2\overline{z_1'}}{\operatorname{Im}(\tau)},$$

$$\alpha^{(2)}(m + n\tau, m' + n'\tau) = (-1)^{mn'+m'n}.$$

We claim that the line bundle $L(H^{(2)}, \alpha^{(2)})$ is isomorphic to the line bundle $\Lambda(\mathcal{O}_T(0)) = \mathcal{O}_T(m^{-1}(0) - p_1^{-1}(0) - p_2^{-1}(0))$ considered earlier. Indeed, let $L = L(H, \alpha)$ be the line bundle on $T$ corresponding to the Hermitian form $H(z_1, z_2) = \frac{z_1 \overline{z_2}}{\mathrm{Im}(\tau)}$ and some compatible homomorphism $\alpha : \Gamma \to U(1)$. Then it is easy to see that $\phi = \phi_L$ (e.g., using Exercise 2 of Chapter 1). The line bundle $L$ has degree 1 (see Exercise 5 of Chapter 3), hence, it is algebraically equivalent to $\mathcal{O}_T(0)$. Now our claim follows from the isomorphism (9.3.5).

## Exercises

1. Let $X$ be a quasiprojective variety, $G$ be a finite group acting on $X$, $H \subset G$ be a subgroup. Show that there is a natural isomorphism
$$X/H \times_{X/G} X \simeq \bigsqcup_{G/H} X.$$

2. Show that a commutative diagram of homomorphisms between abelian varieties
$$\begin{array}{ccc} A & \xrightarrow{f} & B \\ \Big\downarrow{\scriptstyle g} & & \Big\downarrow{\scriptstyle h} \\ C & \xrightarrow{k} & D \end{array} \tag{9.6.1}$$
is cartesian if and only if the sequence
$$0 \to A \overset{(f,g)}{\to} B \times C \overset{(h,-k)}{\to} D \to 0$$
is exact.

3. Let $G$ be a group scheme acting on a scheme $X$, $a : G \times X \to X$ be the morphism defining the action. Let $\mathcal{F}$ be a coherent sheaf on $X$.
   (a) Give a definition of an action of $G$ on a sheaf $\mathcal{F}$ compatible with the action of $G$ on $X$ (informally it should be given by a system of isomorphisms $g : \mathcal{F}_x \widetilde{\to} \mathcal{F}_{gx}$, where $g \in G$, $x \in X$, satisfying a cocycle condition). Sheaves equipped with such an action are called *G-equivariant sheaves.*
   (b) Let $Y$ be another scheme with $G$-action and let $f : X \to Y$ be a $G$-morphism. Show that an action of $G$ on $\mathcal{F}$ induces an action of $G$ on $f_*(\mathcal{F})$.

(c) Construct the canonical action of $\mathbb{G}_m$ on any coherent sheaf $\mathcal{F}$ over $X$, where $\mathbb{G}_m$ acts trivially on $X$ such that $t \in k^* = \mathbb{G}_m(k)$ acts on $\mathcal{F}$ via the embedding $k^* \subset \mathcal{O}_X$.

4. Let $f : A \to B$ be an isogeny of abelian varieties, i.e., a surjective morphism with finite kernel. Show that there exists an isogeny $g : B \to A$ such that $f \circ g = [n]_B$, $g \circ f = [n]_A$ for some $n > 0$.

# 10

# Extensions, Biextensions, and Duality

In this chapter we study the relation between the duality of abelian varieties and Cartier duality of finite commutative group schemes (the latter is induced by the natural notion of duality for finite-dimensional commutative and cocommutative Hopf algebras). The main result is that the kernels of dual isogenies of abelian varieties are Cartier dual. The proof is based on the interpretation of the functor $A \mapsto \hat{A}$ on abelian varieties in terms of $\mathrm{Ext}^1(?, \mathbb{G}_m)$. Similarly, Cartier duality functor on finite commutative group schemes can be interpreted in terms of $\mathrm{Hom}(?, \mathbb{G}_m)$. Then the above result follows from the consideration of the long exact sequence connecting $\mathrm{Ext}^*(?, \mathbb{G}_m)$. As a corollary we get a canonical perfect pairing $A_n \times \hat{A}_n \to \mathbb{G}_m$ between groups of points of order $n$ in an abelian variety $A$ and the dual abelian variety $\hat{A}$ called the *Weil pairing*. Another corollary is that the canonical map $\mathrm{can}_A : A \to \hat{\hat{A}}$ is an isomorphism (for the proof one has to choose a symmetric isogeny $f : A \to \hat{A}$ and use the fact that the orders of finite group schemes $\ker(f)$ and $\ker(\hat{f})$ are equal).

A convenient tool in the theory of duality of abelian varieties is the notion of *biextension* which we consider as a categorification of the notion of bilinear pairing. A *biextension* of $G_1 \times G_2$ by $\mathbb{G}_m$, where $G_1$ and $G_2$ are commutative group schemes, is a line bundle $\mathcal{B}$ on $G_1 \times G_2$ equipped with isomorphisms $\mathcal{B}_{x+x',y} \simeq \mathcal{B}_{x,y} \otimes \mathcal{B}_{x',y}$ and $\mathcal{B}_{x,y+y'} \simeq \mathcal{B}_{x,y} \otimes \mathcal{B}_{x,y'}$ satisfying some natural compatibilities. For example, the theorem of the cube proved in Chapter 8 implies that for a line bundle $L$ on an abelian variety $A$, the line bundle $\Lambda(L) = m^*L \otimes p_1^*L^{-1} \otimes p_2^*L^{-1}$ is a biextension on $A \times A$. This notion is related to duality as follows: a biextension on $A \times B$, where $A$ and $B$ are abelian varieties, is the same as a morphism $A \to \hat{B}$. On the other hand, every biextension $\mathcal{B}$ on $A \times B$ has the left and right kernels which are the maximal subgroup schemes $\ker^l(\mathcal{B}) \subset A$ and $\ker^r(\mathcal{B}) \subset B$, such that $\mathcal{B}$ restricts to trivial biextensions of $\ker^l(\mathcal{B}) \times B$ and $A \times \ker^r(\mathcal{B})$. The difference between trivializations of these two biextensions gives a pairing $\ker^l(\mathcal{B}) \times \ker^r(\mathcal{B}) \to \mathbb{G}_m$. In the case when $\mathcal{B}$ corresponds to an isogeny

$A \to \hat{B}$, this pairing induces Cartier duality between the left and right kernels (as kernels of dual isogenies) discussed above. If we further specialize to the case $\mathcal{B} = \Lambda(L)$, where $L$ is a line bundle on $A$ (and $B = A$), then both left and right kernels coincide with the subgroup scheme $K(L) \subset A$ considered in the previous chapter. The corresponding pairing $K(L) \times K(L) \to \mathbb{G}_m$ is relevant for the problem of descent. Namely, if $K \subset A$ is a subgroup scheme, then in order for $L$ to be a pull-back of a line bundle on $A/K$, $K$ should be a subgroup of $K(L)$, isotropic with respect to the above pairing. This result will be useful in the context of algebraic theory of Heisenberg groups considered in Chapter 12.

## 10.1. Cartier Duality

Let $G$ be a finite group scheme over $k$. Then $A = \mathcal{O}(G)$ is a finite dimensional commutative algebra over $k$. The data consisting of the group law $m : G \times G \to G$, the inversion morphism $i : G \to G$, and the neutral element $e \in G$, translate into algebra homomorphisms $\Delta : A \to A \otimes A$, $S : A \to A$, and $\varepsilon : A \to k$. The axioms of a group can be spelled in terms of the data $(\Delta, S, \varepsilon)$ (an algebra $A$ equipped with these data is called a *Hopf algebra*). In particular, they imply that $\Delta^* : A^* \otimes A^* \to A^*$ is an associative product with the unit $\varepsilon^* : k \to A^*$. If $G$ is reduced then the algebra $A^*$ is just the group algebra of $G(k)$. For this reason, we sometimes denote $A^*$ as $k[G]$ and call it the group algebra of $G$. If $G$ is commutative then the algebra $A^*$ is commutative, so we can consider the scheme $\widehat{G} = \text{Spec}(A^*)$. Now the map $A^* \to A^* \otimes A^*$, dual to the multiplication in $A$ gives rise to a morphism $\hat{m} : \widehat{G} \times \widehat{G} \to \widehat{G}$. Also, the unit in $A$ gives rise to an element $\hat{e} \in \widehat{G}$, while the map $S^* : A^* \to A^*$ produces a morphism $\hat{i} : \widehat{G} \to \widehat{G}$. It is easy to check that $\widehat{G}$ with the data $(\hat{m}, \hat{i}, \hat{e})$ is a commutative finite group scheme over $k$. It is called *Cartier dual* to $G$. The definition immediately implies that the double Cartier dual to $G$ coincides with $G$. The main property of this duality is the following natural isomorphism of functors on the category of $k$-schemes $S$ with values in commutative groups:

$$\widehat{G}(S) \simeq \text{Hom}_S(G_S, \mathbb{G}_{m,S}), \tag{10.1.1}$$

where $G_S = G \times_k S$, $\text{Hom}_S$ denotes the set of homomorphisms of group schemes over $S$. It suffices to define the isomorphism (10.1.1) for affine schemes $S$. Let $S = \text{Spec}(R)$ be such a scheme, where $R$ is a $k$-algebra. A morphism of $S$-schemes $f : G_S \to \mathbb{G}_{m,S}$ corresponds to a homomorphism of $R$-algebras $R[t, t^{-1}] \to A \otimes_k R$, i.e., to an invertible element $\alpha \in A \otimes_k R$.

We can view $\alpha$ as a morphism of $R$-modules $\alpha' : A^* \otimes_k R \to R$. It is easy to check that $f$ is a homomorphism of group schemes if and only if $\alpha'$ is a homomorphism of algebras, i.e., corresponds to an $S$-point of $\widehat{G}$.

**Examples.**

1. Let us consider the discrete group $\mathbb{Z}/n\mathbb{Z}$ (as a scheme it is the disjoint union of $n$ copies of Spec($k$)). We claim that the Cartier dual group to it is $\mu_n := \ker([n] : \mathbb{G}_m \to \mathbb{G}_m)$. Indeed, for any commutative group scheme $G$ we have a functorial isomorphism of groups $\mathrm{Hom}_S((\mathbb{Z}/nZ)_S, G_S) = \ker([n] : G(S) \to G(S))$ where $S$ is a scheme over $k$. Applying this to $G = \mathbb{G}_m$ and using (10.1.1) we derive an isomorphism $\widehat{\mathbb{Z}/nZ}(S) \simeq \mu_n(S)$ for all $S$. Note that the group $\mu_n$ is not reduced if the characteristic of $k$ divides $n$.
2. Assume that characteristic of $k$ is $p > 0$ and let us denote $\alpha_p := \ker([p] : \mathbb{G}_a \to \mathbb{G}_a)$. Then $\alpha_p$ is a finite group scheme of order $p$ with the function ring $k[x]/(x^p)$. It is easy to see that $\alpha_p$ is Cartier dual to itself. The corresponding bilinear pairing
$$\alpha_p \times \alpha_p \to \mathbb{G}_m$$
is given by the truncated exponent $\exp_p(xy) \in (k[x, y]/(x^p, y^p))^*$, where $\exp_p(t) = 1 + t + t^2/2 + \cdots + t^{p-1}/(p-1)!$.

In fact, the category of commutative finite group schemes over $k$ is abelian (this result is due to Grothendieck) and the Cartier duality is an exact functor (cf. [103]). If the characteristic of $k$ is $p > 0$ then all simple objects in this category are: $\mathbb{Z}/l\mathbb{Z}$, where $l$ is a prime different from $p$, $\mathbb{Z}/p\mathbb{Z}$, $\mu_p$ and $\alpha_p$ (this is due to Gabriel [47], the proof can also be found in [103]). The above examples show that $\mathbb{Z}/l\mathbb{Z}$ and $\alpha_p$ are self-dual, while $\mathbb{Z}/p\mathbb{Z}$ is dual to $\mu_p$.

## 10.2. Central Extensions

Let us recall some general nonsense concerning central extensions (borrowed from [58], exp.VII, see also Section 1 of [28]). Below we identify line bundles and the corresponding $\mathbb{G}_m$-torsors. In particular, we denote the sum of $\mathbb{G}_m$-torsors as tensor product. A central extension of a group scheme $G$ by $\mathbb{G}_m$ is equivalent to the following data:

(i) for every point $g \in G(S)$, a line bundle $L_g$ over $S$ (where $S$ is a $k$-scheme),

(ii) for every pair of points $g_1, g_2 \in G(S)$, an isomorphism $a_{g_1,g_2} : L_{g_1} \otimes L_{g_2} \to L_{g_1 g_2}$,

(iii) for every morphism $\phi : S \to S'$ and a point $g \in G(S)$, an isomorphism $b_{g,\phi} : L_g \to \phi^* L_{\phi(g)}$.

The data (ii) should satisfy the following cocycle condition for every three points $g_1, g_2, g_3 \in G(S)$:

$$a_{g_1 g_2, g_3} \circ (a_{g_1,g_2} \otimes \mathrm{id}) = a_{g_1, g_2 g_3} \circ (\mathrm{id} \otimes a_{g_2,g_3}). \tag{10.2.1}$$

The data (iii) should be transitive with respect to composition of morphisms $S \to S' \to S''$ and compatible with data (ii) in the obvious sense.

Using functoriality, we can rewrite these data in terms of a single line bundle $L$ over $G$ corresponding to the canonical point $\mathrm{id} \in G(G)$. Then the data (ii) becomes an isomorphism $m^*L \simeq p_1^*L \otimes p_2^*L$ of line bundles on $G \times G$, where $m : G \times G \to G$ is the group law. The cocycle condition becomes an equality of isomorphisms between line bundles on $G \times G \times G$. Note that the above definition works also for a group scheme over any base scheme. If $G$ is commutative one can define commutative extensions of $G$ by $\mathbb{G}_m$ in a similar way by adding the condition that $a_{g_1,g_2} = a_{g_2,g_1}$. For example, a line bundle $L \in \mathrm{Pic}^0(A)$ defines a commutative extension of $A$ by $\mathbb{G}_m$: if we trivialize $L$ at zero then there is a canonical isomorphism

$$m^*L \simeq p_1^*L \otimes p_2^*L$$

on $G \times G$ compatible with the trivialization of $L$ at zero, which induces the above data $a_{x,y}$. More generally, for every $k$-scheme $S$ one has an isomorphism of groups (functorial in $S$)

$$\hat{A}(S) = \mathrm{Ext}^1_S(A_S, \mathbb{G}_{m,S}),$$

where $\mathrm{Ext}^1_S$ is the group of extensions in the category of commutative groups schemes over $S$, $A_S = A \times_k S$. This should be compared with the property (10.1.1) for finite commutative groups schemes. Using this description of $\hat{A}$ we can interpret the canonical morphism $\mathrm{can}_A : A \to \hat{\hat{A}} = \mathrm{Ext}^1(\hat{A}, \mathbb{G}_m)$ considered in Section 9.3 as follows: $\mathrm{can}_A(x)$ is the $\mathbb{G}_m$-torsor over $\hat{A}$ consisting of isomorphism classes of triples $(L, \alpha_0, \alpha_x)$ where $L$ is a line bundle in $\mathrm{Pic}^0(A)$, $\alpha_0$ (resp., $\alpha_x$) is a trivialization of $L|_0$ (resp., $L|_x$). One can similarly define the map on $S$-points.

### 10.3. Biextensions

Let $A$ and $B$ be abelian varieties and let $\mathcal{B}$ be a line bundle on $A \times B$ that has trivial restrictions to $A \times 0$ and $0 \times B$. Using the theorem of the cube it is easy to see that we have the following isomorphisms of line bundles on $A \times A \times B$ and $A \times B \times B$:

$$a_{x_1,x_2;y} : \mathcal{B}_{x_1+x_2,y} \widetilde{\to} \mathcal{B}_{x_1,y} \otimes \mathcal{B}_{x_2,y},$$
$$a_{x;y_1,y_2} : \mathcal{B}_{x,y_1+y_2} \widetilde{\to} \mathcal{B}_{x,y_1} \otimes \mathcal{B}_{x,y_2}. \qquad (10.3.1)$$

Let us choose a trivialization of the fiber $\mathcal{B}_{0,0}$. Then we can choose isomorphisms (10.3.1) uniquely, so that they restrict to identity maps under this trivialization. The obtained isomorphisms satisfy the following cocycle conditions:

(i) the equation (10.2.1) for $a_{x_1,x_2;y}$ in $(x_1, x_2)$,
(ii) the equation (10.2.1) for $a_{x;y_1,y_2}$ in $(y_1, y_2)$,
(iii) $(a_{x_1,x_2;y_1} \otimes a_{x_1,x_2;y_2})a_{x_1+x_2;y_1,y_2} = (a_{x_1;y_1,y_2} \otimes a_{x_2;y_1,y_2})a_{x_1,x_2;y_1+y_2}$.

In general, if $G_1$ and $G_2$ are groups schemes, $\mathcal{B}$ is a line bundle over $G_1 \times G_2$ equipped with isomorphisms (10.3.1) which satisfy the conditions (i)–(iii), then we say that $\mathcal{B}$ is a *biextension* of $G_1 \times G_2$ by $\mathbb{G}_m$ (or simply that $\mathcal{B}$ is a biextension on $G_1 \times G_2$). The reason for this term is that $\mathcal{B}$ defines the structure of commutative extension by $\mathbb{G}_m$ of both $G_1 \times G_2$ considered as a $G_1$-scheme and $G_1 \times G_2$ considered as a $G_2$-scheme. One can also rewrite the definition of biextension using $S$-points for all schemes $S$ (similar to the definition of central extension we gave above).

Clearly, the Poincaré bundle $\mathcal{P}$ on $A \times \hat{A}$ is a biextension. More generally, if $A$ and $B$ are abelian varieties then any homomorphism $f : B \to \hat{A}$ gives rise to a biextension $(\mathrm{id} \times f)^*\mathcal{P}$ on $A \times B$. It is easy to see that all biextensions of $A \times B$ by $\mathbb{G}_m$ arise in this way (see Exercise 3). Switching the roles of $A$ and $B$ in this correspondence, we obtain a natural involutive isomorphism

$$\mathcal{D} : \mathrm{Hom}(A, \hat{B}) \widetilde{\to} \mathrm{Hom}(B, \hat{A}).$$

It is easy to see that this isomorphism sends $f \in \mathrm{Hom}(A, \hat{B})$ to $\hat{f} \circ \mathrm{can}_B$, where $\mathrm{can}_B : B \to \hat{\hat{B}}$ is the canonical morphism.

For example, for every line bundle $L$ on $A$ (trivialized at 0) we have a biextension $\Lambda(L) = m^*L \otimes p_1^*L^{-1} \otimes p_2^*L^{-1}$ of $A \times A$ by $\mathbb{G}_m$ which corresponds to the morphism $\phi_L : A \to \hat{A}$. This biextension is *symmetric* (with respect to switching two factors of $A \times A$) which corresponds to the fact that the morphism $\phi_L$ is self-dual with respect to the duality $\mathcal{D}$ above: $\mathcal{D}(\phi_L) = \phi_L$. Equivalently, the morphism $\phi_L$ is symmetric (see Section 9.3).

The notion of a biextension can be considered as a categorification of the notion of a bilinear pairing: the basic identities from the definition of bilinear form get replaced by isomorphisms (10.3.1) which leads to necessity of adding higher constraints. On the other hand, the theorem of the cube shows that in some sense line bundles on abelian varieties behave like quadratic forms. The relation between a line bundle $L$ and a biextension $\Lambda(L)$ is similar to the relation between a quadratic form $q$ and the associated symmetric bilinear pairing $b_q(x, y) = q(x + y) - q(x) - q(y)$.

## 10.4. Double Dual and the Weil Pairing

Let $f : A \to B$ be an *isogeny* of abelian varieties, i.e., a surjective homomorphism with finite kernel, and let $\hat{f} : \hat{B} \to \hat{A}$ be the dual homomorphism.

**Theorem 10.1.** *The homomorphism $\hat{f}$ is also an isogeny. The finite group schemes* $\ker(f)$ *and* $\ker(\hat{f})$ *are Cartier dual to each other.*

*Proof.* Consider the exact sequence of group schemes

$$0 \to \ker(f) \to A \to B \to 0.$$

The induced sequence of Ext-groups is

$$\cdots \to \mathrm{Hom}(A, \mathbb{G}_m) \to \mathrm{Hom}(\ker(f), \mathbb{G}_m) \to \mathrm{Ext}^1(B, \mathbb{G}_m) \to \mathrm{Ext}^1(A, \mathbb{G}_m)$$

(where $\mathrm{Ext}^i(G, G')(S) = \mathrm{Ext}^i_S(G_S, G'_S)$ for $i \leq 1$). Since $\mathrm{Hom}(A, \mathbb{G}_m) = 0$ we get the exact sequence

$$0 \to \widehat{\ker(f)} \to \hat{B} \xrightarrow{\hat{f}} \hat{A}.$$

Since $\hat{f}$ has a finite kernel and $\dim A = \dim B$, $\hat{f}$ is also surjective. □

**Corollary 10.2.** *The canonical map* $\mathrm{can}_A : A \to \hat{\hat{A}}$ *is an isomorphism.*

*Proof.* Let $\phi : A \to \hat{A}$ be a symmetric isogeny with kernel $K$ (e.g., take $\phi = \phi_L$ for some ample line bundle $L$ on $A$). Then we have a morphism of commutative diagrams

$$\begin{array}{ccccccccc} 0 & \longrightarrow & K & \longrightarrow & A & \xrightarrow{\phi} & \hat{A} & \longrightarrow & 0 \\ & & \big\downarrow & & \big\downarrow{\scriptstyle \mathrm{can}_A} & & \big\downarrow{\scriptstyle \mathrm{id}} & & \\ 0 & \longrightarrow & \widehat{K} & \longrightarrow & \widehat{\hat{A}} & \xrightarrow{\hat{\phi}} & \hat{A} & \longrightarrow & 0 \end{array} \qquad (10.4.1)$$

It follows that $\mathrm{can}_A : A \to \widehat{\hat{A}}$ is an isogeny, in particular, it is surjective. Therefore, the morphism $K \to \widehat{K}$ is surjective. Since these are finite group schemes of the same order it follows that the morphism $K \to \widehat{K}$ is an isomorphism, hence $\mathrm{can}_A : A \to \widehat{\hat{A}}$ is an isomorphism. □

Henceforward, we will always identify $A$ with $\widehat{\hat{A}}$ using $\mathrm{can}_A$. Note that under this identification the Poincaré line bundles on $A \times \hat{A}$ and on $\widehat{\hat{A}} \times \hat{A}$ coincide (by the definition of $\mathrm{can}_A$).

Now we are going to interpret the definition of the pairing between $\ker(f)$ and $\ker(\hat{f})$ constructed in Theorem 10.1, in terms of biextensions (the abelian variety $B$ is denoted below as $\hat{B}$).

Let $\mathcal{B}$ be the biextension on $A \times B$, where $A$ and $B$ are abelian varieties. Let us define the *left kernel* $\ker^l(\mathcal{B}) \subset A$ (resp., *right kernel* $\ker^r(\mathcal{B}) \subset B$) of $\mathcal{B}$ to be the kernel of the corresponding homomorphism $f : A \to \hat{B}$ (resp., $\hat{f} : B \to \hat{A}$). The restrictions $\mathcal{B}|_{\ker^l(\mathcal{B}) \times B}$ and $\mathcal{B}|_{A \times \ker^r(\mathcal{B})}$ are canonically trivialized, but these trivializations are not necessarily compatible over $\ker^l(\mathcal{B}) \times \ker^r(\mathcal{B})$. The difference between them gives a canonical pairing

$$e_f = e_{\mathcal{B}} : \ker^l(\mathcal{B}) \times \ker^r(\mathcal{B}) \to \mathbb{G}_m$$

In the case when $f$ is isogeny, both kernels are finite and $e_f$ is exactly the pairing of Theorem 10.1. Therefore, in this case $e_f$ is nondegenerate.

Let $e_{\hat{f}} : \ker^r(\mathcal{B}) \times \ker^l(\mathcal{B}) \to \mathbb{G}_m$ be the similar pairing associated with $\mathcal{B}$ considered as a biextension on $B \times A$. It is clear from the above description that $e_f(x, y) = e_{\hat{f}}(y, x)^{-1}$. In particular, if $f : A \to \hat{A}$ is a symmetric isogeny then the induced pairing on $\ker(f) = \ker(\hat{f})$ is skew-symmetric. In fact, the following strong form of the skew-symmetry holds.

**Proposition 10.3.** *Let $f : A \to \hat{A}$ be a symmetric isogeny, $e_f$ be the corresponding bilinear form on* $\ker(f)$ *with values in $\mathbb{G}_m$. Then* $e_f(x, x) \equiv 1$.

*Proof.* Let $\mathcal{B}$ be a symmetric biextension of $A \times A$ corresponding to $f$. The symmetry isomorphism $\sigma : \mathcal{B}_{x,y} \simeq \mathcal{B}_{y,x}$ restricts to the identity on $\mathcal{B}_{0,0}$.

Therefore, the restriction of $\sigma$ to the diagonal in $A \times A$ is equal to the identity morphism. Now our statement follows from the fact that the trivialization of $\mathcal{B}|_{\ker(f)\times A}$ is obtained from the trivialization of $\mathcal{B}|_{A\times\ker(f)}$ by applying $\sigma$. $\square$

Applying the above construction to the $n$th power of the Poincaré line bundle on $A \times \hat{A}$ (which corresponds to the isogeny $[n]_A : A \to A$) we obtain the canonical perfect pairing

$$e_n : A_n \times \hat{A}_n \to \mathbb{G}_m$$

called the *Weil pairing*.

## 10.5. Descent and Biextensions

Let $\mathcal{B}$ be a biextension on $A \times B$, where $A$ and $B$ are abelian varieties. Let $A \to A'$ (resp., $B \to B'$) be a surjective homomorphism of abelian varieties with kernel $K_A \subset A$ (resp., $K_B \subset B$). We want to find a necessary and sufficient condition for $\mathcal{B}$ to be the pull-back of a biextension on $A' \times B'$. Clearly, it is necessary that $K_A \subset \ker^l(\mathcal{B})$ and $K_B \subset \ker^r(\mathcal{B})$.

**Proposition 10.4.** *A biextension $\mathcal{B}$ on $A \times B$ descends to a biextension on $A' \times B'$ if and only if $K_A \subset \ker^l(\mathcal{B})$, $K_B \subset \ker^r(\mathcal{B})$ and the restriction of the canonical pairing $e_{\mathcal{B}}$ to $K_A \times K_B$ is trivial.*

*Proof.* Clearly, the condition $K_A \subset \ker^l(\mathcal{B})$ is equivalent to the condition that $\mathcal{B}$ descends to a biextension on $A' \times B$. The corresponding descent data on $\mathcal{B}$ are given by a trivialization of $\mathcal{B}|_{K_A\times B}$. Now descent data on $\mathcal{B}$ for the projection $A \times B \to A' \times B'$ amount to commuting descent data for the projections $A \times B \to A' \times B$ and $A \times B \to A \times B'$. In other words, we should have trivializations of $\mathcal{B}$ over $K_A \times B$ and over $A \times K_B$, which are compatible on $K_A \times K_B$. This compatibility is equivalent to the condition that $K_A$ and $K_B$ are orthogonal with respect to $e_{\mathcal{B}}$. $\square$

Note that in the situation of the above proposition, a biextension $\overline{\mathcal{B}}$ on $A' \times B'$, such that $\mathcal{B}$ is a pull-back of $\overline{\mathcal{B}}$, is unique.

Now let $L$ be a line bundle on an abelian variety $A$, and let $\pi : A \to A'$ be a surjective morphism of abelian varieties with kernel $K \subset A$. We want to find a criterion for $L$ to be a pull-back of a line bundle on $A'$. To find such a criterion we can use the analogy between line bundles and quadratic forms. Indeed, let $q$ is a quadratic form on an abelian group $G$, $H \subset G$ be a subgroup. Then $q$ descends to a quadratic form on $G/H$ if and only if two conditions are satisfied: 1) the restriction $q|_H$ is trivial, 2) the restriction of the bilinear map $b_q(g, g') = q(g+g') - q(g) - q(g')$ to $H \times G$ is trivial. In the case of line bundles we have to require $L|_K$ to be trivial and the biextension $\Lambda(L)|_{K\times A}$ to be trivial. Note that the condition of triviality of a biextension $\Lambda(L)$ on $K \times A$ just means that $K \subset K(L)$. Moreover, in this case there exists a unique trivialization of the biextension $\Lambda(L)|_{K\times A}$ (because there are no bilinear maps $K \times A \to \mathbb{G}_m$). However, if $K$ is not connected, the choice of a trivialization of $L|_K$ is an additional structure, so we have to be more careful. Here is the precise statement.

**Theorem 10.5.** *$L$ is isomorphic to the line bundle of the form $p^*L'$ for some $L'$ on $A'$ if and only if $K \subset K(L)$, $K$ is isotropic with respect to the canonical pairing $e_{\phi_L} = e_{\Lambda(L)} : K(L) \times K(L) \to \mathbb{G}_m$ and the line bundle $L|_K$ is trivial. More precisely, a choice of $L'$ corresponds to a choice of trivialization of $L|_K$, such that the induced trivialization of $\Lambda(L)|_{K\times K}$ coincides with the restriction of the trivialization of the biextension $\Lambda(L)|_{K(L)\times A}$.*

*Proof.* It is clear that an isomorphism $L \simeq p^*L'$ gives rise to the described structure. Conversely, assume that $K$ is an isotropic subgroup of $K(L)$ and we have a trivialization of $L|_K$ as in the theorem. Then we can define the descent data on $L$ for the projection $A \to A' = A/K$ as follows. Let us identify $A \times_{A'} A$ with $K \times A$, so that the descent data should be an isomorphism $\alpha : L_{k+x} \simeq L_x$ on $K \times A$ satisfying the natural cocycle condition on $K \times K \times A$. Let us define $\alpha$ as the composition of the canonical isomorphism $L_{k+x} \simeq L_x \otimes L_k \otimes \Lambda(L)_{k,x}$ with the trivializations of $L_k$ and of $\Lambda(L)_{k,x}$. It is easy to see that the cocycle condition is equivalent to the compatibility of these two trivializations. □

## 10.6. Transcendental Computation of the Weil Pairing

Let us compute the Weil pairing in the case of complex abelian varieties. Let $A = V/\Gamma$ be such a variety where $V$ is complex vector space, $\Gamma \subset V$ is a lattice, $\hat{A} = \overline{V}^{\vee}/\Gamma^{\vee}$ be the dual variety. Then the Poincaré line bundle on $A \times \hat{A}$ can be identified with the line bundle $L(H_{\text{univ}}, \alpha_{\text{univ}})$ described in Section 1.4. According to the definition, the value of the Weil pairing at a point $(x, y) \in A_n \times \hat{A}_n$ measures the difference between two trivializations of the fiber $L(nH_{\text{univ}}, \alpha^n_{\text{univ}})_{(x,y)}$: one coming from the isomorphism

$$i_1 : L(nH_{\text{univ}}, \alpha^n_{\text{univ}}) \widetilde{\to} ([n] \times \mathrm{id})^* L(H_{\text{univ}}, \alpha_{\text{univ}})$$

and the trivialization of $L(H_{\text{univ}}, \alpha_{\text{univ}})$ at $(0, y)$, and another coming from the isomorphism

$$i_2 : L\big(nH_{\text{univ}}, \alpha^n_{\text{univ}}\big) \widetilde{\to} (\text{id} \times [n])^* L(H_{\text{univ}}, \alpha_{\text{univ}})$$

and the trivialization of $L(H_{\text{univ}}, \alpha_{\text{univ}})$ at $(x, 0)$. To compute these isomorphisms we note that a lifting of a point $u \in A \times \hat{A}$ to a point $\widetilde{u} \in V \oplus \overline{V}^{\vee}$ gives a canonical nonzero element $s_{\widetilde{u}}$ of the fiber $L(H_{\text{univ}}, \alpha_{\text{univ}})_u$. A different lifting $\widetilde{u} + \lambda$ where $\lambda \in \Gamma \oplus \Gamma^{\vee}$ leads to the trivialization

$$s_{\widetilde{u}+\lambda} = \alpha_{\text{univ}}(\lambda) \exp\left(\pi H_{\text{univ}}(\widetilde{u}, \lambda) + \frac{\pi}{2} H_{\text{univ}}(\lambda, \lambda)\right) \cdot s_{\widetilde{u}}. \quad (10.6.1)$$

Now let $(x, y)$ be a point in $A_n \times \hat{A}_n$. We can lift it to a point of the form $(\gamma/n, \gamma^{\vee}/n)$ in $V \oplus \overline{V}^{\vee}$. Then it is easy to see that

$$i_1\big(s_{(\gamma/n, \gamma^{\vee}/n)}\big) = s_{(\gamma, \gamma^{\vee}/n)},$$
$$i_2\big(s_{(\gamma/n, \gamma^{\vee}/n)}\big) = s_{(\gamma/n, \gamma^{\vee})}.$$

Now applying (10.6.1) to $\lambda = (\gamma, 0)$ and to $\lambda = (0, \gamma^{\vee})$ we obtain

$$s_{(\gamma, \gamma^{\vee}/n)} = \exp(\pi H_{\text{univ}}(\gamma^{\vee}/n, \gamma)) s_{(0, \gamma^{\vee}/n)},$$
$$s_{(\gamma/n, \gamma^{\vee})} = \exp(\pi H_{\text{univ}}(\gamma/n, \gamma^{\vee})) s_{(\gamma/n, 0)}.$$

Since the elements $s_{(0,\gamma^{\vee}/n)}$ and $s_{(\gamma/n,0)}$ are compatible with the trivializations of $L(H_{\text{univ}}, \alpha_{\text{univ}})$ at $0 \times \hat{A}$ and $A \times 0$, the value of the Weil pairing at $(\gamma/n, \gamma^{\vee}/n)$ is equal to

$$\exp\left(\frac{\pi}{n}(H_{\text{univ}}(\gamma^{\vee}, \gamma) - H_{\text{univ}}(\gamma, \gamma^{\vee}))\right) = \exp\left(\frac{2\pi i}{n}\langle \gamma^{\vee}, \gamma \rangle\right).$$

## Exercises

1. Let $0 \to A \to B \to C \to 0$ be an exact sequence of abelian varieties. Show that the dual abelian varieties also form an exact sequence $0 \to \hat{C} \to \hat{B} \to \hat{A} \to 0$.
2. (a) Let $f : A \to B$ and $g : B \to C$ be isogenies of abelian varieties. Show that the natural exact sequences

$$0 \to \ker(f) \to \ker(gf) \to \ker(g) \to 0$$

and

$$0 \to \ker(\hat{g}) \to \ker(\hat{f}\hat{g}) \to \ker(\hat{f}) \to 0$$

are dual to each other.

(b) Let $f : A \to \hat{A}$ be a symmetric homomorphism. Show that the restriction of the pairing $e_{nf}$ to $A_n$ is given by

$$e_{nf}(x, y) = e_n(x, f(y)),$$

where $x, y \in A_n$.

(c) Show that the restriction of the Weil pairing $e_{nk}$ to $A_n \times \hat{A}_n \subset A_{nk} \times \hat{A}_{nk}$ is equal to $e_n^k$.

3. Let $\mathcal{B}$ be a biextension of $G_1 \times G_2$ by $\mathbb{G}_m$. Then we have a canonical map $G_1(S) \to \operatorname{Ext}^1(G_2(S), \mathbb{G}_{m,S})$. Show that in the case of abelian varieties this gives a morphism $\phi : G_1 \to \hat{G}_2$. Check that $\mathcal{B}$ is isomorphic to $(\phi \times \mathrm{id})^*\mathcal{P}$ as a biextension.
4. Let $L$ be a line bundle on an abelian variety $A$, $B$ be an abelian subvariety such that $B \subset K(L)$, $p : A \to A/B$ be the quotient morphism.
   (a) Show that there exists a point $x \in \hat{A}$ such that $L \simeq p^*L' \otimes \mathcal{P}_x$ for some line bundle $L'$ on $A/B$.
   (b) Show that all line bundles $L|_{B+a}$ on $B$, where $a \in A$, are isomorphic.
   (c) Show that if $\dim B > 0$ then there exists a point $x \in \hat{A}$ such that $Rp_*(L \otimes \mathcal{P}_x) = 0$.
5. Let $K$ be a finite commutative group scheme which is annihilated by 2.
   (a) Let $0 \to \mu_2 \to K' \to K \to 0$ be an exact sequence of finite commutative group schemes. Show that this sequence splits if and only if $K'$ is annihilated by 2. [*Hint*: Use Cartier duality.]
   (b) Let $e : K \times K \to \mu_2$ be a bilinear pairing such that $e(x, x) \equiv 1$. Show that there exists a morphism $\phi : K \to \mu_2$ such that $e(x, y) = \phi(x + y)\phi(x)\phi(y)$. [*Hint*: Construct an extension $K'$ of $K$ by $\mu_2$ using $e$.]
6. Let $L$ be a line bundle on an abelian variety $A$, $S$ be a symmetric integer-valued $n \times n$ matrix. Let us define a line bundle $L^S$ on $A^n = A \times \cdots \times A$ ($n$ times) by the formula

$$L^S = \bigotimes_{i=1}^{n} p_i^* L^{s_{ii}} \otimes \bigotimes_{i<j} p_{ij}^* \Lambda(L)^{s_{ij}},$$

where $S = (s_{ij})$. For every $n \times m$ matrix $M = (m_{ij})$ we have a homomorphism

$$[M]_A : A^m \to A^n : (x_j)_{j=1,\dots,m} \mapsto \left( \sum_i m_{ij} x_j \right)_{i=1,\dots,n}.$$

Show that $[M]_A^* L^S$ is algebraically equivalent to $L^{{}^tMSM}$. Show that if

$L$ is symmetric (i.e., $[-1]_A^* L \simeq L$) then these two line bundles are isomorphic.

7. Let $I \subset A_n$ be a subgroup, $I^\perp \subset \hat{A}_n$ be the orthogonal complement to $I$ with respect to the Weil pairing.
   (a) Construct an isomorphism $\widehat{A/I} \simeq \hat{A}/I^\perp$.
   (b) Assume that there exists a symmetric isomorphism $\phi : A \widetilde{\rightarrow} \hat{A}$ such that $I$ is Lagrangian with respect to the induced pairing on $A_n$. Prove that $\phi$ induces a symmetric isomorphism $A/I \widetilde{\rightarrow} \widehat{A/I}$.
8. Let $K_A \subset A$, $K_A \subset B$ be finite subgroup schemes in abelian varieties, $\mathcal{B}$ be a biextension on $A \times B$ which is the pull-back of a biextension $\overline{\mathcal{B}}$ on $A/K_A \times B/K_B$. Prove that $\ker^l(\overline{\mathcal{B}}) = K_B^\perp/K_A$, $\ker^r(\overline{\mathcal{B}}) = K_A^\perp/K_B$, where $K_B^\perp \subset \ker^l(\mathcal{B})$ is the orthogonal complement to $K_B$ with respect to $e_{\mathcal{B}}$, etc.

# 11

# Fourier–Mukai Transform

The beginning of this chapter is devoted to the proof of Theorem 9.4 (stating that the dual abelian variety represents a certain functor). The proof is quite technical, however, the basic idea is simple. Every object $K$ of derived category on the product $X \times Y$ gives rise to a functor $\Phi_{K,X\to Y}$ from the derived category of quasi-coherent sheaves on $X$ to the similar category on $Y$. The definition mimics that of integral transform with a kernel: to apply $\Phi_{K,X\to Y}$ to a sheaf on $X$, one has to pull this sheaf back to $X \times Y$, then take a tensor product with $K$ and finally apply the push-forward to $Y$. The Fourier–Mukai transform is the functor $\Phi_{\mathcal{P},A\to\hat{A}}$ corresponing to the Poincaré bundle $\mathcal{P}$ on $A \times \hat{A}$. The main ingredient of the proof of Theorem 9.4 is the fact that the composition $\Phi_{\mathcal{P},A\to\hat{A}} \circ \Phi_{\mathcal{P}^{-1}[g],\hat{A}\to A}$ is isomorphic to the identity. This gives a nice way to recover a point $\xi \in \mathrm{Pic}^0(A)$ from the corresponding line bundle $\mathcal{P}_\xi$ on $A$: the transform $\Phi_{\mathcal{P},A\to\hat{A}}(\mathcal{P}_\xi)$ is isomorphic to the structure sheaf of the point $-\xi \in \hat{A}$ (up to a shift of degree). To prove Theorem 9.4 we apply the similar operation to a family of line bundles in $\mathrm{Pic}^0(A)$.

Other parts of this chapter are devoted to the study of the Fourier–Mukai transform and some of its applications. Aside from the properties one expects by analogy with the usual Fourier transform, we show that this transform can be used in studying cohomology of *nondegenerate line bundles* on abelian variety (line bundles $L$ with finite $K(L)$). Combining properties of the Fourier–Mukai transform with the theorem of the cube we prove that the transform of such a line bundle $L$ is a vector bundle concentrated in one degree $i(L)$ (called the *index* of a line bundle) and of rank $|K(L)|^{\frac{1}{2}}$. This easily implies the similar statement about cohomology of $L$, which is an algebraic analogue of Theorem 7.2. In a different direction we show that the functor of tensoring with a line bundle $L$, such that $K(L)$ is trivial, together with the Fourier–Mukai transform generate an action of a central extension of $\mathrm{SL}_2(\mathbb{Z})$ on the derived category of $A$. This action can be considered as a (partial) categorical analogue of the Weil representation constructed in the context of representation theory of the Heisenberg group (see Chapter 15 for more about this analogy). It will be used in Chapter 14 for the study of vector bundles on elliptic curves.

## 11.1. Functors between Derived Categories of Coherent Sheaves

Let $X$ be a scheme of finite type over an algebraically closed field $k$. By $D^*_{qc}(X)$ we denote the derived category of quasi-coherent sheaves on $X$ (where $* \in \{+, -, b\}$), it has a full subcategory $D^*(X)$ consisting of complexes with coherent cohomology. If $X$ and $Y$ are two such schemes and $K$ is an object of $D^-_{qc}(X \times Y)$ then we consider the corresponding exact functor

$$\Phi_K = \Phi_{K,X\to Y} : D^-_{qc}(X) \to D^-_{qc}(Y) : F \mapsto Rp_{2*}(p_1^* F \otimes^{\mathbb{L}} K),$$

where $p_1$ and $p_2$ are projections of $X \times Y$ to $X$ and $Y$. We will say that $K$ is the *kernel* defining the functor $\Phi_K$. If $K$ has finite Tor-dimension then the functor $\Phi_{K,X\to Y}$ sends $D^b_{qc}(X)$ to $D^b_{qc}(Y)$. If $X$ is proper and $K \in D^-(X \times Y)$ then $\Phi_{K,X\to Y}$ sends $D^-(X)$ to $D^-(Y)$. Hypothetically, in this way one can obtain all exact functors between derived categories of coherent sheaves (at least in the case of smooth projective varieties). The theorem of D. Orlov (see Appendix C, or [104]) implies that this is the case for exact equivalences between such categories (for smooth projective varieties).

**Proposition 11.1.** *For any $K \in D^-_{qc}(X \times Y)$ and $L \in D^-_{qc}(Y \times Z)$ one has a natural isomorphism of functors*

$$\Phi_{L,Y\to Z} \circ \Phi_{K,X\to Y} \simeq \Phi_{K*L,X\to Z},$$

*where*

$$K * L = Rp_{13*}(p_{12}^* K \otimes^{\mathbb{L}} p_{23}^* L) \in D^-_{qc}(X \times Z),$$

*$p_{ij}$ are projections from $X \times Y \times Z$ to pairwise products.*

*Proof.* Applying the flat base change ([61], Proposition 5.12) to the Cartesian square

$$\begin{array}{ccc} X \times Y \times Z & \xrightarrow{p_{12}} & X \times Y \\ \downarrow{\scriptstyle p_{23}} & & \downarrow \\ Y \times Z & \longrightarrow & Y \end{array} \tag{11.1.1}$$

we obtain

$$\begin{aligned}\Phi_{L,Y\to Z}\circ\Phi_{K,X\to Y}(F) &\simeq Rp_{Z*}(Rp_{23*}p_{12}^*(p_X^*F\otimes^{\mathbb{L}}K)\otimes^{\mathbb{L}}L)\\ &\simeq Rp_{Z*}(Rp_{23*}(p_1^*F\otimes^{\mathbb{L}}p_{12}^*K)\otimes^{\mathbb{L}}L),\end{aligned}$$

where $p_X : X\times Y\to X$, $p_Z : Y\times Z\to Z$, $p_1 : X\times Y\times Z\to X$ are the natural projections. Now the projection formula ([61], Proposition 5.6) for the morphism $p_{23}$ gives an isomorphism

$$\begin{aligned}Rp_{Z*}(Rp_{23*}(p_1^*F\otimes^{\mathbb{L}}p_{12}^*K)\otimes^{\mathbb{L}}L) &\simeq Rp_{Z*}Rp_{23*}(p_1^*F\otimes^{\mathbb{L}}p_{12}^*K\otimes^{\mathbb{L}}p_{23}^*L)\\ &\simeq Rp_{3*}(p_1^*F\otimes^{\mathbb{L}}p_{12}^*K\otimes^{\mathbb{L}}p_{23}^*L),\end{aligned}$$

where $p_3 : X\times Y\times Z\to X\times Z$ is the projection. Similarly, applying the projection formula to the morphism $p_{13} : X\times Y\times Z\to X\times Z$, we get

$$\Phi_{K*L,X\to Z}(F)\simeq Rp_{3*}(p_1^*F\otimes^{\mathbb{L}}p_{12}^*K\otimes^{\mathbb{L}}p_{23}^*L).$$ □

**Definition.** The operation $K*L$ defined in the above proposition will be called *convolution* of the kernels $K$ and $L$.

More generally, an object $K\in D_{qc}^-(X\times Y)$ defines a family of functors

$$\Phi_{K,X\times S\to Y\times S} : D_{qc}^-(X\times S)\to D_{qc}^-(Y\times S) : F\mapsto Rp_{S,2*}(p_{S,1}^*F\otimes^{\mathbb{L}}K),$$

where $S$ is a scheme, $p_{S,1} : X\times Y\times S\to X\times S$ and $p_{S,2} : X\times Y\times S\to Y\times S$ are the projections. We still have a natural isomorphism of functors

$$\Phi_{L,Y\times S\to Z\times S}\circ\Phi_{K,X\times S\to Y\times S}\simeq\Phi_{K*L,X\times S\to Z\times S}. \tag{11.1.2}$$

Using the base change of a flat morphism (see Appendix C), one can easily check that the above functors commute with (derived) pull-back functors associated to morphisms $S\to S'$.

## 11.2. Proof of Theorem 9.4

Let $A$ and $B$ be abelian varieties of the same dimension $g$, and $\mathcal{B}$ be a biextension of $A\times B$. We will deduce Theorem 9.4 from the following result.

**Theorem 11.2.** *Assume that the following conditions are satisfied:*

*(i) the maximal subscheme $S\subset B$ such that $\mathcal{B}|_{A\times S}$ is trivial coincides with $e_B\subset B$;*

*(ii) the map $B(k)\to \mathrm{Pic}^0(A)$ induced by $\mathcal{B}$ is surjective;*

*(iii) the maximal subscheme $S\subset A$ such that $\mathcal{B}|_{S\times B}$ is trivial, is finite.*

*Then the pair $(B,\mathcal{B})$ represents the functor from Theorem 9.4.*

**Lemma 11.3.** *For every scheme $S$ the functor $\Phi_{\mathcal{B}^{-1}[g],B\times S\to A\times S}$ is left-adjoint to $\Phi_{\mathcal{B},A\times S\to B\times S}$.*

*Proof.* For every $F \in D^-(B \times S)$, $G \in D^-(A \times S)$ we have

$$\begin{aligned}\mathrm{Hom}(F, \Phi_{\mathcal{B},A\times S\to B\times S}(G)) &\simeq \mathrm{Hom}(p_2^*F, p_1^*G \otimes \mathcal{B})\\ &\simeq \mathrm{Hom}(p_2^*F \otimes \mathcal{B}^{-1}, p_1^!G[-g])\\ &\simeq \mathrm{Hom}(Rp_{1*}(p_2^*F \otimes \mathcal{B}^{-1})[g], G)\\ &= \mathrm{Hom}(\Phi_{\mathcal{B}^{-1}[g],B\times S\to A\times S}(F), G),\end{aligned}$$

where we used the isomorphism $p_1^!G \simeq p_1^*G[g]$ which follows from the triviality of the canonical bundle $\omega_B$. □

**Theorem 11.4.** *Assume that the condition (i) of Theorem 11.2 is satisfied. Let us choose a trivialization of $\omega_B$. Then the canonical adjunction morphism*

$$\mathrm{Id} \to \Phi_{\mathcal{B},A\times S\to B\times S} \circ \Phi_{\mathcal{B}^{-1}[g],B\times S\to A\times S}$$

*is an isomorphism.*

**Lemma 11.5.** *Let $R$ be a regular local ring of dimension $g$, $P_0 \to P_1 \to \ldots \to P_n \to \ldots$ be a complex of finitely generated free $R$-modules. Assume that all modules $H^i(P_\bullet)$ have finite length. Then $H^i(P_\bullet) = 0$ for $i < g$.*

*Proof.* We use induction in $g$. For $g = 0$ the statement is trivial. Assume that the result is true for $g - 1$. Choose an element $x \in \mathfrak{m} \setminus \mathfrak{m}^2$. Then the ring $R/xR$ is regular of dimension $g - 1$ and $P_\bullet/xP_\bullet$ is a complex of finitely generated free $R/xR$-modules. The long exact sequence

$$\cdots \to H^{i-1}(P_\bullet) \to H^{i-1}(P_\bullet/xP_\bullet) \to H^i(P_\bullet) \xrightarrow{x} H^i(P_\bullet) \to \cdots.$$

shows that cohomology modules of $P_\bullet/xP_\bullet$ are of finite length. Thus, by induction assumption we get that $H^i(P_\bullet/xP_\bullet) = 0$ for $i < g - 1$. Now the same exact sequence implies that the multiplication by $x$ on $H^i(P_\bullet)$ is injective for $i < g$. Since some power of $x$ annihilates $H^i(P_\bullet)$ this is possible only if $H^i(P_\bullet) = 0$ for $i < g$. □

*Proof of Theorem 11.4.* Using (11.1.2) and the fact that $\mathcal{B}$ is a biextension, we can easily reduce the proof to showing that $Rp_{2*}(\mathcal{B}) \simeq \mathcal{O}_{e_B}[-g]$. Note that from Theorem 8.8 we immediately get that $Rp_{2*}(\mathcal{B})$ is supported on $e_B \subset B$. Thus, to compute $Rp_{2*}(\mathcal{B})$ we can replace $B$ by $\mathrm{Spec}(R)$ where $R$ is the local ring of $B$ at $e_B$. By the base change theorem (see Appendix C), $Rp_{2*}(\mathcal{B})|_{\mathrm{Spec}(R)}$

is represented by a complex $P_0 \to P_1 \to \cdots P_g$ of finitely generated free $R$-modules. Since cohomology groups of this complex are finitely generated $R$-modules supported on $e_B$ and $R$ is a regular ring of dimension $g$, applying Lemma 11.5 we obtain that $H^i(P_\bullet) = 0$ for $i < g$. It follows that the complex $P_g^\vee \to \cdots P_1^\vee \to P_0^\vee$ is a resolution of $M = \operatorname{coker}(P_1^\vee \to P_0^\vee)$. As we have shown in the proof of Proposition 9.3, the $R$-module $M$ has form $A/I$, where $I$ is the ideal of the maximal subscheme in $\operatorname{Spec}(R)$ over which $\mathcal{B}$ is trivial. Therefore, $I = \mathfrak{m}$ is the maximal ideal corresponding to $e_B$, so $M \simeq k$. Thus, $P_\bullet^\vee$ is a free resolution of $k$. It follows that $H^g(P_\bullet) \simeq \operatorname{Ext}_R^g(k, A) \simeq k$, i.e., $P_\bullet$ is quasi-isomorphic to $k[-g]$. □

*Proof of Theorem 11.2.* Let $S$ be a scheme, $L$ be a line bundle on $A \times S$ such that $L|_{A\times\{s\}} \in \operatorname{Pic}^0(A)$ for every $s \in S$, $L|_{e_A \times S} \simeq \mathcal{O}_S$. We want to construct a morphism $f : S \to B$ such that $L \simeq (\operatorname{id} \times f)^*\mathcal{B}$. Let us consider the object $\Phi_{\mathcal{B}}(L)$ in the derived category of sheaves on $B \times S$ (here and below $\Phi_{\mathcal{B}}$ is short for $\Phi_{\mathcal{B}, A\times S\to B\times S}$). The condition (ii) implies that for every point $s \in S$ the line bundle $L|_{A\times\{s\}}$ has form $\mathcal{B}^{-1}_{A\times b}$ for some point $b \in B$. On the other hand, by Theorem 11.4 we have

$$\Phi_{\mathcal{B}, A\to B}(\mathcal{B}^{-1}|_{A\times b}) \simeq \mathcal{O}_b[-g].$$

It follows that

$$\Phi_{\mathcal{B}}(L)|_{B\times\{s\}} \simeq \Phi_{\mathcal{B}, A\to B}(L|_{A\times\{s\}}) \simeq \mathcal{O}_b[-g].$$

Hence, $\Phi_{\mathcal{B}}(L) \simeq F[-g]$ for some coherent sheaf $F$ on $B \times S$, such that for every point $s \in S$ the restriction of $F$ to $B \times \{s\}$ is the structure sheaf of a point on $B$. In particular, $F$ is locally finitely generated as $\mathcal{O}_S$-module, We claim that $F$ is also flat over $S$. Indeed, it suffices to prove this when $S$ is affine. By Lemma 11.3 we have

$$\operatorname{Hom}(\mathcal{O}_{B\times S}, F) \simeq \operatorname{Hom}(\Phi_{\mathcal{B}^{-1}}(\mathcal{O}_{B\times S}), L).$$

But

$$\Phi_{\mathcal{B}^{-1}}(\mathcal{O}_{B\times S}) \simeq p_1^{AS*} R p_{1*}^{AB}(\mathcal{B}),$$

where $p_1^{AS} : A \times S \to A$, $p_1^{AB} : A \times B \to A$ are projections. The assumption (iii) implies that $Rp_{1*}^{AB}(\mathcal{B})$ has finite support. Furthermore, the same argument as earlier shows that $Rp_{1*}^{AB}(\mathcal{B}) \simeq M[-g]$, where $M$ is a coherent sheaf on $A$ with finite support. Therefore,

$$\operatorname{Hom}(\mathcal{O}_{B\times S}, F) \simeq \operatorname{Ext}^g\left(p_1^{AS*}(M), L\right) \simeq \operatorname{Ext}^g\left(M, p_{1*}^{AS}(L)\right).$$

Since $M$ has finite support we have

$$\mathrm{Ext}^g\left(M, p_{1*}^{AS}(L)\right) \simeq H^0\left(A, \underline{\mathrm{Ext}}^g\left(M, p_{1*}^{AS}(L)\right)\right).$$

Furthermore, since the $\mathcal{O}_A$-module $p_{1*}^{AS}(L)$ is flat, we have

$$\underline{\mathrm{Ext}}^g\left(M, p_{1*}^{AS}(L)\right) \simeq \underline{\mathrm{Ext}}^g(M, \mathcal{O}_A) \otimes_{\mathcal{O}_A} p_{1*}^{AS}(L).$$

Our claim follows easily from the fact that the coherent sheaf $\underline{\mathrm{Ext}}^g(M, \mathcal{O}_A)$ has finite support. Now we can finish the proof as follows. Since $F$ is a finitely generated flat $\mathcal{O}_S$-module, it is actually a locally free $\mathcal{O}_S$-module. Since the restriction of $F$ to $B \times \{s\}$ is a structure sheaf of a point on $B$, we conclude that $F$ is a line bundle supported on the graph of a morphism $\widetilde{f} : S \to B$. Finally, we set $f = -\widetilde{f} := [-1]_B \circ \widetilde{f}$. Then we have

$$\Phi_{\mathcal{B}^{-1}}(F) \simeq (\mathrm{id} \times f)^*\mathcal{B} \otimes p_2^*K$$

for some line bundle $K$ on $S$. On the other hand, the canonical adjunction morphism

$$\Phi_{\mathcal{B}^{-1}}(F) \simeq \Phi_{\mathcal{B}^{-1}[g]}\Phi_{\mathcal{B}}(L) \to L$$

is an isomorphism, since it is an isomorphism on every fiber $A \times \{s\}$. Therefore, $L \simeq (\mathrm{id} \times f)^*\mathcal{B} \otimes p_2^*K$. Comparing restrictions to $e_A \times S$ we conclude that $K \simeq \mathcal{O}_S$. □

*Proof of Theorem 9.4.* We have to check that the conditions (i)–(iii) of Theorem 11.2 are satisfied for $B = A/K(L)$ and for the line bundle $\mathcal{P}$ obtained by the descent from $\Lambda(L)$. If $S \subset B$ is a subscheme such that $\mathcal{P}|_{A \times S}$ is trivial, then $\Lambda(L)|_{A \times \pi^{-1}(S)}$ is trivial where $\pi : A \to B$ is the canonical projection. Therefore, $\pi^{-1}(S)$ is contained in $K(L)$ which implies that $S = e_B$. On the other hand, by Theorem 9.1 the homomorphism $\phi_L : A(k) \to \mathrm{Pic}^0(A)$ is surjective. Since it is equal to the composition of the projection $\pi : A(k) \to B(k)$ and of a homomorphism $B(k) \to \mathrm{Pic}^0(A)$, the latter homomorphism is also surjective. Finally, if $S \subset A$ is a subscheme such that $\mathcal{P}|_{S \times B}$ is trivial then $\Lambda(L)|_{S \times A}$ is also trivial. Hence, $S$ is contained in $K(L)$, so it is finite. □

## 11.3. Definition and Some Properties of the Fourier–Mukai Transform

Let $A$ be an abelian variety, $\hat{A}$ be the dual abelian variety, and $\mathcal{P}$ be the Poincaré line bundle on $A \times \hat{A}$. The *Fourier–Mukai transform* is the functor

$$\mathcal{S} = \mathcal{S}_A = \Phi_{\mathcal{P}} : D^b(A) \to D^b(\hat{A}).$$

Using Theorem 11.4 and the isomorphism $A \simeq \hat{\hat{A}}$ we get the following result due to S. Mukai [91].

**Theorem 11.6.** *One has isomorphisms of functors*

$$\mathcal{S}_{\hat{A}} \circ \mathcal{S}_A \simeq [-1]_A^*[-g],$$
$$\mathcal{S}_A \circ \mathcal{S}_{\hat{A}} \simeq [-1]_{\hat{A}}^*[-g],$$

*where* $g = \dim A$.

**Remark.** The same statement is true for the functors $\Phi_{\mathcal{P}, A\times S\to\hat{A}\times S}$, etc. (as follows from Theorem 11.4). Moreover, one can generalize this to the case of nontrivial families of abelian varieties. On the other hand, there is a similar Fourier transform between derived categories of coherent sheaves on dual complex tori.

**Example.** Let $\mathcal{O}_x$ be the structure sheaf of a point $x \in A$. Then $\mathcal{S}(\mathcal{O}_x) = \mathcal{P}_x$. Using the above theorem, we derive that $\mathcal{S}(\mathcal{P}_\xi) \simeq \mathcal{O}_{-\xi}[-g]$ for any $\xi \in \hat{A}$.

**Corollary 11.7.** *One has an isomorphism of vector spaces* $H^*(A, \mathcal{O}) \simeq \bigwedge^*(H^1(A, \mathcal{O}))$ *and* $\dim H^1(A, \mathcal{O}) = g$.

*Proof.* We have $\mathcal{S}(\mathcal{O}) = \mathcal{O}_e[-g]$. Therefore, we can compute $H^*(A, \mathcal{O})$ as the (derived functor of) restriction of $\mathcal{O}_e[-g]$ to $e \subset \hat{A}$. Locally $e \subset \hat{A}$ is given by $g$ equations, so considering the corresponding Koszul complex we immediately deduce the result. □

As is well known, the usual Fourier transform interchanges operators of translation and multiplication by a character. We leave for the reader to establish the similar property of the Fourier–Mukai transform stating that there are canonical isomorphisms

$$\mathcal{S}(t_x^* F) \simeq \mathcal{P}_{-x} \otimes \mathcal{S}(F), \tag{11.3.1}$$
$$\mathcal{S}(F \otimes \mathcal{P}_\xi) \simeq t_\xi^* \mathcal{S}(F) \tag{11.3.2}$$

for $x \in A$, $\xi \in \hat{A}$, functorial in $F \in D^b(A)$.

**Definition.** A vector bundle $E$ on an abelian variety $A$ is called *homogeneous* if $t_x^* E \simeq E$ for every $x \in A$.

**Proposition 11.8.** *The Fourier transform induces an equivalence between the category of coherent sheaves on A with finite support and the category of homogeneous bundles on $\hat{A}$.*

*Proof.* If $F$ has finite support, then $F \otimes \mathcal{P}_\xi \simeq F$ for every $\xi \in \hat{A}$. Applying (11.3.2) we get that $\mathcal{S}(F)$ is homogeneous. Conversely, let $E$ be a homogeneous bundle. Then for $F_i = H^i\mathcal{S}(E)$ we have $F_i \otimes \mathcal{P}_\xi \simeq F_i$. We claim that this implies that the support of $F_i$ is finite. Indeed, assume that $\mathrm{supp}(F_i)$ contains an irreducible curve $C$. Let $f : C \to A$ be the embedding (we consider $C$ with the reduced scheme structure). Then denoting $G = f^*F_i$, we get $G \otimes f^*\mathcal{P}_\xi \simeq G$ for every $\xi \in \hat{A}$. Let $\overline{G}$ be the quotient of $G$ by its torsion subsheaf. By assumption $\overline{G}$ has a nonzero rank $r$. Now we have $\overline{G} \otimes f^*\mathcal{P}_\xi \simeq \overline{G}$ for every $\xi \in \hat{A}$. Passing to determinants, we obtain that $f^*\mathcal{P}_\xi^r$ is trivial for every $\xi \in \hat{A}$. In other words, the map $f' = [r]_A \circ f : C \to A$ induces a trivial homomorphism from $\mathrm{Pic}^0(A)$ to $\mathrm{Pic}(C)$. Since $C$ is proper, this implies that the pull-back of the (normalized) Poincaré bundle by the morphism $f' \times \mathrm{id} : C \times \hat{A} \to A \times \hat{A}$ is trivial. On the other hand, $(f' \times \mathrm{id})^*\mathcal{P}$ is the family of line bundles on $\hat{A}$ that corresponds to the morphism $\mathrm{can} \circ f' : C \to \hat{\hat{A}}$. Therefore, the morphism $f'$ is constant, which is a contradiction. Thus, all the sheaves $F_i$ have finite support. Since $\mathcal{S}(\mathcal{S}(E))$ is concentrated in degree $g$ we obtain that $F_i = 0$ for $i \neq g$. □

The Fourier–Mukai transform also interchanges the operations of tensor product and convolution. Namely, as the reader can easily check, there is a canonical isomorphism:

$$\mathcal{S}(F) \otimes \mathcal{S}(G) \simeq \mathcal{S}(F * G),$$

where $F, G \in D^b(A)$, $F * G$ is the (derived) push-forward of $F \boxtimes G$ by the group law morphism $m : A \times A \to A$.

For any homomorphism $f : A \to B$ of abelian varieties one has the following canonical isomorphisms:

$$\mathcal{S}_B \circ Rf_* \simeq L\hat{f}^* \circ \mathcal{S}_A, \tag{11.3.3}$$

$$\mathcal{S}_A \circ f^! \simeq R\hat{f}_* \circ \mathcal{S}_B. \tag{11.3.4}$$

Here $f^! : D^b(B) \to D^b(A)$ is the right adjoint functor to $Rf_*$ (cf. [61]). The first isomorphism is easy to establish using the projection formula and the flat base change (cf. [61], Propositions 5.6 and 5.12). The second isomorphism follows from the first, from the adjointness of the functors $(Rf_*, f^!)$ and from the fact that $\mathcal{S}_A$ and $\mathcal{S}_B$ are equivalences. Since canonical bundles of $A$ and $B$

are trivial, one has a (noncanonical) isomorphism $f^! \simeq Lf^*[\dim A - \dim B]$. In particular, if $f$ is an isogeny then we have a (noncanonical) isomorphism

$$\mathcal{S}_A \circ f^* \simeq R\hat{f}_* \circ \mathcal{S}_B. \qquad (11.3.5)$$

The particular case of (11.3.3) is the isomorphism

$$Le^* \circ \mathcal{S}_A \simeq R\pi_*, \qquad (11.3.6)$$

where $e : \mathrm{Spec}(k) \to A$ is the neutral element, $\pi : A \to \mathrm{Spec}(k)$ is the projection. Using the fact that $\mathrm{rk}(F) = \mathrm{rk}(Le^*F)$ for every $F \in D^b(A)$ (which follows from the existence of a finite complex of vector bundles quasi-isomorphic to $F$), we obtain the relations

$$\mathrm{rk}(\mathcal{S}(F)) = \chi(F), \ \chi(\mathcal{S}(F)) = (-1)^g \,\mathrm{rk}(F). \qquad (11.3.7)$$

## 11.4. Fourier–Mukai Transform and Line Bundles

**Definition.** We say that a line bundle $L$ on an abelian variety $A$ is *nondegenerate* if $K(L)$ is finite. Equivalently, $L$ is nondegenerate if the morphism $\phi_L : A \to \hat{A}$ is an isogeny.

**Proposition 11.9.** *Let $L$ be a nondegenerate line bundle on $A$ trivialized at zero. Then one has a canonical isomorphism*

$$\phi_L^* \mathcal{S}_A(L) \simeq \pi^* R\pi_* L \otimes L^{-1},$$

*where $\pi : A \to \mathrm{Spec}(k)$ is the projection to the point.*

*Proof.* Making the flat base change $\phi_L : A \to \hat{A}$ of the projection $p_2 : A \times \hat{A} \to \hat{A}$ we can write

$$\phi_L^* \mathcal{S}_A(L) \simeq \phi_L^* Rp_{2*}(p_1^* L \otimes \mathcal{P}) \simeq Rp_{2*}(p_1^* L \otimes (\mathrm{id}_A, \phi_L)^* \mathcal{P}),$$

where in the latter expression $p_2$ denotes the projection of the product $A \times A$ on the second factor. But we have an isomorphism

$$m^* L \simeq p_1^* L \otimes p_2^* L \otimes (\mathrm{id}_A, \phi_L)^* \mathcal{P}.$$

Hence,

$$\phi_L^* \mathcal{S}_A(L) \simeq Rp_{2*}(p_2^* L^{-1} \otimes m^* L) \simeq L^{-1} \otimes Rp_{2*} m^*(L) \simeq L^{-1} \otimes \pi^* R\pi_* L$$

as required. □

**Corollary 11.10.** *For a nondegenerate line bundle $L$ one has*

$$\chi(L)^2 = |K(L)|,$$

*where $\chi$ is the Euler-Poincaré characteristic.*

*Proof.* We have from the above proposition that

$$\chi(\phi_L^* \mathcal{S}(L)) = \chi(L) \cdot \chi(L^{-1}).$$

Using the Grothendieck–Riemann-Roch theorem it is easy to deduce that for any coherent sheaf $F$ on $\hat{A}$ one has $\chi(\phi_L^* F) = \deg(\phi_L) \cdot \chi(F)$ (see Exercise 5). Thus, the LHS is equal to $\deg(\phi_L) \cdot \chi(\mathcal{S}(L))$. It remains to use the equalities $\deg(\phi_L) = |K(L)|$, $\chi(\mathcal{S}(L)) = (-1)^g \operatorname{rk}(L) = (-1)^g$ (by (11.3.7)), and $\chi(L^{-1}) = (-1)^g \chi(L)$ (by the Serre duality). □

It turns out that the Fourier–Mukai transform of a nondegenerate line bundle has only one nontrivial cohomology. More precisely, the following result holds.

**Theorem 11.11.** *Let $L$ be a nondegenerate line bundle. Then there exists an integer $i(L)$, $0 \leq i(L) \leq g$, and a vector bundle $E$ on $\hat{A}$ such that $\mathcal{S}(L) \simeq E[-i(L)]$.*

*Proof.* The proof is based on the formula

$$\mathcal{S}(L) \otimes \mathcal{S}([-1]^* L^{-1}) \simeq \mathcal{S}(L * [-1]^* L^{-1}),$$

where $*$ denotes the convolution of sheaves on $A$. Using the isomorphism

$$p_1^* L \simeq m^* L \otimes (-p_2)^* L \otimes (m, -\phi_L \circ p_2)^* \mathcal{P}$$

on $A \times A$, we obtain

$$L * [-1]^* L^{-1} \simeq L \otimes m_*(m, -\phi_L \circ p_2)^* \mathcal{P}.$$

Making the change of variables $x' = x + y$, $y' = -y$ on $A \times A$ we can rewrite this as

$$L \otimes Rp_{1*}(\mathrm{id} \times \phi_L)^* \mathcal{P} \simeq L \otimes Rp_{1*}(\Lambda(L)).$$

Using the symmetry of $\Lambda(L)$ we can write

$$Rp_{1*}(\Lambda(L)) \simeq Rp_{1*}(\phi_L \times \mathrm{id})^* \mathcal{P} \simeq \phi_L^* Rp_{1*}(\mathcal{P}).$$

Finally, using the fact that $Rp_{1*}(\mathcal{P}) \simeq \mathcal{O}_e[-g]$ we obtain the isomorphism

$$L * [-1]^* L^{-1} \simeq L|_{K(L)}[-g].$$

It follows that

$$\mathcal{S}(L) \otimes \mathcal{S}([-1]^* L^{-1}) \simeq \mathcal{S}(L|_{K(L)})[-g].$$

Note that $\mathcal{S}(L|_{K(L)})$ is the vector bundle of rank $|K(L)|$ on $\hat{A}$. According to Proposition 11.9, the pull-back of the LHS under $\phi_L$ is isomorphic to $\pi^*(R\pi_*(L) \otimes R\pi_*(L^{-1})) \otimes \mathcal{O}$. It follows that $R\pi_*(L)$ is concentrated in one degree and applying again Proposition 11.9 we derive that $\mathcal{S}(L)$ is locally free. □

**Corollary 11.12.** *If $L$ is a nondegenerate line bundle on an abelian variety $A$ then $H^i(A, L) = 0$ for $i \neq i(L)$ while $H^{i(L)}(A, L)$ has dimension $|K(L)|^{\frac{1}{2}}$.*

*Proof.* This follows from the above theorem and from (11.3.6). □

**Remark.** As we have seen in Chapter 7, in the case when $(A, L)$ is defined over $\mathbb{C}$, the number $i(L)$ is equal to the number of negative eigenvalues of the Hermitian form associated with $L$. In the case of an arbitrary ground field $k$, it is equal to the number of positive roots of the polynomial $P$, defined by $P(n) = \chi(L(n))$, where $\mathcal{O}_A(1)$ is some ample line bundle on $A$ (see [95]). Although we do not prove this fact here, we will obtain some information on $i(L)$ in Sections 11.6 and 15.4.

## 11.5. Action of $\mathbf{SL}_2(\mathbb{Z})$

The following theorem gives a relation between the Fourier–Mukai transform and the functor of tensoring by a nondegenerate line bundle.

**Theorem 11.13.** *Let $L$ be a nondegenerate line bundle on $A$. Let $T_L$ be the functor of tensor multiplication by $L$, $S_L = \phi_L^* \circ \mathcal{S}$. Then there is an isomorphism of functors from $D^b(A)$ to itself:*

$$T_L S_L T_L S_L T_L \simeq R\Gamma(A, L) \otimes [-1]^* \circ S_L.$$

*Proof.* Let $\langle x, y\rangle_L$ denotes the biextension $\Lambda(L) = m^*L \otimes p_1^* L^{-1} \otimes p_2^* L^{-1}$ on $A \times A$. The functor $T_L$ is given by the kernel $\Delta_* L$, where $\Delta : A \to A \times A$ is the diagonal. The functor $S_L$ is given by the kernel $\Lambda(L)$, hence the composition

$T_L S_L$ corresponds to $\Lambda(L) * \Delta_* L \simeq \Lambda(L) \otimes p_2^* L$. Thus, the functor $(T_L S_L)^2$ is given by the kernel

$$K_{x,z} = \int_y \langle x, y\rangle_L L_y \langle y, z\rangle_L L_z \simeq \int_y L_y \langle y, x+z\rangle_L L_z \simeq \int_y L_{x+y+z} L_{x+z}^{-1} L_z,$$

where $\int_y F_{x,y,z}$ denotes the derived push-forward with respect to the projection $p_{13} : A^3 \to A^2$ applied to $F \in D^b(A^3)$. Making the change of variable $y \mapsto y + x + z$, we can rewrite this as

$$\left(\int_y L_y\right) L_{x+z}^{-1} L_z \simeq \left(\int L\right) \langle x, z\rangle_L^{-1} L_x^{-1}.$$

The latter kernel represents the functor $(\int L) \otimes [-1]^* S_L T_L^{-1}$ as required. □

**Corollary 11.14.** *Let $L$ be a line bundle on $A$ with $|\chi(L)| = 1$. Then*

$$(T_L S_L)^3 \simeq \mathrm{id}[-i(L) - g].$$

Recall that the group $\mathrm{SL}_2(\mathbb{Z})$ has the following presentation: it is generated by $S = \left(\begin{smallmatrix} 0 & 1 \\ -1 & 0 \end{smallmatrix}\right)$ and $T = \left(\begin{smallmatrix} 1 & 1 \\ 0 & 1 \end{smallmatrix}\right)$ with the relations $S^4 = (TS)^3 = 1$.

**Corollary 11.15.** *Let $L$ be a line bundle on $A$ with $|\chi(L)| = 1$. Then the functors $S_L$ and $T_L$ generate the action of a central extension of $\mathrm{SL}_2(\mathbb{Z})$ by $\mathbb{Z}$ on $D^b(A)$.*

*Proof.* In view of the previous corollary, we only have to compute $S_L^4$. We have

$$S_L^2 = \phi_L^* \circ \mathcal{S} \circ \phi_L^* \circ \mathcal{S} \simeq \phi_L^* \circ (\phi_L)_* \circ \mathcal{S} \circ \mathcal{S} \simeq [-1]_A^*[-g]$$

(here we used the isomorphism (11.3.4), the self-duality of $\phi_L$, and Theorem 11.6). Hence, $S_L^4 \simeq \mathrm{id}[-2g]$. □

**Remark.** In this corollary, by an action of a group $G$ on a category $\mathcal{C}$ we mean a homomorphism from $G$ to the group of autoequivalences of $\mathcal{C}$ considered up to an isomorphism. There is a stronger notion of an action of a group $G$ on a category defined in [128]. One can show that in the above situation there is an action of a central extension of $\mathrm{SL}_2(\mathbb{Z})$ by $\mathbb{Z}$ on $D^b(A)$ in the strong sense (see [106]).

## 11.6. Index of Nondegenerate Line Bundles

In this section we establish some simple properties of the index $i(L)$.

**Lemma 11.16.** *If $L$ and $L'$ are algebraically equivalent nondegenerate line bundles on an abelian variety, then $i(L) = i(L')$.*

*Proof.* Since $\phi_L : A \to \hat{A}$ is a surjection, we have $L' \simeq t_x^* L$ for some $x \in A$. Therefore, $H^*(L) \simeq H^*(L')$. □

**Proposition 11.17.** *Let $f : A \to B$ be an isogeny of abelian varieties, $L$ a nondegenerate line bundle on $B$. Then $f^*L$ is also nondegenerate and $i(f^*L) = i(L)$.*

*Proof.* Since $\phi_{f^*L} = \hat{f} \circ \phi_L \circ f$, we immediately see that $f^*L$ is nondegenerate. Now using (11.3.5) we obtain

$$\mathcal{S}_A f^* L \simeq R\hat{f}_* \mathcal{S}_B(L).$$

Note that the functor $\hat{f}_*$ is exact, since the morphism $\hat{f}$ is finite. It remains to apply Theorem 11.11. □

**Proposition 11.18.** *Let $L$ be a nondegenerate line bundle on $A$. Then for every positive integer $n$ one has $i(L) = i(L^n)$.*

*Proof.* By Proposition 11.17, for every $m > 0$ one has $i([m]_A^* L) = i(L)$. Since $[m]_A^* L$ is algebraically equivalent to $L^{m^2}$ (see Exercise 7 of Chapter 8), by Lemma 11.16 we have $i(L^{m^2}) = i(L)$. This gives a proof in the case $n = m^2$. In general we can use the fact that every $n > 0$ can be represented as the sum of four squares: $n = a^2 + b^2 + c^2 + d^2$. Consider the morphism $f : A^4 \to A^4$ given by the $4 \times 4$-matrix $M$ with integer coefficients representing the multiplication by the quaternion $a + ib + jc + kd$. Note that $M^t \cdot M = nI_4$. It follows that $f^*(L^{\boxtimes 4})$ is algebraically equivalent to $(L^n)^{\boxtimes 4}$ (see Exercise 6 of Chapter 10). It remains to use the fact that $i(N^{\boxtimes 4}) = 4i(N)$ for a nondegenerate line bundle $N$. □

**Corollary 11.19.** *If $L$ is an ample line bundle on an abelian variety $A$ then $H^i(A, L) = 0$ for $i > 0$ while $\dim H^0(A, L) = |K(L)|^{\frac{1}{2}}$. In particular, if $L$ defines a principal polarization then $H^0(A, L)$ is 1-dimensional.*

## 11.7. Fourier Transform on the Chow Groups and on the Cohomology Groups

We refer to Fulton [46] for the notions related to algebraic cycles that are used below.

Being an exact equivalence of triangulated categories, the Fourier transform induces an isomorphism

$$\mathcal{S} : K_0(A) \xrightarrow{\sim} K_0(\hat{A})$$

of the Grothendieck groups. The Chern character induces an isomorphism $\mathrm{ch} : K_0(X)_{\mathbb{Q}} \to \mathrm{CH}^*(X)_{\mathbb{Q}}$, hence, we obtain the transform

$$\mathcal{S} : \mathrm{CH}^*(A)_{\mathbb{Q}} \xrightarrow{\sim} \mathrm{CH}^*(\hat{A})_{\mathbb{Q}}.$$

Applying the Grothendieck–Riemann-Roch theorem, one can easily write this transform explicitly: for $\alpha \in \mathrm{CH}^*(A)_{\mathbb{Q}}$ one has

$$\mathcal{S}(\alpha) = p_{2*}(p_1^* \cdot \exp(\mathrm{ch}(\mathcal{P}))), \tag{11.7.1}$$

where $p_{2*}$ denotes the push-forward of cycles under the projection $p_2 : A \times \hat{A} \to \hat{A}$, $p_1^*$ denotes the pull-back of cycles under $p_1 : A \times \hat{A} \to A$. Most of the properties of the Fourier–Mukai transform described in Section 11.3 have obvious analogues for the Fourier transform on algebraic cycles. For example, $\mathcal{S}^2 = (-1)^g[-1]_A^*$.

It is clear that one can define similar transform on groups of cycles modulo algebraic (resp., numerical, or homological) equivalence. Also, one can consider the cohomological Fourier transform $\mathcal{S} : H^*(A, \mathbb{Q}) \xrightarrow{\sim} H^*(\hat{A}, \mathbb{Q})$ given by the same formula (11.7.1), where the Chern charater is replaced by the topological Chern character (with values in cohomology).

**Proposition 11.20.** *The cohomological Fourier transform sends $H^*(A, \mathbb{Z})$ into $H^*(\hat{A}, \mathbb{Z})$. More precisely, the restriction of $\mathcal{S}$ to $H^i(A, \mathbb{Z})$ is equal to $(-1)^{i(i+1)/2+g}$ times the standard isomorphism*

$$H^i(A, \mathbb{Z}) \to H^{2g-i}(A, \mathbb{Z})^\vee \simeq H^{2g-i}(\hat{A}, \mathbb{Z})$$

*induced by the Poincaré duality (where $g = \dim A$).*

*Proof.* For every $i$ let us denote by $\delta_i \in H^i(A, \mathbb{Z}) \otimes H^i(\hat{A}, \mathbb{Z})$ the tensor $\sum_j f_j \otimes f_j^*$ for some dual bases $(f_j)$ and $(f_j^*)$ of $H^i(A, \mathbb{Z})$ and $H^i(\hat{A}, \mathbb{Z})$. Künneth isomorphism identifies $H^*(A \times \hat{A}, \mathbb{Z})$ with $H^*(A, \mathbb{Z}) \otimes H^*(\hat{A}, \mathbb{Z})$. Under this identification we have $c_1(\mathcal{P}) = \delta_1$ (see Exercise 8 of Chapter 1). It is easy to check that the $n$-fold wedge power of $\delta_1$ is equal to $(-1)^{n(n+1)/2} n! \delta_n$. Hence,

$$\operatorname{ch}(\mathcal{P}) = \exp(c_1(\mathcal{P})) = \sum_{n=0}^{2g} (-1)^{n(n+1)/2} \delta_n.$$

Now the result follows easily from the definition of $\mathcal{S}$. □

## Exercises

1. Let us define the bilinear form on the Grothendieck group $K_0(A)$ by the formula

$$\chi([F],[G]) = \sum_i (-1)^i \dim \operatorname{Ext}^i(F, G).$$

The Fourier transform defines a homomorphism $\mathcal{S} : K_0(A) \to K_0(\hat{A})$. Show that

$$\chi(\mathcal{S}(x), \mathcal{S}(y)) = \chi(x, y).$$

2. Let $L$ be a line bundle on an abelian variety $A$.
   (a) Assume that $K(L)$ is infinite. Let $B$ be the connected component of zero in $K(L)$, $p : A \to A/B$ be the quotient morphism. Show that the support of $\mathcal{S}(L)$ is of the form $\hat{p}(\widehat{A/B}) + x$ for some $x \in \hat{A}$. [*Hint*: Use Exercise 4 of Chapter 10.]
   (b) Prove that $L$ is nondegenerate if and only if $\chi(L) \neq 0$.
   (c) Assume that $H^0(A, L) \neq 0$ and $H^i(A, L) = 0$ for $i > 0$. Prove that $L$ is ample.
3. Let

$$\begin{array}{ccc} A & \xrightarrow{\;f\;} & B \\ {\scriptstyle g}\downarrow & & \downarrow{\scriptstyle h} \\ C & \xrightarrow{\;k\;} & D \end{array} \qquad (11.7.2)$$

be a cartesian diagram in the category of abelian varieties and their homomorphisms. Show that for any $F \in D^b_{qc}(B)$ one has

$$\mathcal{S}_C(Rg_* f^! F) \simeq R\hat{k}_* L\hat{h}^* \mathcal{S}_B(F).$$

4. Let $f : X \to A$ be a morphism from a proper connected variety to an abelian variety.
   (a) Show that if the induced homomorphism $f^* : \mathrm{Pic}^0(A) \to \mathrm{Pic}(X)$ is trivial, then $f(X)$ is a point.
   (b) Show that $f$ is determined by $f^*$ uniquely up to translation (i.e., up to replacing $f$ with $t_x^* \circ f$ for some $x \in A$).
5. The Grothendieck–Riemann-Roch theorem for a projective morphism of smooth varieties $f : X \to Y$ states that for any coherent sheaf $F$ on $X$ one has $\mathrm{ch}(Rf_* F) = f_*(\mathrm{ch}(F) \cdot \mathrm{Td}_f)$ where ch is the Chern character, $\mathrm{Td}_f$ is a relative Todd class. In the case when $f$ is a morphism between abelian varieties the relative Todd class is trivial since the tangent bundles of abelian varieties are trivial. Check that if $f$ is an isogeny between abelian varieties then $\mathrm{ch}(Rf_* f^* F) = \deg(f) \cdot \mathrm{ch}(F)$. Deduce that $\chi(f^* F) = \deg(f) \cdot \chi(F)$.
6. Let $L$ be an ample line bundle on an abelian variety $A$.
   (a) Show that $\chi(L^n) = n^g \chi(L)$ for $n > 0$, where $g = \dim A$. [*Hint*: Use Riemann-Roch.]
   (b) Prove that for every $n > 0$ the degree of the morphism $[n] : A \to A$ is equal to $n^{2g}$. Equivalently, the finite group scheme $A_n$ has order $n^{2g}$. [*Hint*: Use the previous Exercise.]
7. Let $E$ be an elliptic curve. Show that
$$\deg \mathcal{S}(F) = -\operatorname{rk} F,$$
$$\operatorname{rk} \mathcal{S}(F) = \deg F.$$
   [*Hint*: Use formulas 11.3.7.] Show that for $L = \mathcal{O}_E(e)$ the action of the functors $S_L[1]$ and $T_L$ on $D^b(E)$ induces by passing to invariants (deg, rk) the standard action of the matrices $S$ and $T$ on $\mathbb{Z}^2$.
8. Identify the group of isomorphism classes of central extensions of $\mathrm{SL}_2(\mathbb{Z})$ by $\mathbb{Z}$ with $\mathbb{Z}/12\mathbb{Z}$. Show that under this identification the class of the central extension arising in Corollary 11.15 corresponds to $\pm(2g - 4i(L)) \in \mathbb{Z}/12\mathbb{Z}$. This means that when $g \equiv 2i(L) \bmod (6)$ we can correct functors $S_L$ and $T_L$ by shifts of degree to get an action of $\mathrm{SL}_2(\mathbb{Z})$ on $D^b(A)$. For example, we can do this when $A$ is an abelian surface and $i(L) = 1$.

# 12

# Mumford Group and Riemann's Quartic Theta Relation

In this chapter we develop an algebraic version of the theory of finite Heisenberg groups. Namely, we consider central extensions of finite commutative groups schemes by $\mathbb{G}_m$, such that the corresponing commutator pairing is perfect in the sense of Cartier duality (we call them *finite Heisenberg group schemes*). The most important example of such an extension is the Mumford group $G(L)$ acting on the cohomology of a non-degenerate line bundle $L$ on an abelian variety. By the definition, $G(L)$ is a central extension of $K(L)$ by $\mathbb{G}_m$, such that the corresponding $\mathbb{G}_m$-torsor over $K(L)$ is obtained from the line bundle $L|_{K(L)}$ (the structure of central extension on this $\mathbb{G}_m$-torsor comes from the theorem of the cube). The commutator pairing $K(L) \times K(L) \to \mathbb{G}_m$ coming from this central extension coincides with the perfect pairing considered in Chapter 10. Other examples of finite Heisenberg group schemes are obtained by considering the restriction of the central extension $G(L) \to K(L)$ to appropriate subgroups of $K(L)$. As in Chapter 2, we consider representations of a finite Heisenberg group scheme on which $\mathbb{G}_m$ acts in the standard way. We prove that there is a unique such irreducible representation. In the case $G = G(L)$ we show that the natural representation of $G(L)$ on $H^{i(L)}(L)$ is irreducible (recall that $i(L)$ is the unique degree in which cohomology does not vanish).

We consider two applications of this theory: to the construction of simple vector bundles on abelian varieties and to the proof of Riemann's quartic theta relation (more precisely, we prove an analogue of this identity valid in arbitrary characteristic $\neq 2$). The former construction associates to a line bundle $L$ on an abelian variety $A$ and to a finite subgroup scheme $K \subset K(L)$ such that the restriction of the canonical pairing $K(L) \times K(L) \to \mathbb{G}_m$ to $K$ is nondegenerate, a simple vector bundle on $A/K$. Riemann's quartic theta relation essentially expresses the theorem of the cube in terms of theta functions. Namely, according to this theorem, for $\theta \in H^0(A, L)$ the expressions $\theta(x+y+z+t)\theta(x+t)\theta(y+t)\theta(z+t)$ and $\theta(t)\theta(x+y+t)\theta(x+z+t)\theta(y+z+t)$ are sections of the same line bundle $M$ on $A^4$. Riemann's quartic relation states that in the case when $K(L) = 0$, one is obtained from another by applying the projectors associated with some explicit Lagrangian subgroups in the Mumford group $G(M)$.

In the next chapter we will consider some other applications of the Mumford group.

In this chapter we will often identify line bundles with the corresponding $\mathbb{G}_m$-torsors (and denote them by the same letter).

## 12.1. Algebraic Theory of Heisenberg Groups

The notion of a finite Heisenberg group considered in Chapter 2 has a natural algebraic counterpart (in arbitrary characteristic). Namely, one can consider central extensions of the form

$$0 \to \mathbb{G}_m \to G \to K \to 0$$

in the category of group schemes over $k$, where $K$ is a finite commutative group scheme. Then the commutator pairing gives a bihomomorphism $e : K \times K \to \mathbb{G}_m$. We say that $G$ is a *finite Heisenberg group scheme* if $e$ induces an isomorphism of $K$ with the Cartier dual group scheme $\widehat{K}$.

By a representation of $G$ we mean a vector space $V$ over $k$ equipped with an action of $G$ such that $\mathbb{G}_m$ acts in the standard way ($t \in \mathbb{G}_m(k) = k^*$ is represented by $t \cdot \mathrm{id}_V$). In other words, this structure is given by an isomorphism

$$G \otimes_{\mathcal{O}_K} \pi^* V \widetilde{\to} \pi^* V$$

on $K$, satisfying the natural cocycle condition, where $G$ is considered as a line bundle on $K$, $\pi : K \to \mathrm{Spec}(k)$ is the projection.

We are going to prove the following analogue of Stone-von Neumann theorem for this situation.

**Theorem 12.1.** *There exists a unique (up to isomorphism) representation $V_0$ of $G$ of dimension $|K|^{\frac{1}{2}}$. The functor $V \mapsto V \otimes V_0$ is an equivalence between categories of $k$-vector spaces and $G$-representations.*

The unique representation of $G$ of dimension $|K|^{\frac{1}{2}}$ will be called *Schrödinger representation* of $G$.

**Lemma 12.2.** *If $I \subset K$ is an isotropic subgroup scheme, then the restriction $G|_I$ of the central extension $G \to K$ to $I$ is trivial (as a central extension).*

*Proof.* Assume that $I$ is annihilated by $N > 0$. Then the isomorphism $G_0 = G_{Nx} \simeq G_x^N$ gives a trivialization of $G^N|_I$. It is easy to see that this trivialization of is compatible with the structure of central extension (because $G|_I$ is commutative). Therefore, $G|_I$ is isomorphic to the push-out of a commutative extension

$$0 \to \mu_N \to G_f \to I \to 0$$

under the canonical embedding $\mu_N \hookrightarrow \mathbb{G}_m$. It suffices to prove that the push-out of this extension under the canonical embedding $\mu_N \hookrightarrow \mu_{MN}$ splits for some $M > 0$. Passing to Cartier dual groups we get an exact sequence

$$0 \to \widehat{I} \to \widehat{G_f} \to \mathbb{Z}/N\mathbb{Z} \to 0.$$

Let $\xi \in \widehat{G_f}(k)$ be an element mapping to the generator of $\mathbb{Z}/N\mathbb{Z}$. Choose $M > 0$ such that $MN\xi = 0$. Then there exists a homomorphism $\mathbb{Z}/MN\mathbb{Z} \to \widehat{G_f}$ mapping the generator to $\xi$. The dual homomorphism $G_f \to \mu_{MN}$ gives the required splitting. □

**Definition.** We say that a subgroup scheme $I \subset K$ is Lagrangian if $I$ is isotropic with respect to the commutator pairing $e$ and the morphism $K/I \to \widehat{I}$ induced by $e$ is an isomorphism.

**Lemma 12.3.** *For any finite Heisenberg group scheme $G \to K$ there exists a Lagrangian subgroup scheme $I \subset K$.*

*Proof.* We argue by induction in $|K|$. If $|K| = 1$ we take $I = K = 1$. Now assume that $|K| > 1$. Let $I_0$ be a subgroup scheme of $K$ of prime order (such a subgroup always exists). We claim that $e|_{I_0 \times I_0} \equiv 1$. If $I_0$ is reduced then this is clear, so we can assume that characteristic of $k$ is $p > 0$. If $I_0 = \mu_p$, then again there are no nontrivial bihomomorphisms $I_0 \times I_0 \to \mathbb{G}_m$, because the dual group scheme to $\mu_p$ is $\mathbb{Z}/p\mathbb{Z}$ and there are no nontrivial homomorphisms $\mu_p \to \mathbb{Z}/p\mathbb{Z}$. It remains to consider the case $I_0 = \alpha_p$. Let us choose a trivialization $G|_{\alpha_p} \simeq \mathbb{G}_m \times \alpha_p$ of the $\mathbb{G}_m$-torsor $G|_{\alpha_p}$, compatible with the natural trivialization over $0 \in \alpha_p$. Then the group structure on $G|_{\alpha_p}$ is given by some morphism $c : \alpha_p \times \alpha_p \to \mathbb{G}_m$ satisfying the cocycle condition

$$c(x, y)c(x + y, z) = c(x, y + z)c(y, z)$$

which should be regarded as an equality of morphisms $\alpha_p \times \alpha_p \times \alpha_p \to \mathbb{G}_m$. Also, we have $c(0, 0) = 1$. The ring of functions on $\alpha_p \times \alpha_p$ is $k[x, y]/(x^p, y^p)$. Since $c(0, 0) = 1$, we have $c(x, y) = \exp_p(C(x, y))$ (where $\exp_p(t) = \sum_{i=0}^{p-1} t^i/i!$). The multiplicative cocycle condition for $c$ is equivalent to the additive cocycle condition for $C$:

$$C(x, y) + C(x + y, z) = C(x, y + z) + C(y, z).$$

Comparing the coefficients with $x^i y^j z^i$ in this identity we conclude that $C(x, y) = C(y, x)$, hence, $c(x, y) = c(y, x)$. Therefore, the group $G|_{\alpha_p}$ is commutative, i.e., $I_0$ is isotropic as claimed. By Lemma 12.2, we can lift $I_0$ to a subgroup of $G$. According to Exercise 1 in this situation we have the Heisenberg group scheme structure on $N_G(I_0)/I_0$. By induction assumption we can choose a Lagrangian subgroup scheme $\overline{I}$ in $I_0^{\perp}/I_0$. Now we define $I$ to be the preimage of $\overline{I}$ under the projection $I_0^{\perp} \to I_0^{\perp}/I_0$. □

*Proof of Theorem 12.1.* Let us choose a Lagrangian subgroup scheme $I \subset K$ and its lifting to a subgroup in $G$ (this is possible by Lemma 12.3 and Lemma 12.2). We can define a representation $V_0$ of $G$ as follows: $V_0$ is the subspace in $H^0(I\backslash G, \mathcal{O})$ consisting of functions $f$ such that $f(\lambda g) = \lambda f(g)$ for $\lambda \in \mathbb{G}_m$ (to intepret this equation consider $f$ as a morphism $I\backslash G \to \mathbb{A}^1$). The action of $G$ on $V_0$ is induced by the right action of $G$ on $I\backslash G$. Trivializing the $\mathbb{G}_m$-torsor $G \to K$ we can define an isomorphism of vector spaces $V_0 \simeq H^0(K/I, \mathcal{O})$. In particular, the dimension of $V_0$ is $|K|^{\frac{1}{2}}$. Now let $V$ be arbitrary representation of $G$. First, we can consider $V$ as a representation of $I$. Recall that the group algebra of $I$ coincides with algebra of functions on $\widehat{I}$. Therefore, $V$ has a structure of a module over the algebra $H^0(\widehat{I}, \mathcal{O})$. This means that $V \simeq H^0(\widehat{I}, \mathcal{F})$ for some quasi-coherent sheaf $\mathcal{F}$ on $\widehat{I}$. The action of $G$ on $V$ induces an action of $G$ on $\mathcal{F}$ compatible with the action of $K$ on $\widehat{I} \simeq K/I$ by translations (where $I$ acts by multiplication with characters and the action of $\mathbb{G}_m$ is standard). In particular, $\mathcal{F}$ is a locally free $\mathcal{O}$-module. Let $\pi : V \to \mathcal{F}|_0$ be the morphism of evaluation at $0 \in \widehat{I}$. We can construct a linear map

$$i : V \to V_0 \otimes \mathcal{F}|_0$$

as follows. We identify $V_0 \otimes \mathcal{F}|_0$ with functions $f$ on $I\backslash G$ with values in $\mathcal{F}|_0$, such that $f(\lambda g) = \lambda f(g)$ for $\lambda \in \mathbb{G}_m$. Then we set $i(v)(g) = \pi(gv)$. It is clear that $i$ is a map of $G$-representations. We claim that $i$ is injective. Indeed, let $p : G/I \to K/I \simeq \widehat{I}$ be the natural morphism. Consider the composition of the embedding $p^* : H^0(\widehat{I}, \mathcal{F}) \hookrightarrow H^0(G/I, p^*\mathcal{F})$ with the isomorphism $H^0(G/I, p^*\mathcal{F}) \simeq H^0(I\backslash G, \mathcal{O}) \otimes \mathcal{F}|_0$ induced by the natural trivialization of $p^*\mathcal{F}$ and by the isomorphism $H^0(G/I, \mathcal{O}) \simeq H^0(I\backslash G, \mathcal{O})$ sending $f(g)$ to $f(g^{-1})$. The obtained embedding $H^0(\widehat{I}, \mathcal{F}) \hookrightarrow H^0(I\backslash G, \mathcal{O}) \otimes \mathcal{F}|_0$ factors through $i$ hence $i$ is injective. If $V$ is finite-dimensional, then we immediately deduce that $i$ is an isomorphism (comparing dimensions).

In general, $V$ is a union of finite-dimensional subrepresentations (since any subspace of $\mathcal{F}|_0$ extends to a $G$-invariant subsheaf of $\mathcal{F}$), so the statement follows by functoriality of our construction. □

**Corollary 12.4.** *For every $G$-representation $V$ the canonical map*

$$V_0 \otimes \mathrm{Hom}_G(V_0, V) \to V$$

*is an isomorphism.*

We denote by $G^{op}$ the opposite group scheme to $G$. Note that $G^{op}$ is also a Heisenberg group scheme and $V_0^*$ is the Schrödinger representation of $G^{op}$.

**Corollary 12.5.** *Consider the space $H^0(K, G)$ where $G$ is regarded as a line bundle over $K$. Then one has a canonical isomorphism $H^0(K, G) \simeq V_0 \otimes V_0^*$ compatible with $G \times G^{op}$-actions (where the action of $G \times G^{op}$ on $H^0(K, G)$ is induced by the left and right actions of $G$ on itself).*

*Proof.* Let us denote by $G * G^{op}$ the quotient of $G \times G^{op}$ by the subgroup $\mathbb{G}_m$ embedded as $t \mapsto (t, t^{-1})$. Then $G * G^{op}$ is a Heisenberg group scheme (a central extension of $K \times K$). Now the $G * G^{op}$-representations $H^0(K, G)$ and $V_0 \otimes V_0^*$ should be isomorphic (by uniqueness of the representation of minimal dimension). We can choose this isomorphism canonically by requiring that it identifies some nonzero $I$-invariant functionals, where $I$ is a Lagrangian subgroup in $K \times K$ equipped with a lifting to $G * G^{op}$. As $I$ we take the subgroup $K$ embedded as $k \mapsto (k, -k)$. It has a natural lifting to a subgroup in $G * G^{op}$, namely, the image of the homomorphism $G \to G * G^{op} : g \mapsto (g, g^{-1})$. The $I$-invariant functional on $H^0(K, G)$ is the evaluation at $0 \in K$. The $I$-invariant functional on $V_0 \otimes V_0^*$ is given by the trace. □

**Remark.** One can immediately generalize Theorem 12.1 to representations of $G$ on coherent sheaves over arbitrary noetherian scheme $S$, where $G$ acts trivially on $S$ and the induced action of $\mathbb{G}_m$ on $\mathcal{F}$ is standard (see Exercise 3 of Chapter 9). Namely, we claim that for every such $G$-equivariant sheaf $\mathcal{F}$ the canonical map

$$V_0 \otimes \mathrm{Hom}_G(V_0, \mathcal{F}) \to \mathcal{F} \tag{12.1.1}$$

is an isomorphism of $G$-equivaraiant sheaves. Indeed, it suffices to prove this in the case when $S$ is affine. Then we can apply Corollary 12.4 to $G$-representation $H^0(S, \mathcal{F})$. This implies that the map (12.1.1) is an isomorphism. Note that if $\mathcal{F}$ is a vector bundle, then $\mathrm{Hom}_G(V_0, \mathcal{F})$ is also a vector bundle (as a direct summand of $\mathcal{F}$).

## 12.2. Mumford Group

Let $L$ be a line bundle $L$ on an abelian variety $A$. As before, we denote by the same letter $L$ the corresponding $\mathbb{G}_m$-torsor over $A$. We are going to construct a central extension

$$0 \to \mathbb{G}_m \to G(L) \to K(L) \to 0. \tag{12.2.1}$$

As $\mathbb{G}_m$-torsor, $G(L)$ is just the restriction of $L$: $G(L) = L|_{K(L)}$. To give it a structure of central extension we have to provide an isomorphism

$$G(L)_{k_1+k_2} \widetilde{\rightarrow} G(L)_{k_1} \otimes G(L)_{k_2}$$

over $K(L) \times K(L)$ satisfying the cocycle condition (see Section 10.2). In other words, we have to trivialize the restriction of the biextension $\Lambda(L)$ to $K(L) \times K(L)$. Actually, we have two such trivializations: one is restricted from the canonical trivialization of $\Lambda(L)$ over $K(L) \times A$ and another is restricted from $A \times K(L)$. We prefer the former one, since then we obtain a canonical *left* action of the group $G(L)$ on $L$ which is compatible with the action of $K(L)$ on $A$ by translations. Indeed, it is easy to see that such an action is given by an isomorphism

$$G(L)_k \otimes L_x \widetilde{\rightarrow} L_{k+x}$$

over $K(L) \times A$ (where $k \in K$, $x \in A$), which corresponds to the trivialization of the biextension $\Lambda(L)$ over $K(L) \times A$.

From the theorem of the cube one can get another description of $G(L)$ as the group scheme of automorphisms of $L$ compatible with some translations on $A$. For example, let us describe the group $G(L)(k)$. Recall that $K(L)(k)$ is the subgroup in $A(k)$ consisting of points $x \in A(k)$ such that $t_x^*L \simeq L$. Now $G(L)(k)$ can be interpreted as the group of pairs $(x, \alpha)$ where $x \in K(L)$, $\alpha : L \to t_x^*L$ is an isomorphism. Indeed, for every $x \in K(L)$ we have a canonical isomorphism $t_x^*L \simeq L_x \otimes L$ induced by the trivialization of $\Lambda(L)$ over $\{x\} \times A$. Thus, isomorphisms $\alpha : L \to t_x^*L$ are in bijective correspondence with $L_x \setminus \{0\}$. It is easy that the group law in this interpretation takes the form

$$(x, \alpha) \cdot (y, \beta) = (x + y, t_y^*\alpha \circ \beta).$$

Henceforward, we assume that $L$ is nondegenerate, so that $K(L)$ is finite. We claim that $G(L)$ is a Heisenberg group scheme, i.e., the commutator pairing

$$e_L : K(L) \times K(L) \to k^*$$

is nondegenerate. Indeed, it is easy to see that $e_L$ coincides with the pairing $e_{\phi_L}$ induced by the biextension $\Lambda(L) = m^*L \otimes p_1^*L^{-1} \otimes p_2^*L^{-1}$ on $A \times A$, which corresponds to the symmetric morphism $\phi_L : A \to \hat{A}$ (see Section 10.4). Therefore, our claim follows from Theorem 10.1.

There is a natural action of $G(L)$ on $H^{i(L)}(A, L)$ (the only nonzero cohomology group of $L$) so that the center $\mathbb{G}_m$ acts in the standard way. We have the following important observation.

**Proposition 12.6.** *The representation* $H^{i(L)}(A, L)$ *of* $G(L)$ *is the Schrödinger representation of* $G(L)$.

*Proof.* Indeed, by Corollary 11.12, the dimension of $H^{i(L)}(A, L)$ is equal to $|K(L)|^{\frac{1}{2}}$, so the statement follows from Theorem 12.1. □

Note that in the case $i(L) = 0$ the above theorem is an algebraic analogue of Proposition 3.1 in Chapter 3. In fact, in the case $k = \mathbb{C}$ these results coincide. Indeed, assume that $A = V/\Gamma$, $L = L(H, \alpha)$, where $V$ is a complex vector space, $\Gamma$ is a lattice in $V$, $H$ is a positive Hermitian form on $V$ such that $E = \operatorname{Im} H$ takes integer values on $\Gamma$, $\alpha : \Gamma \to U(1)$ is a map satisfying (1.2.2). Then we have $K(L) = \Gamma^{\perp}/\Gamma$ (see Section 8.6) and one can easily identify the group $G(L)(\mathbb{C})$ with the push-out of the Heisenberg extension

$$1 \to U(1) \to G(E, \Gamma, \alpha^{-1}) \to \Gamma^{\perp}/\Gamma \to 0$$

defined in Chapter 3, by the standard embeddding $U(1) \hookrightarrow \mathbb{C}^*$. Furthermore, we have an identification of $H^0(A, L)$ with the space of theta functions $T(H, \Gamma, \alpha^{-1})$, so that the action of $G(L)$ on $H^0(A, L)$ coincides with the action of $G(E, \Gamma, \alpha^{-1})$ on $T(H, \Gamma, \alpha^{-1})$.

## 12.3. Descent and Vector Bundles

Let $L$ be a nondegenerate line bundle on an abelian variety $A$, and let $f : A \to B$ be an isogeny with kernel $I$. Theorem 10.5 can be restated as saying that a choice of a line bundle $M$ on $B$ together with an isomorphism $L \simeq f^*M$ is equivalent to the choice of a lifting homomorphism $I \to G(L)$ (so $I$ should be an isotropic subgroup of $K(L)$).

The relation between Mumford groups of $M$ and $L = f^*M$ is the following: $G(M) = N(I)/I$ where $N(I)$ is the normalizer of $I$ in $G(L)$. In particular, $K(M) = I^{\perp}/I$ where $I^{\perp} \subset K(L)$ is the orthogonal complement to $I$ with respect to $e_L$.

Lemma 12.2 implies that a lifting homomorphism $I \to G(L)$ for a subgroup scheme $I \subset K(L)$ exists if and only if $I$ is isotropic with respect to $e_L$. Together with Lemma 12.3 this proves that for every nondegenerate line bundle $L$ on $A$ one can find an isogeny $f : A \to B$ such that $L \simeq f^*M$ and $K(M) = 0$.

Recall that a vector bundle $\mathcal{V}$ is called *simple* if $\mathrm{Hom}(\mathcal{V}, \mathcal{V}) = k$. One can see the interplay between the Mumford group and the descent in the following construction of simple vector bundles on abelian varieties.

**Proposition 12.7.** *Let $L$ be a line bundle on an abelian variety $A$, $K \subset K(L)$ be a finite subgroup scheme, such that the restriction of $e_L$ to $K \times K$ is nondegenerate. Let $G \to K$ be the restriction of the central extension $G(L) \to K(L)$ to $K$, $V$ be the Schrödinger representation of $G$. Then the following three constructions produce isomorphic simple vector bundles $\mathcal{V}_1 \simeq \mathcal{V}_2 \simeq \mathcal{V}_3$ on $A/K$:*

*(i) Pick a Lagrangian subgroup scheme $I \subset K$ and its lifting to $G$. Let $p : A \to A/I$, $q : A/I \to A/K$ be natural projections. Then the subgroup $I$ and its lifting define a line bundle $M$ on $A/I$ such that $p^*M \simeq L$. Set $\mathcal{V}_1 = q_*M$.*

*(ii) We have an action of $G$ on $L$ compatible with the action of $K$ on $A$ by translations. Hence, we get the induced action of $K$ on $L \otimes V^*$, i.e., the descent data on this bundle for the morphism $\pi : A \to A/K$. Set $\mathcal{V}_2$ to be the descended vector bundle, so that $\pi^*\mathcal{V}_2 \simeq L \otimes V^*$.*

*(iii) Consider the vector bundle $\pi_*L$ on $A/K$. It has a natural action of $G$ (compatible with the trivial action of $G$ on $A/K$). Set $\mathcal{V}_3 = \mathrm{Hom}_G(V, \pi_*L)$.*

*Proof. Equivalence of (i) and (iii).* It is easy to see that for $V$ induced from the trivial representation of $I$ we have $\mathrm{Hom}_G(V, \pi_*L) = q_*M$.
*Equivalence of (ii) and (iii).* We have an isomorphism $\pi_*L \simeq V \otimes \mathrm{Hom}_G(V, \pi_*L)$ compatible with $G$-actions (see (12.1.1)). Hence,

$$V \otimes \pi^*\mathcal{V}_3 \simeq \pi^*\pi_*L.$$

Let us consider the Cartesian square

$$\begin{array}{ccc} K \times A & \xrightarrow{m} & A \\ \downarrow{\scriptstyle p_2} & & \downarrow{\scriptstyle \pi} \\ A & \xrightarrow{\pi} & A/K \end{array} \tag{12.3.1}$$

where $m(k, a) = k + a$. By the flat base change ([61], Proposition 5.12), we have $\pi^*\pi_*L \simeq p_{2*}m^*L$. Since $K$ is a subgroup in $K(L)$, we have an isomorphism $m^*L \simeq L|_K \boxtimes L$. Hence, we have

$$\pi^*\pi_*L \simeq H^0(K, L|_K) \otimes L.$$

By Corollary 12.5 we have an isomorphism $H^0(K, L|_K) \simeq V \otimes V^*$. Therefore, we get

$$V \otimes \pi^*\mathcal{V}_3 \simeq V \otimes V^* \otimes L.$$

It is easy to check that this isomorphism is compatible with the action of $G \times K$ (where $G$ acts on $V$, $K$ on $\pi^*\mathcal{V}_3$ and on $V^* \otimes L$). Hence, $\pi^*\mathcal{V}_3$ and $V^* \otimes L$ are isomorphic as sheaves with $K$-action.

It remains to check that the vector bundle $\mathcal{V}_1 \simeq \mathcal{V}_2 \simeq \mathcal{V}_3$ is simple. It is convenient to do this for $\mathcal{V}_2$. Indeed, endomorphisms of $\mathcal{V}_2$ are given by endomorphisms of $L \otimes V^*$ compatible with $K$-action. Clearly, these are just $G$-endomorphisms of $V^*$ that reduce to scalars. □

In Chapter 14 we will see that all simple vector bundles on elliptic curves are obtained by the above construction.

## 12.4. Riemann's Quartic Theta Relation

Let $A$ be an abelian variety, $L$ be an ample line bundle on $A$ with $K(L) = 0$, $s$ be a nonzero global section of $L$ (unique up to a nonzero constant). In what follows we assume that char$(k) \neq 2$. An easy consequence of theorem of the cube is the following canonical isomorphism of line bundles on $A^4$ (see Exercise 4 of Chapter 8):

$$g^*(L^{\boxtimes 4}) \xrightarrow{\tau} f^*(L^{\boxtimes 4}),$$

where $f, g : A^4 \to A^4$ are the following homomorphisms:

$$f(x, y, z, t) = (x + y + z + t, x + t, y + t, z + t),$$
$$g(x, y, z, t) = (t, x + y + t, x + z + t, y + z + t).$$

It is easy to see that there are natural isomorphisms

$$A_2 \xrightarrow{\sim} \ker(f) : x \mapsto (x, x, x, x),$$
$$A_2 \xrightarrow{\sim} \ker(g) : x \mapsto (x, x, x, 0).$$

Let us denote $M := f^*(L^{\boxtimes 4})$. Note that since $K(L) = 0$ we get that $\ker(f)$ and $\ker(g)$ are Lagrangian subgroups in $K(M)$ (see Section 12.3). On the other hand, $\ker(f) \cap \ker(g) = 0$, hence

$$K(M) = \ker(f) \oplus \ker(g).$$

Notice also that the subgroups $\ker(f)$ and $\ker(g)$ are equipped with liftings to subgroups of $G(M)$ (for $\ker(f)$ this is clear, while for $\ker(g)$ the lifting is induced by the isomorphism $\tau$ above), hence they both act on the space $H^0(A^4, M)$. We will denote this action by $x \cdot t$ where $x \in \ker(f)$ (resp., $\ker(g)$), $t \in H^0(A^4, M)$.

**Theorem 12.8.** *One has the following equalities in $H^0(A^4, M)$:*

$$f^*(s^{\boxtimes 4}) = \epsilon \cdot 2^{-\dim A} \cdot \sum_{x \in \ker(f)} x \cdot \tau g^*(s^{\boxtimes 4}),$$
$$\tau g^*(s^{\boxtimes 4}) = \epsilon \cdot 2^{-\dim A} \cdot \sum_{x \in \ker(g)} x \cdot f^*(s^{\boxtimes 4}),$$

*where $\epsilon = \pm 1$.*

*Proof.* By definition, the line in $H^0(A^4, M)$ generated by $s_1 := f^*(s^{\boxtimes 4})$ is stable under the action of $\ker(f)$. Since the subgroup $\ker(f)$ is Lagrangian and the action of $G(M)$ on $H^0(A^4, M)$ is irreducible, the space of invariants of $\ker(f)$ in $H^0(A^4, M)$ is spanned by $s_1$. Similarly, the space of invariants of $\ker(g)$ in $H^0(A^4, M)$ is spanned by $s_2 := \tau g^*(s^{\boxtimes 4})$. It follows that

$$\sum_{x \in \ker(f)} x \cdot s_2 = c s_1 \tag{12.4.1}$$

for some constant $c \in k$. Consider the following involution on $A^4$:

$$i : (x, y, z, t) \mapsto (-z, -y, -x, x + y + z + t).$$

Then we have $f \circ i = g$, $g \circ i = f$. We claim that

$$i^*(\tau) = \tau^{-1} : f^*(L^{\boxtimes 4}) \to g^*(L^{\boxtimes 4}).$$

Indeed, this follows from the fact that both $i^*(\tau)$ and $\tau^{-1}$ restrict to identity on fibers over 0. The involution $i$ induces an involution on the Heisenberg group $G(M)$ which switches the subgroups $\ker(f)$ and $\ker(g)$. Applying $i^*$ to equation (12.4.1) we get

$$\sum_{y \in \ker(g)} y \cdot s_1 = c s_2.$$

Therefore, we have

$$\sum_{y \in \ker(g)} \sum_{x \in \ker(f)} y \cdot x \cdot s_2 = c \cdot \sum_{y \in \ker(g)} y \cdot s_1 = c^2 s_2.$$

On the other hand,

$$\begin{aligned} \sum_{y \in \ker(g), x \in \ker(f)} y \cdot x \cdot s_2 &= \sum_{y \in \ker(g), x \in \ker(f)} [y, x] \cdot x \cdot y \cdot s_2 \\ &= \sum_{y \in \ker(g), x \in \ker(f)} [y, x] \cdot x \cdot s_2. \end{aligned}$$

Now the sum $\sum_{y \in \ker(g)} [y, x]$ is nonzero only when $x = 0$, hence, we get

$$\sum_{y \in \ker(g), x \in \ker(f)} y \cdot x \cdot s_2 = |\ker(g)| \cdot s_2.$$

Thus, we get $c^2 = |\ker(g)| = 2^{2 \dim(A)}$. □

**Remark.** There seems to be no simple algebraic way to find the sign $\epsilon$ in the above identities. Below we will show that in the case when the ground field is $\mathbb{C}$ one always have $\epsilon = 1$. Also, Exercise 4 states that $\epsilon$ does not change if we translate the line bundle $L$, so it depends only on the corresponding *polarization* $\phi_L : A \to \hat{A}$. More generally, $\epsilon$ is constant in connected families of data $(A, L)$. One can derive from this that $\epsilon = 1$ in arbitrary characteristics using the existence and irreducibility of the relevant moduli space defined over $\mathbb{Z}$ (see [38]).

## 12.5. Transcendental Computation

Let $V/\Gamma$ be a complex torus equipped with a line bundle $L = L(H, \alpha)$ such that $H$ is positive, and the restriction of the symplectic form $E = \operatorname{Im} H$ to $\Gamma$ is unimodular. We assume also that there exists a real Lagrangian subspace $U \subset V$ such that $\Gamma \cap U$ is a lattice in $U$ and $\alpha|_{\Gamma\cap U} \equiv 1$. We have the corresponding theta function $\theta = \theta^{\alpha}_{H,\Gamma,U}$ generating the space $T(H, \Gamma, \alpha)$. Now let us consider the maps $f, g : V^{\oplus 4} \to V^{\oplus 4}$ as in Theorem 12.8. Note that both these maps preserve the lattice $\Gamma^{\oplus 4}$ and the subspace $U^{\oplus 4}$. Let us consider the Hermitian form

$$H_q = f^*(H^{\oplus 4}) = g^*(H^{\oplus 4})$$

on $V^{\oplus 4}$. Then $U^{\oplus 4}$ is a Lagrangian subspace with respect to the symplectic form $E_q = \operatorname{Im} H_q$. The lattices $\Gamma_1 = f^{-1}(\Gamma^{\oplus 4})$ and $\Gamma_2 = g^{-1}(\Gamma^{\oplus 4})$ are self-dual with respect to $E_q$ and are equipped with quadratic maps $\alpha_1 = \alpha^{\oplus 4} \circ f$ and $\alpha_2 = \alpha^{\oplus 4} \circ g$ satisfying (1.2.2). Notice also that

$$\Gamma_1 \cap \Gamma_2 = \Gamma^{\oplus 4}$$

and that the maps $\alpha_1$ and $\alpha_2$ coincide on $\Gamma^{\oplus 4}$. Now we have

$$f^*(\theta^{\boxtimes 4}) = \theta^{\alpha_1}_{H_q,\Gamma_1,U^{\oplus 4}},$$

$$g^*(\theta^{\boxtimes 4}) = \theta^{\alpha_2}_{H_q,\Gamma_2,U^{\oplus 4}}.$$

We want to apply the result of Section 5.5 to lattices $\Gamma_1$ and $\Gamma_2$. We claim that

$$(\Gamma_1 + \Gamma_2) \cap U^{\oplus 4} = \Gamma_1 \cap U^{\oplus 4} + \Gamma_2 \cap U^{\oplus 4}.$$

Indeed, this follows easily from the explicit description of our lattices:

$$\Gamma_1 = \left\{(x, y, z, t) \in \frac{1}{2}\Gamma^{\oplus 4} | \, x \equiv y \equiv z \equiv t \bmod \Gamma\right\},$$

$$\Gamma_2 = \left\{(x, y, z, t) \in \frac{1}{2}\Gamma^{\oplus 4} | \, t \in \Gamma, \ x \equiv y \equiv z \bmod \Gamma\right\}.$$

Now applying Corollary 5.6 we obtain

$$f^*\theta^{\boxtimes 4} = 2^{-\dim_{\mathbb{C}} V} \cdot \sum_{x \in \Gamma/2\Gamma} U_{(\alpha(x)^5,(x/2,x/2,x/2,x/2))} g^*\theta^{\boxtimes 4}.$$

Thus, we see that the constant $\epsilon$ in Theorem 12.8 is equal to 1 in the case $k = \mathbb{C}$. The obtained equation can be rewritten more explicitly as follows:

$$\theta(x+y+z+t)\theta(x+t)\theta(y+t)\theta(z+t) = 2^{-\dim_{\mathbb{C}} V}$$
$$\times \sum_{u \in \Gamma/2\Gamma} \alpha(u)^{-5} \cdot \theta_{\frac{u}{2}}(t)\theta_{\frac{u}{2}}(x+y+t)\theta_{\frac{u}{2}}(x+z+t)\theta_{\frac{u}{2}}(y+z+t), \tag{12.5.1}$$

where we denoted $\theta_c = U_{(1,c)}\theta$. Changing variables and assuming that $\alpha^2 \equiv 1$, we get the following more classical form of the Riemann's quartic theta relation:

$$\theta\left(\frac{x+y+z-t}{2}\right)\theta\left(\frac{x+y+t-z}{2}\right)\theta\left(\frac{x+t+z-y}{2}\right)\theta\left(\frac{y+z+t-x}{2}\right)$$
$$= 2^{-\dim_{\mathbb{C}} V} \cdot \sum_{u\in\Gamma/2\Gamma} \alpha(u)\cdot\theta_{\frac{u}{2}}(x)\theta_{\frac{u}{2}}(y)\theta_{\frac{u}{2}}(z)\theta_{\frac{u}{2}}(t).$$

## Exercises

1. Let $G$ be a finite Heisenberg group scheme, $I \subset K$ be an isotropic subgroup equipped with a lifting to a subgroup of $G$. Show that the normalizer of $I$ in $G$, $N_G(I)$ coincides with the preimage of $I^\perp \subset K$, where $I^\perp$ is the orthogonal complement to $I$ with respect to $e$. Prove that $N_G(I)/I$ is a finite Heisenberg group scheme (an extension of $I^\perp/I$ by $\mathbb{G}_m$).
2. Assume that characteristic of the ground field is zero. Let $L$ be an ample line bundle on an abelian variety $A$ with $\chi(L) = 1$. We assume in addition that $L$ is symmetric, i.e., $[-1]_A^* L \simeq L$. For every pair of relatively prime numbers $(n, k)$ with $n > 0$ let us denote by $G_n^k \to A_n$ the restriction to $A_n \subset A_{kn}$ of the Heisenberg extension $G(L^{kn}) \to A_{kn}$.
   (a) Show that $G_n^k$ is a Heisenberg group. Show that the corresponding extension
   $$1 \to \mathbb{G}_m \to G_n^k \to A_n \to 0$$
   is the push-forward of the extension
   $$1 \to \mathbb{G}_m \to G(L^n) \to A_n \to 0$$
   under the homomorphism $[k] : \mathbb{G}_m \to \mathbb{G}_m$ of raising to the $k$th power.
   (b) Let $V_{n,k}$ be the vector bundle on $A$ obtained by applying the construction of Proposition 12.7 to $L^{kn}$ and $A_n \subset A$, so that $[n]_A^* V_{n,k} \simeq V^* \otimes L^{kn}$, where $V$ is the Schrödinger representation of $G_n^k$. Show that $\operatorname{rk} V_{n,k} = n^g$, $\chi(V_{n,k}) = k^g$.
   (c) Prove that $V_{n,k+n} \simeq V_{n,k} \otimes L$.
   (d) Prove that $V_{n,k}^\vee \simeq V_{n,-k}$, where $V_{n,k}^\vee$ is the dual bundle to $V_{n,k}$.
   (e) Let $I \subset A_n$ (resp., $J \subset A_k$) be a Lagrangian subgroup with respect to the commutator pairing of $G(L^n)$ (resp., $G(L^k)$). Let $q : A/I \to A$ and $q' : A/I + J \to A/J$ be morphisms induced by multiplication by $n$, $r : A \to A/J$ and $r' : A/I \to A/I + J$ be the quotient

morphisms. Apply Exercise 3 of Chapter 11 to the Cartesian diagram

$$\begin{array}{ccc} A/I & \xrightarrow{r'} & A/I+J \\ \Big\downarrow q & & \Big\downarrow q' \\ A & \xrightarrow{r} & A/J \end{array} \tag{12.5.2}$$

to prove that for $k > 0$ one has

$$\phi_L^* \mathcal{S}(V_{n,k}) \simeq V_{k,-n}$$

while for $k < 0$ one has

$$\phi_L^* \mathcal{S}(V_{n,k}) \simeq V_{-k,n}[-g].$$

(f) Show that for any sheaf $F$ on $A$ one has

$$\chi(V_{n,k} \otimes F) = \frac{\chi(L^{nk} \otimes [n]_A^* F)}{n^g}.$$

3. This is a sequel to the previous exercise. Recall that the Riemann-Roch theorem for an abelian variety $A$ takes the following simple form:

$$\chi(F) = \int_A \mathrm{ch}(F)_g,$$

where $F$ is a coherent sheaf on $A$, $\mathrm{ch}(F)_i \in H^{2i}(A, \mathbb{Q})$ is the degree $i$ component of the Chern character of $F$, $g = \dim A$.

(a) Using the fact that $H^*(A, \mathbb{Q})$ is the exterior algebra of $H^1(A, \mathbb{Q})$ prove that $\mathrm{ch}([n]_A^* F)_i = n^{2i}\,\mathrm{ch}(F)$. Deduce from this that

$$\chi(V_{n,k} \otimes F) = \sum_{i=0}^{g} \frac{n^i k^{g-i}}{(g-i)!} \int_A c_1(L)^{g-i} \cdot \mathrm{ch}_i(F).$$

(b) Let us denote

$$\chi_i(F) = \frac{1}{(g-i)!} \int_A c_1(L)^{g-i} \cdot \mathrm{ch}_i(F),$$

where $i = 0, 1, \ldots, g$. In particular, $\chi_0(F) = \mathrm{rk}(F)$, $\chi_g = \chi(F)$. Check that these functions on $K_0(A)$ are linearly independent.

(c) For every pair of relatively prime integers $(n, k)$ let us set

$$v_{n,k} = \sum_i n^i k^{g-i} \chi_i.$$

Prove that these vectors generate a subgroup $M_g$ of finite index in $\mathbb{Z}\chi_0 \oplus \cdots \oplus \mathbb{Z}\chi_g$. Prove that there is an action of $\mathrm{SL}_2(\mathbb{Z})$ on $M_g$ such that the standard generators act on vectors $(v_{n,k})$ as follows:

$$T(v_{n,k}) = v_{n,k+n},$$
$$S(v_{n,k}) = v_{k,-n}.$$

Show that $M_3$ is a proper subgroup of $\mathbb{Z}\chi_0 \oplus \mathbb{Z}\chi_1 \oplus \mathbb{Z}\chi_2 \oplus \mathbb{Z}\chi_3$.

4. Show that the constant $\epsilon$ in Theorem 12.8 does not change if $L$ is replaced by $t_x^* L$, where $x \in A$.
5. In the notation of Section 12.5, prove the following identity:

$$\theta_v(x)\theta_w(x)\theta(0)\theta_{v+w}(0) = 2^{-\dim_{\mathbb{C}} V} \cdot \sum_{u \in \Gamma/2\Gamma} \alpha(u)^{-5} \exp(\pi i E(u, v+w))$$
$$\times \theta_{\frac{u}{2}}(x)\theta_{\frac{u}{2}+v+w}(x)\theta_{\frac{u}{2}+v}(0)\theta_{\frac{u}{2}+w}(0).$$

where $v, w, x \in V$. Taking $v, w \in \frac{1}{2}\Gamma/\Gamma$ we get a system of quadratic equations on the functions $\theta_v(x)$, where $v \in \frac{1}{2}\Gamma/\Gamma$.

6. Let

$$\theta_{ab}(x) = \sum_{n \in \mathbb{Z}} \exp(\pi i \tau (n + a/2)^2 + 2\pi i (z + b/2)(a + 1/2)),$$

where $a, b \in \{0, 1\}$, be classical theta-functions (with half-integer characteristics) on elliptic curve (corresponding to some fixed element $\tau$ in the upper half-plane).

(a) Show that $\theta_{ab}(0) = 0$ if and only if $(ab) = (11)$. [*Hint*: Use Exercise 5 of Chapter 3.]

(b) Prove the following identity

$$\theta_{00}(0)^4 = \theta_{10}(0)^4 + \theta_{01}(0)^4.$$

[*Hint*: Use the previous Exercise.]

(c) Show that

$$\theta_{11}(x)\theta_{11}(y-z)\theta_{11}(t)\theta_{11}(y+z-x-t)$$
$$= \theta_{11}(z)\theta_{11}(y-x)\theta_{11}(y+t)\theta_{11}(z-x+t)$$
$$- \theta_{11}(y)\theta_{11}(z-x)\theta_{11}(z+t)\theta_{11}(y-x+t).$$

Check that the Kronecker function $F(x, y)$, introduced in Exercise 7 of Chapter 3, satisfies

$$F(x, y) = -\frac{\theta_{11}'(0)\theta_{11}(x + y)}{\theta_{11}(x)\theta_{11}(y)}.$$

Show that the above identity is equivalent to the following identity for $F$:

$$\begin{aligned} &F(-u', v)F(u + u', v + v') - F(u, v)F(u + u', v') \\ &\quad + F(u', v')F(u, v + v') = 0. \end{aligned}$$

# 13
# More on Line Bundles

In this chapter we study which parts of the description of holomorphic line bundles on complex tori (see Chapter 1) can be transferred to the context of line bundles on abelian varieties (in arbitrary characteristic). Clearly, the symmetric morphism $\phi_L : A \to \hat{A}$ associated with a line bundle $L$ on an abelian variety $A$ can be considered as an analogue of the Hermitian form corresponding to a holomorphic line bundle (see Section 9.6). One result we prove in this chapter is that every symmetric homomorphism $\phi : A \to \hat{A}$ is equal to $\phi_L$ for some line bundle $L$ on $A$. However, we do not have a nice description of the $\hat{A}$-torsor $\hat{A}(\phi)$ consisting of line bundles $L$ on $A$ with $\phi_L = \phi$ (recall that in Chapter 1 we used quadratic maps $\alpha : \Gamma \to U(1)$ for this purpose). In order to retain some connection with quadratic functions we have to look at *symmetric line bundles*, i.e., line bundles $L$ such that $[-1]_A^* L \simeq L$. In the analogy between line bundles and quadratic functions, symmetric line bundles should be considered as analogues of (homogeneous) quadratic forms. As usual, the categorical notion is richer. Thus, the restriction of the symmetry isomorphism $[-1]^* L \widetilde{\to} L$ to points of order 2 gives a quadratic function $q_L : A_2 \to \mu_2$. The corresponding bilinear pairing is given by $e_2(x, \phi_L(y))$, where $e_2 : A_2 \times \hat{A}_2 \to \mu_2$ is the Weil pairing. Using this construction, it is not difficult to see that the $\hat{A}_2$-torsor $\hat{A}_2(\phi)$ consisting of symmetric line bundles with $\phi_L = \phi$ can be canonically identified with the $\hat{A}_2$-torsor consisting of quadratic functions $q : A_2 \to \mu_2$ whose associated bilinear pairing is $e_2(x, \phi_L(y))$. Now the $\hat{A}$-torsor $\hat{A}(\phi)$ is canonically isomorphic to the push-out of $\hat{A}_2(\phi)$ with respect to the embedding $\hat{A}_2 \to \hat{A}$.

If $L$ is a symmetric line bundle with $K(L) = 0$, then the quadratic function $q_L$ is nondegenerate. In the case when characteristic of $k$ is different from 2, the form $q_L$ should be either even or odd (see Section 5.2). We prove that if in addition $L$ is ample, then the nonzero section of $L$ is even (resp., odd) precisely when $q_L$ is even (resp., odd). We also show that this is equivalent to the condition that the divisor of zeroes of the unique section of $L$ has even (resp., odd) multiplicity at zero.

## 13.1. Quadratic Form Associated to a Symmetric Line Bundle

Let $L$ be a symmetric line bundle on an abelian variety $A$. This means that $[-1]^*L \simeq L$. The isomorphism

$$\tau = \tau^L : [-1]^*L \to L$$

can be chosen uniquely in such a way that $\tau_0 : L|_0 \to L|_0$ is the identity, where $0 \in A$ is the neutral element. Then the composition

$$L \stackrel{[-1]^*\tau}{\to} [-1]^*L \stackrel{\tau}{\to} L$$

is the identity. Now given an $S$-point $x \in A_2(S)$ (where $S$ is a $k$-scheme), we can evaluate $\tau$ at $x$ and get an isomorphism $\tau|_x : L|_x = L|_{-x} \to L|_x$. This isomorphism is a multiplication by $q_L(x) \in \mathcal{O}^*(S)$. Using the fact that $\tau \circ [-1]^*\tau = \mathrm{id}$ we derive that $q_L(x)^2 = 1$. Thus, we have defined a morphism

$$q_L : A_2 \to \mu_2$$

such that $q_L(0) = 1$. It is clear that if $L$ and $M$ are symmetric line bundles then

$$q_{L\otimes M}(x) = q_L(x)q_M(x).$$

Also if $f : A \to B$ is a homomorphism of abelian varieties and $M$ is a symmetric line bundle on $B$ then $f^*M$ is a symmetric line bundle on $A$ and one has

$$q_{f^*M}(x) = q_M(f(x))$$

for any $x \in A_2$.

**Proposition 13.1.** *The function $q_L$ satisfies*

$$q_L(x + y) = e_2(x, \phi_L(y))q_L(x)q_L(y),$$

*where $e_2 : A_2 \times \hat{A}_2 \to \mu_2$ is the Weil pairing. For $\xi \in \hat{A}_2$ one has*

$$q_{\mathcal{P}_\xi}(x) = e_2(x, \xi).$$

*Proof.* Let $\Lambda(L) = m^*L \otimes p_1^*L^{-1} \otimes p_2^*L^{-1}$. Then $\Lambda(L)$ is a symmetric line bundle on $A \times A$ and the first identity is equivalent to

$$q_{\Lambda(L)}(x, y) = e_2(x, \phi_L(y)).$$

Since $\Lambda(L) \simeq (\mathrm{id} \times \phi_L)^*\mathcal{P}$ we have

$$q_{\Lambda(L)}(x, y) = q_{\mathcal{P}}(x, \phi_L(y)).$$

It remains to prove the identity

$$q_{\mathcal{P}}(x,\xi) = e_2(x,\xi)$$

for $(x,\xi) \in A_2 \times \hat{A}_2$. Indeed, the symmetry isomorphism for $\mathcal{P}$ is the composition

$$\mathcal{P}_{-x,-\xi} \stackrel{\alpha_x}{\to} \mathcal{P}_{x,-\xi}^{-1} \stackrel{\alpha_\xi^{-1}}{\to} \mathcal{P}_{x,\xi},$$

where $\alpha_x$ and $\alpha_\xi$ are given by the biextension structure on $\mathcal{P}$. For $x \in A_2$ we have $\mathcal{P}_{-x,-\xi} = \mathcal{P}_{x,-\xi}$ and the isomorphism $\alpha_x$ induces the trivialization of $\mathcal{P}^2|_{A_2 \times \hat{A}}$. Similarly, $\alpha_\xi$ corresponds to the trivialization of $\mathcal{P}^2|_{A \times \hat{A}_2}$. Therefore, our claim follows from the definition of the Weil pairing given in Section 10.4. □

Let $L$ be a nondegenerate symmetric line bundle on $A$, and let $I \subset K(L)$ be a subgroup scheme, isotropic with respect to $e_L$. As we have seen before (see Section 12.3), descents of $L$ to a line bundle on $A/I$ correspond to liftings of $I$ to a subgroup in the Mumford group $G(L)$. The following result shows that if we want to descend $L$ to a *symmetric* line bundle on $A/I$, we need triviality of the quadratic form $q_L$ at the points of order 2 in $I$.

**Proposition 13.2.** *Let $p : A \to A/I$ be the natural projection. Then a symmetric line bundle $\overline{L}$ on $A/I$ such that $p^*\overline{L} \simeq L$ exists if and only if $q_L|_{I_2} \equiv 1$.*

*Proof.* We have a natural involution $i : G(L) \to G(L)$ of the Mumford group of $L$, which is equal to the composition of the isomorphism $G(L) \to G([-1]^*L)$ induced by the morphism $[-1] : A \to A$, and the isomorphism $G([-1]^*L) \to G(L)$ induced by the symmetry isomorphism $\tau : [-1]^*L \to L$. A line bundle $\overline{L}$ on $A/I$ obtained by the descent from a lifting $\sigma : I \to G(L)$ is symmetric if and only if $\sigma$ is symmetric in the following sense:

$$i(\sigma(x)) = \sigma(-x),$$

where $x \in I$. In particular, it is necessary that the lifting of $I_2$ is preserved by $i$. The latter condition is equivalent to the triviality of $q_L|_{I_2}$ (see Exercise 6(b)). It remains to check that this condition is also sufficient. Indeed, let $\sigma : I \to G(L)$ be arbitrary lifting. The condition $q_L|_{I_2} \equiv 1$ implies that $i(\sigma(x)) = \sigma(x)$ for $x \in I_2$. Then $\chi = (i \circ \sigma) \cdot \sigma^{-1}$ is a homomorphism from $I$ to $\mathbb{G}_m$ that restricts trivially to $I_2$. It follows that there exists a homomorphism $\chi_0 : I \to \mathbb{G}_m$ such that $\chi = \chi_0^2$. Now $\sigma_0 = \chi_0 \cdot \sigma$ is a symmetric lifting. □

## 13.2. Multiplicities of a Symmetric Divisor at Points of Order 2

Another definition of a sign function on $A_2$ starts with a symmetric effective divisor $D \subset A$ rather than a symmetric line bundle ($D$ is called *symmetric* if it is invariant with respect to the involution $[-1] : A \to A$). Namely, if $D$ is such a divisor, then there is a canonical isomorphism

$$\tau_D : [-1]^*\mathcal{O}_A(D) \to \mathcal{O}_A(D).$$

This isomorphism is characterized by the property that $\tau_D([-1]^*s) = s$, where $s \in H^0(A, \mathcal{O}_A(D))$ is a nonzero section vanishing on $D$. Then one can proceed as above and define the map $\epsilon_D : A_2 \to \mu_2$ by

$$\epsilon_D(x) = \tau_D|_x,$$

where $x \in A_2(S)$.

The difference between the construction of $\epsilon_D$ and that of $q_L$ is that $\epsilon_D(0)$ is not necessarily equal to 1. However, the two constructions are closely related. In fact, if we set $L = \mathcal{O}_A(D)$, then $L$ is a symmetric line bundle and we have $\tau_D = \epsilon_D(0)\tau$, where $\tau = \tau^L$. Hence

$$\epsilon_D(x) = q_L(x)\epsilon_D(0). \tag{13.2.1}$$

Thus, the datum of symmetric divisor adds an additional constant $\epsilon_D(0)$ to the information deduced from the corresponding line bundle. The following proposition gives a geometric interpretation of the corresponding map on $k$-points $\epsilon_D(k) : A_2(k) \to \mu_2(k)$.

**Proposition 13.3.** *For every $x \in A_2(k)$ one has*

$$\epsilon_D(k)(x) = (-1)^{\mathrm{mult}_x D}$$

*Proof.* If $\mathrm{mult}_x D = n$ then the section $s$ defines a nonzero element

$$s_x \bmod \mathfrak{m}_x^{n+1}L \in \mathfrak{m}_x^n L/\mathfrak{m}_x^{n+1} nL,$$

where $\mathfrak{m}_x \subset \mathcal{O}_x$ is the maximal ideal in the local ring of $x$. Now for every $y \in A$ we have an isomorphism

$$\mathfrak{m}_{-y}^n L/\mathfrak{m}_{-y}^{n+1} L \simeq \mathfrak{m}_y^n L/\mathfrak{m}_y^{n+1} L$$

induced by $\tau$. For $y = x \in A_2$ this isomorphism is equal to $(-1)^n\epsilon_D(x)$. On the other hand, it leaves $s_x \bmod \mathfrak{m}_x^{n+1}L$ invariant, hence $(-1)^n\epsilon_D(k)(x) = 1$. □

Let us denote by $\tau_* : H^0(L) \to H^0(L)$ the action of the symmetry isomorphism $\tau$ on sections of $L$: $\tau_*(f) := \tau([-1]^* f)$, where $f \in H^0(L)$.

**Corollary 13.4.** *In the above situation the section $s \in H^0(A, L)$ defining $D$ satisfies $\tau_*(s) = (-1)^{\mathrm{mult}_0 D} s$.*

*Proof.* We have $\tau([-1]^*s) = \epsilon_D(0)\tau_D([-1]^*s) = \epsilon_D(0)s$. It remains to use the above Proposition. □

### 13.3. Case of a Principal Polarization

The most interesting case is when a symmetric line bundle $L$ defines a principal polarization, i.e., when $\phi_L : A \to \hat{A}$ is an isomorphism. According to Proposition 13.1, in this situation $q_L$ is a nondegenerate quadratic form on $A_2$. Assume that the characteristic of $k$ is not equal to 2, so that $A_2 \simeq (\mathbb{Z}/2\mathbb{Z})^{2g}$. Recall that the Arf-invariant of a nondegenerate quadratic form $q$ on $(\mathbb{Z}/2\mathbb{Z})^{2g}$ is defined to be equal to 1 (resp., $-1$) if and only if $q$ is even (resp., odd).

**Theorem 13.5.** *Assume that* $\mathrm{char}(k) \neq 2$. *Let $L$ be a symmetric ample line bundle on $A$ with $K(L) = 0$. Then the operator $\tau_* : H^0(L) \to H^0(L)$ is equal to* $\mathrm{Arf}(q_L) \cdot \mathrm{id}$.

*Proof.* First we claim that it suffices to find a symmetric line bundle $L'$ with $\phi_{L'} = \phi_L$ such that the statement holds for $L'$. Indeed, we have $L \simeq t_x^*L'$ for some $x \in A_2$, hence, we have $\tau = q_{L'}(x)t_x^*(\tau')$. It follows that $\tau_* = q_{L'}(x) \cdot \tau'_*$ under the natural identification $H^0(L) \simeq H^0(L')$. On the other hand, we get $q_L = q_{L'}(x) \cdot t_x^*q_{L'}$ and $\mathrm{Arf}(q_L) = q_{L'}(x)\,\mathrm{Arf}(q_{L'})$ which implies our claim.

Now let us fix a subgroup $I \subset A_2$ which is Lagrangian with respect to the form $e_2(x, \phi_L(y)) = e_{L^2}(x, y)$ on $A_2$. Since $q_{L^2} \equiv 1$, by Proposition 13.2 the line bundle $L^2$ descends to a symmetric line bundle $M$ on $B = A/I$. Let us consider the subgroup $J = A_2/I \subset B_2$. Then we have $q_M|_J \equiv 1$ (since $q_{L^2} \equiv 1$). In particular, $J$ is isotropic with respect to the form $e_2(x, \phi_M(y))$. Now we can apply the same construction to $(B, M, J)$ and descend $M^2$ to a symmetric line bundle $L'$ on $B/J$. It is easy to see that $B/J \simeq A$ and $\phi_{L'} = \phi_L$. As before we have $q_{L'}|_{B_2/J} \equiv 1$. But $B_2/J$ is a Lagrangian subgroup in $A_2$, hence the quadratic form $q_{L'}$ is even (see Exercise 1(c) of Chapter 5). It remains to check that the symmetry isomorphism acts as identity on $H^0(L')$. Let us consider the symmetry automorphism $\tau_*^{M^2} : H^0(M^2) \to H^0(M^2)$. We claim that $\tau_*^{M^2} = \mathrm{id}$. Indeed, it is easy to see that the space $H^0(M^2)$ is spanned by the sections of the form $t_a^*s \cdot t_{-a}^*s$, where $s \in H^0(M)$, $a \in A(k)$ and some isomorphism $t_a^*M \otimes t_{-a}^*M \simeq M^2$ is used. Clearly, the symmetry automorphism preserves these sections, hence $\tau_*^{M^2} = \mathrm{id}$. Alternatively, one can check that $\tau_*^{M^2}$ commutes with the action of the Heisenberg group $G(M^2)$ on $H^0(M^2)$ (see Exercise 6). Since by Proposition 12.6 the representation of $G(M^2)$ on $H^0(M^2)$ is irreducible, we should have $\tau_*^{M^2} = \pm\,\mathrm{id}$. It remains to use the fact that $\tau_*^{M^2}(s^2) = s^2$ for $s \in H^0(M)$. $\square$

Combining this theorem with Corollary 13.4, we get the following result.

**Corollary 13.6.** *Assume that* $\mathrm{char}(k) \neq 2$. *Let $D$ be a symmetric effective divisor such that $K(L) = 0$, where $L = \mathcal{O}_A(D)$. Then* $\mathrm{mult}_0\, D$ *is even if and only if the quadratic form $q_L$ is even.*

**Remark.** It would be interesting to study the case of a symmetric line bundle $L$ with $K(L) = 0$ but not necessarily ample. According to Corollary 11.12, in this case there is an integer $i(L)$, $0 \leq i(L) \leq g$, such that $H^{i(L)}(L) \neq 0$. The symmetry isomorphism acts on $H^{i(L)}(L)$ by $\pm 1$. In the case when the ground field is $\mathbb{C}$ we will show below that this sign is equal to the $(-1)^{i(L)}\,\mathrm{Arf}(q_L)$. It seems plausible that this is also true in the case of arbitrary characteristic $\neq 2$. As in the case $i(L) = 0$, it would be enough to check this equality in the case when $q_L$ is even.

## 13.4. Transcendental Picture

Let $T = V/\Gamma$ be a complex torus, where $\Gamma$ is a lattice in a complex vector space $V$, $L = L(H, \Gamma, \alpha)$ be a holomorphic line bundle associated with a Hermitian form $H$ on $V$, such that $E = \mathrm{Im}\, H|_{\Gamma\times\Gamma}$ is integer-valued, and with a quadratic map $\alpha : \Gamma \to U(1)$ (see Section 1.2). Note that $L$ is symmetric if and only if $\alpha(-\gamma) = \alpha(\gamma)$ for any $\gamma \in \Gamma$, which is equivalent to the condition $\alpha^2 = 1$. Then we have a canonical isomorphism $\tau : [-1]^*L \widetilde{\to} L$ given by $f(v) \mapsto f(-v)$ in the standard trivialization of the pull-back of $L$ to $V$. In particular, for a point $\gamma/2 \in V/\Gamma$, where $\gamma \in \Gamma$, the action of $\tau$ at $\gamma/2$ is computed by using the formula

$$f\left(\frac{\gamma}{2}\right) = \alpha(\gamma) f\left(-\frac{\gamma}{2}\right).$$

Hence, we get $q_L(\gamma/2) = \alpha(\gamma)$.

Assume, in addition, that $H$ is nondegenerate and $\Gamma$ is self-dual with respect to $E$. Then the cohomology of $L$ is concentrated in degree $i(L)$, the number of negative eigenvalues of $H$, and $H^{i(L)}(T, L)$ is 1-dimensional. Let us assume also that $q_L$ is even, or equivalently $\alpha$ is even. Then according to Theorem 5.1, there exists a Lagrangian subspace $U \subset V$ compatible with

$(\Gamma, \alpha)$. Now from the proof of Theorem 7.2 we see that the action of the symmetry $\tau : [-1]^*L \to L$ on $H^{i(L)}(T, L)$ coincides with $(-1)^{i(L)}$.

## 13.5. Line Bundles and Symmetric Homomorphisms

Recall that to every line bundle $L$ on an abelian variety we associated a symmetric homomorphism $\phi_L : A \to \hat{A}$. Now we will show that every symmetric homomorphism from $A$ to $\hat{A}$ appears in this way.

**Theorem 13.7.** *Let $f : A \to \hat{A}$ be a symmetric homomorphism. Then there exists a line bundle $L$ on $A$ such that $f = \phi_L$. Furthermore, the line bundle $L$ can be chosen to be symmetric.*

*Proof.* Let $\mathcal{P}$ be the Poincaré line bundle on $A \times \hat{A}$. Let us consider the line bundle $M$ on $A$ defined by

$$M = (\mathrm{id}, f)^*\mathcal{P}^2.$$

From the symmetry of $f$ it follows that $\phi_M = 4f$. Indeed, denoting the fiber of $\mathcal{P}$ at the point $(a, \xi) \in A \times \hat{A}$ by $\langle a, \xi \rangle$, we have $M_a = \langle a, f(a) \rangle^2$. Therefore,

$$M_{a+a'} \otimes M_a^{-1} \otimes M_{a'}^{-1} \simeq \langle a, f(a') \rangle^2 \otimes \langle a', f(a) \rangle^2 \simeq \langle a, f(a') \rangle^4$$

because $\langle a', f(a) \rangle \simeq \langle a, f(a') \rangle$ by the symmetry of $f$. Now we claim that there exists a line bundle $L$ on $A$ such that $[2]_A^*L \simeq M$. To prove this we have to show that there exists a lifting of $A_2 \subset K(M)$ to a subgroup of the Mumford group $G(M)$ (see Section 12.3). According to Lemma 12.2, it suffices to check that $A_2 \subset K(M) = \ker(4f)$ is isotropic with respect to the pairing $e_{4f}$. But $e_{4f}(x, y) = e_2(x, 2f(y)) = 1$ for $x, y \in A_2$ (see Exercise 2(b) of Chapter 10), so our claim follows. The relation $M \simeq [2]_A^*L$ implies $4\phi_L = \phi_M = 4f$, hence $\phi_L = f$.

Finally, we note that for every line bundle $L$ there exists a symmetric line bundle $L'$ such that $\phi_L = \phi_{L'}$. □

Let $\phi : A \to \hat{A}$ be a symmetric morphism. We define $\hat{A}(\phi)$ as the set of isomorphism classes of line bundles $L$ on $A$ with $\phi_L = \phi$. There is a non-canonical isomorphism $\hat{A}(\phi) \simeq \hat{A}(k)$, however, in the case when $A$ is defined over a smaller subfield, such an isomorphism does not necessarily commute with the action of Galois group. On the other hand, there is a canonical $\hat{A}(k)$-torsor structure on $\hat{A}(\phi)$ given by tensor product. Furthermore, it is clear that $\hat{A}(\phi)$ is the push-out of the $\hat{A}_2(k)$-torsor $\hat{A}_2(\phi)$ consisting of symmetric line bundles $L$ with $\phi_L = \phi$.

**Theorem 13.8.** *The $\hat{A}_2(k)$-torsor $\hat{A}_2(\phi)$ is canonically isomorphic to the $\hat{A}_2(k)$-torsor consisting of morphisms $q : A_2 \to \mu_2$ satisfying*

$$q(x+y) = q(x)q(y)e_2(x, \phi(y)). \tag{13.5.1}$$

*Proof.* Note that homomorphisms $A_2 \to \mu_2$ correspond to $k$-points of $\hat{A}_2$ (by Weil pairing), so the set of maps $q : A_2 \to \mu_2$ satisfying (13.5.1) is indeed a $\hat{A}_2(k)$-torsor. Furthermore, according to Proposition 13.1 the map $L \mapsto q_L$ is a morphism of $\hat{A}_2(k)$-torsors. Since by Theorem 13.7, $\hat{A}_2(\phi)$ is nonempty, this map is an isomorphism. □

## Exercises

1. Let $L$ be a line bundle on an abelian variety $A$.
   (a) Prove that $L \simeq M^2$ for some line bundle $M$ on $A$ if and only if $\phi_L(A_2) = 0$.
   (b) Now assume that $L$ is symmetric. Prove that $L \simeq M^2$ for some *symmetric* line bundle $M$ if and only if $q_L \equiv 1$.
2. Let $\phi : A \xrightarrow{\sim} \hat{A}$ be a principal polarization of an abelian variety $A$.
   (a) Show that all ample line bundles $L$ on $A$ with $\phi = \phi_L$ are translations of each other.
   (b) Prove that there exists a symmetric effective divisor $D \subset A$ such that $\phi = \phi_{\mathcal{O}_A(D)}$.
3. Let $A$ be an abelian variety, $\mathcal{B}$ be a symmetric biextension on $A \times A$. Show that there exists a line bundle $L$ on $A$ such that $\mathcal{B} \simeq \Lambda(L)$.
4. Let $E$ be an elliptic curve over a field of characteristic $\neq 2$. Prove that there exists a unique line bundle $L$ of degree 1 on $E$ such that $q_L$ is odd, namely, $L = \mathcal{O}_E(e)$ where $e \in E$ is the neutral element. Show that the corresponding quadratic form $q$ satisfies $q(x) = -1$ for every nontrivial point $x \in E_2$.
5. Prove that the classical theta function $\theta\left[\begin{smallmatrix} n_1/2 \\ n_2/2 \end{smallmatrix}\right](x, Z)$ for $n_1, n_2 \in \mathbb{Z}^g$, defined in Section 5.6, is an even (resp., odd) function of $x$ if and only if the dot-product $n_1 \cdot n_2$ is even (resp., odd).
6. Let $L$ be an ample symmetric line bundle on an abelian variety $A$. Let $i : G(L) \to G(L)$ be the involution introduced in the proof of Proposition 13.2.

(a) Show that the map $\tau_* : H^0(L) \to H^0(L)$ is compatible with the involution $i$ and with the action of $G(L)$ on $H^0(L)$, i.e., that $\tau_*(gs) = i(g)\tau_*(s)$, where $g \in G(L)$, $s \in H^0(L)$.

(b) Let $p : G(L) \to K(L)$ be the natural projection. Show that the restriction of the involution $i$ to the subgroup $p^{-1}(A_2) \subset G(L)$ is given by $i(g) = q_L(p(g)) \cdot g$. In particular, if $q_L \equiv 1$, then this restriction is equal to the identity.

# 14

# Vector Bundles on Elliptic Curves

In this chapter we study vector bundles on an elliptic curve $E$. The main idea is to combine the $\mathrm{SL}_2(\mathbb{Z})$-action (up to shifts) on the derived category of coherent sheaves on $E$ introduced in Chapter 11, with the notion of stability (resp., semistability) of vector bundles that we recall below in 14.1. It is easy to see that all vector bundles on elliptic curve are direct sums of semistable ones. On the other hand, we prove that the Fourier–Mukai transform (and hence, every functor constituting a part of the $\mathrm{SL}_2(\mathbb{Z})$-action) sends semistable bundles to semistable bundles or to torsion coherent sheaves (up to shift). Since the pair (deg, rk) transforms under the $\mathrm{SL}_2(\mathbb{Z})$-action on $D^b(E)$ according to the standard action of $\mathrm{SL}_2(\mathbb{Z})$ on $\mathbb{Z}^2$, we obtain the equivalence of the category of semistable bundles of given slope (the ratio of degree and rank) with the category of torsion coherent sheaves on $E$. Under this equivalence stable bundles correspond to structure sheaves of points (in particular, their degree and rank are relatively prime). We also sketch an inductive construction of all stable bundles on $E$ as successive extensions. This construction is based on the observation that if $V_1$ and $V_2$ are stable vector bundles of degrees and ranks $(d_i, r_i)$, $i = 1, 2$, such that $d_2r_1 - d_1r_2 = 1$, then there is a unique non-trivial extension of $V_2$ by $V_1$, and it is stable.

## 14.1. Stable and Semistable Bundles

The *slope* of a vector bundle $F$ is the ratio $\mu(F) = \deg F/\mathrm{rk}\, F$. A vector bundle $F$ is called *stable* (respectively, *semistable*), if for every proper nonzero subbundle $F_1 \subset F$ one has $\mu(F_1) < \mu(F)$ (resp., $\mu(F_1) \leq \mu(F)$). Equivalently, for every nontrivial quotient-bundle $F \to F_2$ one should have $\mu(F_2) > \mu(F)$ (resp., $\mu(F_2) \geq \mu(F)$).

**Lemma 14.1.** *For a pair of semistable bundles $F$ and $F'$, one has* $\mathrm{Hom}(F, F') = 0$ *unless* $\mu(F) \leq \mu(F')$.

*Proof.* Let $G \subset F'$ be the image of a nonzero morphism $f : F \to F'$. Then $G$ is a quotient-bundle of $F$, hence $\mu(G) \geq \mu(F)$. On the other hand, even though $G$ is just a subsheaf in $F'$, it is contained in the unique subbundle $G' \subset F'$ of the same rank as $G$. Then we have $\mu(G) \leq \mu(G') \leq \mu(F')$. Therefore, $\mu(F) \leq \mu(F')$. □

It is clear that line bundles are stable. To construct semistable bundles of higher rank, the following lemma is useful.

**Lemma 14.2.** *Let* $0 \to F_1 \to F_2 \to F_3 \to 0$ *be an exact sequence of vector bundles of equal slopes. Then* $F_2$ *is semistable if and only if* $F_1$ *and* $F_3$ *are semistable.*

*Proof.* Clearly, if $F_2$ is semistable, then $F_1$ and $F_3$ are also semistable. Now assume that $F_1$ and $F_3$ are semistable of slope $\mu$. Then for arbitrary subbundle $G \subset F_2$, we have the corresponding subsheaves $G \cap F_1 \subset F_1$ and $G/G \cap F_1 \subset F_3$. By semistability of $F_1$ and $F_3$ this implies that $\mu(G \cap F_1) \leq \mu$ and $\mu(G/G \cap F_1) \leq \mu$. Hence, $\mu(G) \leq \mu$. □

**Lemma 14.3.** *A stable bundle is simple.*

*Proof.* Let $F$ be a stable bundle. Using the natural map

$$\det : \mathrm{Hom}(F, F) \to \mathrm{Hom}(\det F, \det F) \simeq H^0(E, \mathcal{O}) = k$$

we can define for every endomorphism $T : F \to F$ its characteristic polynomial $\det(T - \lambda\, \mathrm{id})$. Let $\lambda$ be a root of this polynomial. Then the rank of the image of $T - \lambda\, \mathrm{id}$ is smaller than the rank of $F$. We claim that $T - \lambda\, \mathrm{id} = 0$. Indeed, assume that the image $G$ of $T - \lambda\, \mathrm{id}$ has positive rank. Then since $G$ is simultaneously a nontrivial subbundle of $F$ and a nontrivial quotient-bundle of $F$, we obtain contradicting inequalities $\mu(G) < \mu(F)$ and $\mu(F) > \mu(G)$. □

The filtration introduced in the part (i) of the following lemma is called the *Harder–Narasimhan filtration.* (It was introduced in [60].)

**Lemma 14.4.** *(i) Every bundle* $F$ *has a unique filtration* $0 \subset F_1 \subset \cdots \subset F_n = F$ *such that associated graded quotients* $F_i/F_{i-1}$ *are semistable bundles and* $\mu(F_i/F_{i-1}) > \mu(F_{i+1}/F_i)$.

*(ii) Every semistable bundle $F$ of slope $\mu$ has a filtration whose associated graded quotients are stable bundles of slope $\mu$.*

*Proof.* (i) Before starting the proof, let us observe that slopes of non-zero subbundles of a given bundle $F$ are bounded from above. Indeed, we can realize $F$ as a subbundle of $L^{\oplus n}$, where $L$ is a sufficiently ample bundle. Now our claim follows from semistability of $L^{\oplus n}$ (see Lemma 14.2). Thus, there exists a subbundle of maximal slope in $F$. Clearly, any such subbundle is semistable. Furthermore, if $F_1$ and $F_1'$ are two subbundles of the maximal slope in $F$, then $F_1 + F_1'$ is also a subbundle of the maximal slope. Indeed, this follows from the fact that $F_1 + F_1'$ is a quotient-bundle of the semistable bundle $F_1 \oplus F_1'$. Therefore, we can take $F_1$ to be the maximal subbundle of maximal slope in $F$. Then every non-zero subbundle of $F/F_1$ has slope $< \mu(F_1)$. Now we define $F_2$ to be the preimage in $F$ of the maximal subbundle of maximal slope in $F/F_1$, and so on. In this way we obtain the required filtration. The uniqueness follows immediately from the definition.

(ii) This can be easily proven by induction in the rank of $F$. □

**Lemma 14.5.** *Let $F$ be an indecomposable bundle on an elliptic curve. Then $F$ is semistable.*

*Proof.* Let $0 \subset F_1 \subset \cdots \subset F_n = F$ be the Harder–Narasimhan filtration, that is $G_i = F_i/F_{i-1}$ is semistable for all $i$ and the slope $\mu(G_i)$ of $G_i$ is strictly decreasing function of $i$. Now we observe that for every pair of semistable bundles $G$ and $G'$ such that $\mu(G) < \mu(G')$, one has $\mathrm{Ext}^1(G, G') \simeq \mathrm{Hom}(G', G)^* = 0$. Hence, the above filtration splits. □

## 14.2. Categories of Semistable Bundles

We identify $E$ with the dual elliptic curve $\hat{E}$ using the isomorphism $\phi_L$, where $L$ is a line bundle of degree 1 on $E$ ($\phi_L$ does not depend on such $L$). Then the Fourier–Mukai transform $\mathcal{S}$ can be considered as an autoequivalence of $D^b(E)$.

**Lemma 14.6.** *Let $F$ be a semistable bundle with a negative slope $\mu$. Then the Fourier transform $\mathcal{S}(F)[1]$ is a semistable bundle with the slope $-\mu^{-1}$.*

*Proof.* We can assume that $F$ is indecomposable. Note that for every $L \in \mathrm{Pic}^0(E)$ one has $H^0(F \otimes L) = \mathrm{Hom}(L^{-1}, F) = 0$ by semistability. Therefore, by the base change theorem (see [62], Theorem 12.11) $\mathcal{S}(F)[1]$ is a vector bundle. Using the equation (11.3.7) we find that its rank is equal to $-\deg F$ while its degree is equal to $\mathrm{rk}\, F$. It remains to apply Lemma 14.5. □

Now we can prove the following theorem.

**Theorem 14.7.** *(i) Let $\mathcal{B}_\mu$ be the category of the semistable bundles of the slope $\mu$ on $E$. Then there is a canonical equivalence*

$$T : \mathcal{B}_\mu \simeq \mathrm{Sh}_{\mathrm{tors}},$$

*where $\mathrm{Sh}_{\mathrm{tors}}$ is the category of sheaves with finite support on $E$.*

*(ii) For a semistable bundle $F \in \mathcal{B}_\mu$ one has* $\mathrm{length}(T(F)) = gcd(\deg F, \mathrm{rk}\, F)$.

*Proof.* As we have noted above if $\mu < 0$ then the Fourier transform induces the equivalence $\mathcal{B}_\mu \simeq \mathcal{B}_{-\mu^{-1}}$. Also twisting by $\mathcal{O}(e)$ gives the equivalence $\mathcal{B}_{\mu-1} \simeq \mathcal{B}_\mu$. Applying these two operations by Euclid algorithm we obtain the equivalence $\mathcal{B}_\mu \simeq \mathcal{B}_0$. Now we claim that the Fourier transform $\mathcal{S}[1]$ induces the equivalence $\mathcal{B}_0 \simeq \mathrm{Sh}_{\mathrm{tors}}$. To prove this, we have to check that for a semistable bundle $F$ of degree 0 the object of the derived category $\mathcal{S}(F)[1]$ is a sheaf with finite support. Assume first that there exists a line bundle $L$ of degree zero and a nonzero morphism $L \to F$. Then $L$ is a subbundle in $F$ and the quotient $F/L$ is also a semistable bundle of degree zero. Since $\mathcal{S}(L)[1]$ is a structure sheaf of a point, it suffices to prove the statement for $F/L$. Iterating this argument we reduce to the case when $H^0(F \otimes L) = 0$ for every $L \in \mathrm{Pic}^0(E)$. In this case by the base change theorem ([62], Theorem 12.11) we get that $\mathcal{S}(F)[1]$ is a vector bundle. By formula (11.3.7) its rank is equal to $\deg F = 0$, hence, $\mathcal{S}(F)[1] = 0$. Thus, we have a functor $\mathcal{S}[1] : \mathcal{B}_0 \to \mathrm{Sh}_{\mathrm{tors}}$. Clearly, $[-1]^*\mathcal{S}$ is the inverse functor. The proof of (ii) follows easily from the fact that $gcd(\deg F, \mathrm{rk}\, F)$ is preserved by our equivalences between categories $\mathcal{B}_\mu$. □

**Corollary 14.8.** *Let $F$ be an indecomposable vector bundle of rank $r$ and degree $d$ on $E$. The following conditions are equivalent*

*(i) $F$ is stable;*

*(ii) $F$ is simple;*

*(iii) $d$ and $r$ are relatively prime.*

*Proof.* This follows from the fact that stable bundles (resp., structure sheaves of points) are simple objects of the category $\mathcal{B}_\mu$ (respectively, $\mathrm{Sh}_{\mathrm{tors}}$). □

**Corollary 14.9.** *Every simple vector bundle of rank $r$ and degree $d$ on $E$ is obtained by the construction of Proposition 12.7 applied to some line bundle of degree $rd$ and the subgroup $E_r \subset E$.*

*Proof.* This follows from the proof of Theorem 14.7 and from Exercise 2 of Chapter 12. □

**Remark.** Theorem 14.7 holds also in the case when the base field $k$ is not algebraically closed. So the isomorphism classes of indecomposable bundles of rank $r$ and degree $d$ are still in bijective correspondence with the indecomposable torsion sheaves of degree $d_0 = gcd(r, d)$ on $C$ up to isomorphism which are in turn classified by the pairs: a decomposition $d_0 = d_1 d_2$ and an orbit of cardinality $d_1$ for the action of the Galois group $\mathrm{Gal}(\overline{k}/k)$ on $E(\overline{k})$ where $\overline{k}$ is an algebraic closure of $k$.

## 14.3. Stable Bundles on Elliptic Curves and Rational Numbers

Let us fix a point $x \in E$. Theorem 14.7 implies that there is a bijective correspondence between stable vector bundles on $E$ with determinant of the form $\mathcal{O}_E(dx)$, where $d \in \mathbb{Z}$, and rational numbers. Namely, for every rational number $\mu = \frac{d}{r}$, where $r > 0$, $(r, d) = 1$, there exists a unique stable vector bundle $V_\mu$ of rank $r$ such that $\det V_\mu \simeq \mathcal{O}_E(dx)$ (see Exercise 3).

Let us call two rational numbers $\mu_1 = \frac{d_1}{r_1}$ and $\mu_2 = \frac{d_2}{r_2}$ (where $r_i > 0$, $(r_i, d_i) = 1$) *close* if $|d_1 r_2 - d_2 r_1| = 1$. The reason is that if $\mu_1 < \mu_2$ are close then for every rational number $\mu = \frac{d}{r}$ such that $\mu_1 < \mu < \mu_2$ one should have $r > r_1$ and $r > r_2$. This condition has a simple interpretation for vector bundles. Indeed, assume that $\mu_1 < \mu_2$. Then we have

$$\dim \mathrm{Hom}\left(V_{\mu_1}, V_{\mu_2}\right) = \dim \mathrm{Ext}^1\left(V_{\mu_2}, V_{\mu_1}\right) = d_2 r_1 - d_1 r_2.$$

Thus, $\mu_1$ and $\mu_2$ (such that $\mu_1 < \mu_2$) are close if and only if $\mathrm{Hom}(V_{\mu_1}, V_{\mu_2})$ is 1-dimensional. By Serre duality, in this case there is a unique non-trivial extension

$$0 \to V_{\mu_1} \to V \to V_{\mu_2} \to 0.$$

**Theorem 14.10.** *$V$ is stable.*

*Proof.* Indeed, assume that $U \subset V$ is a stable subbundle such that $\mu(U) \geq \mu(V)$. Then $\mu(U) > \mu_1$, hence $U$ cannot be contained in $V_{\mu_1}$. Therefore, the induced map from $U$ to $V_{\mu_2}$ is nonzero which implies that $\mu(U) \leq \mu_2$. If $\mu(U) = \mu_2$ then $U$ maps to $V_{\mu_2}$ isomorphically which is impossible since the extension does not split. Thus, we have $\mu(V) \leq \mu(U) < \mu_2$. But the numbers $\mu(V) = \frac{d_1+d_2}{r_1+r_2}$ and $\mu_2$ are close and the denominator of $\mu(U)$ is less than the denominator of $\mu(V)$, so this is impossible. □

**Remark.** It was not important for the proof that $\det V_{\mu_i} \simeq \mathcal{O}_E(d_i x)$. The theorem still holds for every pair of stable bundles with slopes $\mu_1$ and $\mu_2$.

Here is an algorithm for obtaining all pairs of close rational numbers between 0 and 1. Start with one close pair $(0/1, 1/1)$. Then every step of the algorithm produces a new generation of close pairs by the following recipe: every close pair $(\frac{d_1}{r_1}, \frac{d_2}{r_2})$ from the previous generation produces two new close pairs, namely, $(\frac{d_1}{r_1}, \frac{d_1+d_2}{r_1+r_2})$ and $(\frac{d_1+d_2}{r_1+r_2}, \frac{d_2}{r_2})$. Combining this algorithm with the theorem above we obtain the algorithm for constructing all stable bundles (up to translation) with slope between 0 and 1. This construction gives the following result.

**Corollary 14.11.** *Assume that $\mu_1 < \mu_2$ are close and $\mu_2 - \mu_1 < 1$. Then the unique (up to scalar) morphism $V_{\mu_1} \to V_{\mu_2}$ is a surjection if $r_1 > r_2$ and is an embedding as a subbundle if $r_1 < r_2$.*

*Proof.* Shifting $\mu_i$ by an integer (which corresponds to tensoring $V_{\mu_i}$ by a power of $\mathcal{O}_E(x)$) we can assume that $(\mu_1, \mu_2)$ is obtained from the above algorithm and $(\mu_1, \mu_2) \neq (0, 1)$. □

## Exercises

1. Let $F$ be a stable bundle on an elliptic curve $E$, $d = \deg(F)$, $r = \mathrm{rk}(F)$. Prove that for every $x \in E$ one has

$$t_{rx}^* F \simeq \mathcal{P}_{dx} \otimes F,$$

   where we use the identification $E \simeq \hat{E}$ given by $\phi_{\mathcal{O}_E(e)}$. [*Hint*: This statement is invariant under the action of $\mathrm{SL}_2(\mathbb{Z})$; in the case of integer slope a stable bundle has rank 1.]
2. Assume that $d \neq 0$. Then all stable bundles of degree $d$ and rank $r$ are obtained from one by translations. [*Hint*: By the previous exercise it suffices to prove that all stable bundles are obtained from one by translations

and tensorings with $\mathrm{Pic}^0(E)$. Now use $\mathrm{SL}_2(\mathbb{Z})$-action to reduce to the case of line bundles.]

3. A stable bundle $F$ is uniquely determined by its rank and by the line bundle $\det F$. [*Hint*: By the previous Exercise it suffices to prove that if $\det(t_x^*F) = \det(F)$ then $t_x^*F \simeq F$. The assumption means that $dx = 0$ where $d = \deg(F)$. Since $r = \mathrm{rk}(F)$ and $d$ are relatively prime, there exists a point $x'$ such that $x = rx'$. Now use Exercise 1.]
4. There exists a sequence of semistable indecomposable bundles $F_1 = \mathcal{O}$, $F_2, F_3, \ldots,$ such that $F_n$ is the unique non-trivial extension of $F_{n-1}$ by $\mathcal{O}$.
5. Assume that characteristic of $k$ is zero. Then every indecomposable bundle on elliptic curve has form $F_n \otimes V$ where $V$ is stable, $F_n$ is one of the bundles defined in the previous exercise. [*Hint*: It suffices to prove that $F_n \otimes V$ is a non-trivial extension of $F_{n-1} \otimes V$ by $V$.]
6. Let $\pi : E' \to E$ be an isogeny of elliptic curves of degree $r$ that is relatively prime to the characteristic.
   (a) Prove that for a line bundle $L$ on $E'$ of degree $d$ such that $(r, d) = 1$, the vector bundle $\pi_* L$ on $E$ is simple. [*Hint*: It suffices to prove that $\pi_* L$ is indecomposable. But

$$\mathrm{Hom}(\pi_*L, \pi_*L) \simeq \mathrm{Hom}(\pi^*\pi_*L, L) \simeq \oplus_{x\in\ker(\pi)} \mathrm{Hom}(t_x^*L, L).$$

   But for all $x \in \ker(\pi)$ we have $rx = 0$, while $K(L) = E'_d$.]
   (b) Prove that for every line bundle $L$ on $E'$ the vector bundle $\pi_* L$ is a direct sum of stable bundles.
   (c) Prove that for a stable bundle $V$ on $E$ the bundle $\pi^* V$ is a direct sum of stable bundles.
   (d) Prove that every stable bundle of rank $r$ on $E$ is isomorphic to $\pi_* L$ for some line bundle $L$ on $E'$.
7. Let $L$ be a line bundle of positive degree $d$ on $E$. Consider the space of extension classes $\mathrm{Ext}^1(L, \mathcal{O}_E)$. For every extension class $e$ let $V_e$ be the corresponding rank 2 vector bundle sitting in the exact sequence

$$0 \to \mathcal{O}_E \to V_e \to L \to 0.$$

   (a) Assume that $d$ is odd. Prove that for generic $e$ the bundle $V_e$ is stable.
   (b) Assume that $d$ is even. Prove that for generic $e$ the bundle $V_e$ is semistable.
8. Below we use the notations of Section 14.3. Prove that for every pair of positive rational numbers $\mu < \mu'$ there exists a sequence of rational

numbers

$$\mu = \mu_{-m} < \cdots < \mu_{-1} < \mu_0 < \mu_1 < \cdots < \mu_n = \mu'$$

such that all the pairs $(\mu_i, \mu_{i+1})$ are close, the sequence of denominators of $\mu_{-m}, \ldots, \mu_0$ is strictly decreasing, and the sequence of denominators of $\mu_0, \ldots, \mu_n$ is (non-strictly) increasing. Show that if $\mu' - \mu < 1$ then the second sequence is strictly increasing. Deduce that if $\mu < \mu'$, then there exists a morphism $V_\mu \to V_{\mu'}$ such that its image is a stable bundle. If $\mu' - \mu < 1$ then in addition show that the image is a subbundle of $V_{\mu'}$.

# 15

# Equivalences between Derived Categories of Coherent Sheaves on Abelian Varieties

In this chapter we study equivalences between derived categories of coherent sheaves on abelian varieties. In Section 15.1 we present a construction of such equivalences based on the Fourier–Mukai transform. The idea is to consider a finite group scheme $K$ equipped with embeddings into an abelian variety $A$ and into the dual abelian variety $\hat{A}$. In this situation one can try to "descend" the Fourier–Mukai transform to an equivalence between derived categories of $A/K$ and $\hat{A}/K$. We show that this is possible provided that there exist a line bundle $L$ on $A$ and a line bundle $L'$ on $\hat{A}$, such that homomorphisms $\phi_L : A \to \hat{A}$ and $\phi_{L'} : \hat{A} \to A$ restrict to the identity on $K$, and there exists an isomorphism $L|_K \simeq L'|_K$ compatible with the additional data on these line bundles coming from the theorem of the cube. Then in Section 15.2 we describe a general framework for studying equivalences between derived categories of coherent sheaves on abelian varieties. The idea is that $D^b(A)$ looks similar to the Schrödinger representation of the Heisenberg group: There are two kinds of autoequivalences of this category, translations and tensoring with line bundles in $\mathrm{Pic}^0(A)$, which are similar to the operators of the Schrödinger representation. We introduce the corresponding notion of the *Heisenberg groupoid* **H** and show that $D^b(A)$ is a *representation* of **H** in appropriate sense. Then we work out the analogue of the classical theory from Chapters 2 and 4 for this situation. The Heisenberg groupoid is naturally attached to the "symplectic" data constructed from $A$, namely, the abelian variety $X = A \times \hat{A}$ equipped with the biextension $\mathcal{E}$ on $X \times X$ defined by $\mathcal{E} = p_{14}^*\mathcal{P} \otimes p_{23}^*\mathcal{P}^{-1}$ ($\mathcal{E}$ should be considered as an analogue of the symplectic form). There is a natural notion of Lagrangian abelian subvariety in $X$ (with respect to $\mathcal{E}$) and every such subvariety $Y$ gives rise to a representation of **H**. In the case when the projection $X \to X/Y$ admits a section, the underlying category of this representation is the derived category of coherent sheaves on $X/Y$. *Intertwining functors* (which are appropriate analogues of intertwining operators) between these representations can be constructed similarly to Chapter 4. In the case when Lagrangian subvarieties $Y_1, Y_2 \subset X$ have finite intersection, the corresponding intertwining functor is an equivalence of the kind considered in Section 15.1.

The main application of this formalism is the proof of the fact that the derived categories of coherent sheaves on abelian varieties $A$ and $A'$ are equivalent provided that the corresponding symplectic data $(X, \mathcal{E})$ and $(X', \mathcal{E}')$ are isomorphic. In fact, the converse statement is also true (see [105] for the proof). Developing the above analogy a little further, we define analogues of constants $c(L_1, L_2, L_3) \in U(1)$ from Chapter 4. These are integers attached to triples of Lagrangian subvarieties, which measure shifts of degree appearing when composing the above intertwining operators. Recall that when $L_1, L_2, L_3$ are Lagrangian subspaces of the symplectic vector space, the constant $c(L_1, L_2, L_3)$ can be expressed in terms of the signature of some quadratic form (see Chapter 4). We prove that the integers arising in the categorical setup are equal to indexes of some nondegenerate line bundles on abelian varieties (see Section 11.6). We use this fact to derive a nontrivial relation between these indexes in Proposition 15.8.

## 15.1. Construction of Equivalences

Let $A$ be an abelian variety, $\hat{A}$ the dual abelian variety, $\mathcal{P}$ the Poincaré line bundle on $A \times \hat{A}$. Let $K \subset A$ be a finite subgroup scheme, $\phi : K \to \hat{A}$ a homomorphism, $G \to K$ a line bundle over $K$, equipped with an isomorphism

$$i_G : G_{k_1+k_2} \widetilde{\to} G_{k_1} \otimes G_{k_2} \otimes \mathcal{P}_{k_1,\phi(k_2)}$$

over $K \times K$ satisfying the natural cocycle condition on $K \times K \times K$. Consider the abelian category $\mathrm{Coh}_{G,\phi}(A)$ whose objects are pairs $(F, i)$, where $F$ is a coherent sheaf on $A$, $i$ is an isomorphism

$$i_{x,k} : F_{x+k} \widetilde{\to} G_k \otimes \mathcal{P}_{x,\phi(k)} \otimes F_x \tag{15.1.1}$$

on $K \times X$, such that the following diagram on $K \times K \times X$ is commutative:

$$\begin{array}{ccc} F_{x+k_1+k_2} & \xrightarrow{i_{x+k_1,k_2}} & G_{k_2} \otimes \mathcal{P}_{x+k_1,\phi(k_2)} \otimes F_{x+k_1} \\ \downarrow{\scriptstyle i_{x,k_1+k_2}} & & \downarrow{\scriptstyle i_{x,k_1}} \\ G_{k_1+k_2} \otimes \mathcal{P}_{x,\phi(k_1+k_2)} \otimes F_x & \xrightarrow{i_G} & G_{k_1} \otimes G_{k_2} \otimes \mathcal{P}_{k_1,\phi(k_2)} \otimes \mathcal{P}_{x,\phi(k_1+k_2)} \otimes F_x \end{array} \tag{15.1.2}$$

The morphisms $(F, i) \to (F', i')$ in the category $\mathrm{Coh}_{G,\phi}(A)$ are morphisms $F \to F'$ between the corresponding coherent sheaves commuting with the isomorphisms $i$ and $i'$.

The category $\mathrm{Coh}_{G,\phi}(A)$ should be considered as a "twist" of the category $\mathrm{Coh}(A/K)$ of coherent sheaves on $A$. Namely, for $\phi = 0$ and trivial $G$ we have a canonical equivalence $\mathrm{Coh}_{\mathcal{O}_K,0} \simeq \mathrm{Coh}(A/K)$ by the descent theory (see Appendix C). More generally, assume that there exists a line bundle $L$ on $A$ such that $\phi = \phi_L|_K$, $G = L|_K$ and $i_G$ is the restriction of the similar isomorphism on $A \times A$ (corresponding to some trivialization of $L|_0$). Then the functor $A \mapsto A \otimes L^{-1}$ extends to an equivalence of categories $\mathrm{Coh}_{G,\phi}(A) \simeq \mathrm{Coh}_{\mathcal{O}_K,0}(A)$, hence in this case the category $\mathrm{Coh}_{G,\phi}(A)$ is also equivalent to $\mathrm{Coh}(A/K)$.

Our interest to categories $\mathrm{Coh}_{G,\phi}(A)$ is due to the fact that the Fourier–Mukai transform $\mathcal{S} : D^b(A) \to D^b(\hat{A})$ is likely to induce an equivalence of the derived category of such a category on $A$ with a similar derived category on $\hat{A}$ (since $\mathcal{S}$ switches translations with tensor multiplications by $\mathrm{Pic}^0$). Indeed, let $F$ be an object in $D^b(A)$ equipped with an isomorphism (15.1.1). For simplicity let us assume that $K$ is reduced (this assumption will not be used in the proof of Theorem 15.2). We can rewrite (15.1.1) as a collection of isomorphisms

$$t_k^* F \simeq G_k \otimes \mathcal{P}_{\phi(k)} \otimes F,$$

where $k \in K$, $\mathcal{P}_\xi = \mathcal{P}|_{A \times \xi}$ for $\xi \in \hat{A}$. Using the isomorphisms (11.3.1) and (11.3.2) we get

$$t_{\phi(k)}^* \mathcal{S}(F) \simeq G_k^{-1} \otimes \mathcal{P}_{-k} \otimes \mathcal{S}(F),$$

where $\mathcal{P}_x = \mathcal{P}|_{x \times \hat{A}}$ for $x \in A$. Let us assume that $\phi$ is an isomorphism of $K$ onto a subgroup $K'$ of $\hat{A}$. Then the obtained collection of isomorphisms for $\mathcal{S}(F)$ is of the same kind as for an object of $\mathrm{Coh}_{\phi_* G^{-1}, -\phi^{-1}}(\hat{A})$. Thus, we can expect that there is an equivalence

$$D^b(\mathrm{Coh}_{G,\phi}(A)) \simeq D^b(\mathrm{Coh}_{\phi_* G^{-1}, -\phi^{-1}}(\hat{A})).$$

We will not prove this in full generality but rather restrict ourselves to the situation when both categories in question can be "untwisted." In other words, we assume that there exist line bundles $L$ on $A$ and $M$ on $\hat{A}$ trivialized at zero, such that $\phi = \phi_L|_K$, $-\phi^{-1} = \phi_M|_{K'}$, and there exist isomorphisms $G \simeq L|_K$, $G^{-1} \simeq \phi^* M|_{K'}$, compatible with $i_G$. Then the functor $F \mapsto F \otimes L^{-1}$ (resp., $F \mapsto F \otimes M^{-1}$) gives an equivalence $\mathrm{Coh}_{G,\phi}(A) \widetilde{\to} \mathrm{Coh}_{\mathcal{O}_K,0}(A)$ (resp., $\mathrm{Coh}_{\phi_* G^{-1}, -\phi^{-1}}(\hat{A}) \widetilde{\to} \mathrm{Coh}_{\mathcal{O}_{K'},0}(\hat{A})$). In other words, in this case we expect to find an equivalence of the categories $D^b(A/K)$ and $D^b(\hat{A}/K')$. The naive way to define it would be to pretend that the descent theorem (see Appendix C) for the morphism $\pi : A \to A/K$ (resp., $\pi' : \hat{A} \to \hat{A}/K'$) extends to objects of derived category. The above argument shows that the functor

$$\Phi : D^b(A) \to D^b(\hat{A}) : F \mapsto \mathcal{S}(L \otimes F) \otimes M^{-1}$$

transforms the descent data for $\pi$ to the descent data for $\pi'$, so it can be considered as an equivalence of $D^b(A/K)$ and $D^b(\hat{A}/K')$. Unfortunately, the descent theorem for a finite flat morphism $f$ extends to objects of derived category only when the degree of $f$ does not divide the characteristic. To circumvent this difficulty we note that the functor $\Phi \circ \pi^* : D^b(A/K) \to D^b(\hat{A})$ is given by the kernel

$$\widetilde{\mathcal{V}} = (\pi \times \mathrm{id}_{\hat{A}})_*(\mathcal{P} \otimes (L \boxtimes M^{-1}))$$

on $A/K \times \hat{A}$, which is actually a vector bundle. Furthermore, $\widetilde{\mathcal{V}}$ is equipped with the descent data for the morphism $\mathrm{id}_{A/K} \times \pi'$, constructed using the isomorphism

$$\begin{aligned}\mathcal{P}_{x,\xi+\phi(k)} \otimes L_x \otimes M^{-1}_{\xi+\phi(k)} &\simeq \mathcal{P}_{x+k,\xi} \otimes \mathcal{P}_{x,\phi(k)} \otimes L_x \otimes M^{-1}_{\phi(k)} \otimes M^{-1}_{\xi} \\ &\simeq \mathcal{P}_{x+k,\xi} \otimes L_{x+k} \otimes M^{-1}_{\xi} \qquad (15.1.3)\end{aligned}$$

on $K \times A \times \hat{A}$ (here we used our assumptions on $L$ and $M$). These descent data defines a vector bundle $\mathcal{V}$ on $A/K \times \hat{A}/K'$. Clearly, the functor

$$\Phi_{\mathcal{V},A/K \to \hat{A}/K'} : D^b(A/K) \to D^b(\hat{A}/K')$$

(here we use the notation of Section 11.1) is compatible with the above functor $\Phi : D^b(A) \to D^b(\hat{A})$ and the pull-back functors corresponding to $\pi$ and $\pi'$. We are going to prove that $\Phi_{\mathcal{V}}$ is an equivalence of categories. First, we are going to present $\mathcal{V}$ in a more symmetric form.

**Lemma 15.1.** *Let $\Gamma_{-\phi} \subset K \times K'$ be the image of the embedding $(\mathrm{id}_K, -\phi) : K \to A \times \hat{A}$. Then the line bundle $\mathcal{P} \otimes (L \boxtimes M^{-1})$ has a natural descent data for the projection $A \times \hat{A} \to (A \times \hat{A})/\Gamma_{-\phi}$. Let $P$ be the corresponding line bundle on $(A \times \hat{A})/\Gamma_{-\phi}$. Then*

$$\mathcal{V} \simeq \rho_* P,$$

*where $\rho : (A \times \hat{A})/\Gamma_{-\phi} \to A/K \times \hat{A}/K'$ is the natural projection.*

*Proof.* The descent data in question is derived immediately from the isomorphism (15.1.3). Applying the flat base change ([61], Proposition 5.12) to the

Cartesian square

$$\begin{array}{ccc} A \times \hat{A} & \longrightarrow & (A \times \hat{A})/\Gamma_{-\phi} \\ \Big\downarrow & & \Big\downarrow \rho \\ A/K \times \hat{A} & \longrightarrow & A/K \times \hat{A}/K' \end{array} \qquad (15.1.4)$$

we obtain an isomorphism

$$\left(\mathrm{id}_{A/K} \times \pi'\right)^* \rho_* P \simeq \widetilde{\mathcal{V}}.$$

It is easy to check that it is compatible with the descent data for the morphism $\mathrm{id}_{A/K} \times \pi'$. □

Similarly, we define a vector bundle $\mathcal{V}'$ on $A/K \times \hat{A}/K'$ by setting

$$\mathcal{V}' = \rho_* P^{-1}.$$

**Theorem 15.2.** *The functors*

$$\Phi_{\mathcal{V}, A/K \to \hat{A}/K'} : D^b(A/K) \to D^b(\hat{A}/K')$$

*and*

$$\Phi_{\mathcal{V}'[g], \hat{A}/K' \to A/K} : D^b(\hat{A}/K') \to D^b(A/K)$$

*are mutually inverse equivalences of categories.*

*Proof.* Let us prove that the composition $\Phi_{\mathcal{V}'[g], \hat{A}/K' \to A/K} \circ \Phi_{\mathcal{V}, A/K \to \hat{A}/K'}$ is isomorphic to the identity functor on $D^b(A/K)$. The composition in the different order is computed similarly. It suffices to prove that

$$Rp_{12*}(p_{13}^* \mathcal{V} \otimes p_{23}^* \mathcal{V}') \simeq \mathcal{O}_{\Delta_{A/K}}[-g], \qquad (15.1.5)$$

where $\Delta_{A/K}$ is the diagonal in $A/K \times A/K$, $p_{ij}$ are projections of $A/K \times A/K \times \hat{A}/K'$ onto products of factors.

The main problem is to express the vector bundle $p_{13}^* \mathcal{V} \otimes p_{23}^* \mathcal{V}'$ in the more convenient form (for the computation of $Rp_{12*}$). Let us define the subgroups $K_1$ and $K_2$ in $A \times A \times \hat{A}$ as the images of the embeddings

$$K \times K \to A \times A \times \hat{A} : (k_1, k_2) \mapsto (k_1, k_2, -\phi(k_1))$$

and

$$K \times K \to A \times A \times \hat{A} : (k_1, k_2) \mapsto (k_1, k_2, -\phi(k_2)),$$

respectively. Let us denote by $\rho_1 : (A \times A \times \hat{A})/K_1 \to A/K \times A/K \times \hat{A}/K'$ and $\rho_2 : (A \times A \times \hat{A})/K_2 \to A/K \times A/K \times \hat{A}/K'$ the natural projections. Then we have

$$p_{13}^* \mathcal{V} \simeq \rho_{1*}(P_{13}),$$

where $P_{13}$ is the pull-back of $P$ under the natural map

$$(A \times A \times \hat{A})/K_1 \to (A \times \hat{A})/\Gamma_{-\phi} : (x_1, x_2, \xi) \mapsto (x_1, \xi).$$

Similarly,

$$p_{23}^* \mathcal{V}' \simeq \rho_{2*}\big(P_{23}^{-1}\big),$$

where $P_{23}$ is the pull-back of $P$ under the natural map

$$(A \times A \times \hat{A})/K_2 \to (A \times \hat{A})/\Gamma_{-\phi} : (x_1, x_2, \xi) \mapsto (x_2, \xi).$$

Let $K_{12} \subset A \times A \times \hat{A}$ be the intersection of $K_1$ and $K_2$, so that $K_{12}$ is the image of the embedding

$$K \mapsto A \times A \times \hat{A} : (k, k, -\phi(k)).$$

Then we have the Cartesian square of finite flat morphisms:

$$\begin{array}{ccc} (A \times A \times \hat{A})/K_{12} & \longrightarrow & (A \times A \times \hat{A})/K_2 \\ \big\downarrow & & \big\downarrow \rho_2 \\ (A \times A \times \hat{A})/K_1 & \xrightarrow{\rho_1} & A/K \times A/K \times \hat{A}/K' \end{array} \qquad (15.1.6)$$

By the flat base change and the projection formula ([61], Propositions 5.12 and 5.6), we obtain that

$$p_{13}^* \mathcal{V} \otimes p_{23}^* \mathcal{V}' \simeq \rho_{1*}(P_{13}) \otimes \rho_{2*}\big(P_{23}^{-1}\big) \simeq \rho_{12*}\big(\widetilde{P}_{13} \otimes \widetilde{P}_{23}^{-1}\big), \qquad (15.1.7)$$

where $\rho_{12} : (A \times A \times \hat{A})/K_{12} \to A/K \times A/K \times \hat{A}/K'$ is the natural projection, $\widetilde{P}_{13}$ and $\widetilde{P}_{23}$ are defined similarly to $P_{12}$ and $P_{23}$ using two natural projections of $(A \times A \times \hat{A})/K_{12}$ to $(A \times \hat{A})/\Gamma_{-\phi}$. Now we claim that

$$\widetilde{P}_{13} \otimes \widetilde{P}_{23}^{-1} \simeq f^*(\mathcal{P} \otimes p_1^* L), \qquad (15.1.8)$$

where

$$f : (A \times A \times \hat{A})/K_{12} \to A \times \hat{A} : (x_1, x_2, \xi) \mapsto (x_1 - x_2, \xi + \phi_L(x_2)),$$

$p_1 : A \times \hat{A} \to A$ is the projection to the first factor. Indeed, let $\pi_{K_{12}} : A \times A \times \hat{A} \to (A \times A \times \hat{A})/K_{12}$ be the natural projection. It suffices to construct an isomorphism

$$\pi_{K_{12}}^* \widetilde{P}_{13} \otimes \widetilde{P}_{23}^{-1} \simeq \pi_{K_{12}}^* f^*(\mathcal{P} \otimes p_1^* L),$$

compatible with the descent data of both sides with respect to $\pi_{K_{12}}$. This isomorphism is obtained as follows:

$$\begin{aligned} \pi_{K_{12}}^* \big(\widetilde{P}_{13} \otimes \widetilde{P}_{23}^{-1}\big)_{(x_1,x_2,\xi)} &\simeq \mathcal{P}_{x_1,\xi} \otimes L_{x_1} \otimes \mathcal{P}_{x_2,\xi}^{-1} \otimes L_{x_2}^{-1} \\ &\simeq \mathcal{P}_{x_1-x_2,\xi+\phi_L(x_2)} \otimes L_{x_1-x_2} \\ &\simeq f^*(\mathcal{P} \otimes p_1^* L)_{(x_1,x_2,\xi)}, \end{aligned}$$

where $(x_1, x_2, \xi) \in A \times A \times \hat{A}$ (here we used the isomorphism $L_{x_1} \simeq L_{x_1-x_2} \otimes L_{x_2} \otimes \mathcal{P}_{x_1-x_2,\phi_L(x_2)}$). We leave for the reader to check that this isomorphism is compatible with the descent data. Combining (15.1.7) and (15.1.8) we get

$$p_{13}^* \mathcal{V} \otimes p_{23}^* \mathcal{V}' \simeq \rho_{12*} f^*(\mathcal{P} \otimes p_1^* L).$$

It follows that $Rp_{12*}(p_{13}^* \mathcal{V} \otimes p_{23}^* \mathcal{V}')$ is isomorphic to the (derived) push-forward of $f^*(\mathcal{P} \otimes p_1^* L)$ with respect to the natural projection $(A \times A \times \hat{A})/K_{12} \to A/K \times A/K$. This morphism is equal to the following composition:

$$(A \times A \times \hat{A})/K_{12} \to (A \times A)/\Delta(K) \to A/K \times A/K,$$

where $\Delta(K)$ is the image of $K \subset A$ under the diagonal embedding $\Delta : A \hookrightarrow A \times A$. Thus, we can compute the push-forward of $f^*(\mathcal{P} \otimes p_1^* L)$ in two steps: first, find the push-forward to $(A \times A)/\Delta(K)$, and then the push-forward of the result to $A/K \times A/K$. In the first step we use the following Cartesian square:

$$\begin{array}{ccc} (A \times A \times \hat{A})/K_{12} & \longrightarrow & (A \times A)/\Delta(K) \\ \Big\downarrow f & & \Big\downarrow d \\ A \times \hat{A} & \xrightarrow{p_1} & A \end{array} \qquad (15.1.9)$$

where $d : (A \times A)/\Delta(K) \to A$ is the difference map sending $(x_1, x_2)$ to $x_1 - x_2$. By the base change we obtain that the push-forward of $f^*(\mathcal{P} \otimes p_1^* L)$ to $(A \times A)/\Delta(K)$ is isomorphic to

$$d^* R p_{1*}(\mathcal{P} \otimes p_1^* L) \simeq d^* \mathcal{O}_e[-g],$$

where $e \in A$ is the neutral element. It remains to note that $d^{-1}(e)$ coincides with the image of the natural map

$$\widetilde{\Delta} : A/K \to (A \times A)/\Delta(K) : x \mapsto (x, x).$$

Since the composition of $\widetilde{\Delta}$ with the projection to $A/K \times A/K$ is the diagonal embedding of $A/K$, the push-forward of $d^* \mathcal{O}_e$ to $A/K \times A/K$ is isomorphic to $\mathcal{O}_{\Delta_{A/K}}$. □

## 15.2. Heisenberg Groupoid and Its Representations

Let $A$ be an abelian variety. Then the abelian variety $X = A \times \hat{A}$ has an additional structure similar to the symplectic structure on the vector space of the form $V \oplus V^*$. Namely, there is a natural biextension $\mathcal{E}$ on $X \times X = A \times \hat{A} \times A \times \hat{A}$ given by

$$\mathcal{E} = p_{14}^* \mathcal{P} \otimes p_{23}^* \mathcal{P}^{-1},$$

where $\mathcal{P}$ is the Poincaré bundle on $A \times \hat{A}$ (and on $\hat{A} \times A$). The fact that $\mathcal{E}$ is a biextension of $X \times X$ follows immediately from the fact that $\mathcal{P}$ is a biextension of $A \times \hat{A}$. The interesting features of $\mathcal{E}$ is that it is *skew-symmetric*; i.e., there is a natural isomorphism $\mathcal{E}_{x',x} \simeq \mathcal{E}_{x,x'}^{-1}$, and nondegenerate; i.e., the morphism $X \to \hat{X}$ given by $x \mapsto \mathcal{E}|_{\{x\} \times X}$ is an isomorphism.

We consider the pair $(X, \mathcal{E})$ as an analogue of the symplectic vector space and we want to construct an analogue of the Heisenberg group and its Schrödinger representation in this situation. The natural idea is to look at the action of functors of translation $t_a^*$ by points $a \in A$ and tensoring by line bundles $\mathcal{P}_\xi$, where $\xi \in \hat{A}$, on the derived category $D^b(A)$ of coherent sheaves on $A$ (at this point one could also work with the abelian category of coherent sheaves, however, the use of derived categories will become crucial in Section 15.3). All these functors commute with each other (since the line bundles $\mathcal{P}_\xi$ are translation-invariant), so it may seem at first that this is not an adequate analogue of the Heisenberg group. However, isomorphisms of functors

$$t_a^* \circ (\otimes \mathcal{P}_\xi) \widetilde{\to} (\otimes \mathcal{P}_\xi) \circ t_a^*$$

are not canonical. In order to choose such an isomorphism one has to trivialize the 1-dimensional space $\langle \xi, a \rangle := \mathcal{P}|_{a,\xi}$. Although one can do this for a given pair $(\xi, a)$, it is in general impossible to do this simultaneously for a family of pairs $(\xi, a)$ parametrized by a scheme. Thus, the correct commutation relation is

$$t_a^* \circ (\otimes \mathcal{P}_\xi) \simeq \langle \xi, a \rangle \otimes (\otimes \mathcal{P}_\xi) \circ t_a^*.$$

More generally, for every $x = (a, \xi) \in X$ we introduce the functors

$$T_x = (\otimes \mathcal{P}_\xi) \circ t_a^* : D^b(A) \to D^b(A).$$

Then from the above commutation relation we get

$$T_x \circ T_{x'} \simeq \mathcal{B}_{x,x'} \otimes T_{x+x'},$$

where for $x = (a, \xi)$, $x' = (a', \xi')$ we set $\mathcal{B}_{x,x'} = \langle \xi', a \rangle$. In other words, $\mathcal{B}_{x,x'}$ is the fiber at $(x, x')$ of the biextension $\mathcal{B} = p_{14}^* \mathcal{P}$.

Let us denote by $\mathbf{H}(k)$ the category of pairs $(L, x)$, where $L$ is a 1-dimensional vector space, $x \in X(k)$. The morphisms $(L, x) \to (L', x')$ exist only if $x = x'$ and are given by the $k$-linear isomorphisms $L \to L'$. All morphisms in $\mathbf{H}(k)$ are isomorphisms, so it is a *groupoid*. There is a natural monoidal structure on $\mathbf{H}(k)$ given by the functor $+ : \mathbf{H}(k) \times \mathbf{H}(k) \to \mathbf{H}(k)$ such that

$$(L, x) + (L', x') = (L \otimes L' \otimes \mathcal{B}_{x,x'}, x + x').$$

Similarly, for every $k$-scheme $S$ one can define the groupoid $\mathbf{H}(S)$ by replacing one-dimensional vector spaces with line bundles on $S$ and $k$-points of $X$ with $S$-points. We will call the corresponding functor $\mathbf{H}$ from $k$-schemes to groupoids *the Heisenberg groupoid* associated with $A$.

Naively, one can try to define a representation of $\mathbf{H}(k)$ on a $k$-linear category $\mathcal{C}$ as a monoidal functor from $\mathbf{H}(k)$ to the category of $k$-linear functors from $\mathcal{C}$ to itself, such that an object $(L, 0)$ corresponds to the functor of tensoring with $k$. However, this definition does not take into account correctly "the nontriviality in families" of the 1-dimensional spaces appearing in the above commutation relations. Namely, since a representation of $\mathbf{H}(k)$ is a collection of functors depending on $x \in X(k)$, we would like to add the condition that these functors depend on $x$ "algebraically." The first difficulty one encounters here is how to define a notion of an algebraic family of functors. Instead of trying to give a general definition we restrict ourselves to the case when $\mathcal{C} = D^b(T)$ for some $k$-variety $T$ and the functors are given by kernels on $T \times T$ as in Section 11.1. The composition of functors gets replaced by the convolution of kernels defined in Proposition 11.1. Then the family of such kernels parametrized by a scheme $S$ is simply an object of $D^b(T \times T \times S)$, such that restricting to $T \times T \times \{s\}$ we recover the kernel corresponding to a point $s \in S$. With this in mind, we can give a definition of a representation of the Heisenberg groupoid $\mathbf{H}$.

**Definition.** Let $T$ be a smooth projective variety over $k$. A *representation* of $\mathbf{H}$ on $D^b(T)$ is an object $K \in D^b(T \times T \times X)$ and an isomorphism

$$Rp_{1345,*}(Lp_{124}^*K \otimes^L Lp_{235}^*K) \simeq \pi_{34}^*\mathcal{B} \otimes L(\pi_{12}, \pi_3 + \pi_4)^*K$$

on $T \times T \times X \times X$, where $p_I$ (resp., $\pi_I$) for $I \subset [1, 5]$ (resp. $I \subset [1, 4]$) are projections of $T \times T \times T \times X \times X$ (resp. $T \times T \times X \times X$) to the corresponding product of factors. Note that the LHS of the above isomorphism is a relative convolution of sheaves on $T \times T \times X \times X$ considered as kernels on $T \times T$ parametrized by $X \times X$. Let us write this isomorphism symbolically as

$$i_{x,x'} : K_x * K_{x'} \widetilde{\rightarrow} \mathcal{B}_{x,x'} \otimes K_{x+x'}$$

(if we set $K_x = K|_{T \times T \times \{x\}}$ then we get precisely this isomorphism of kernels on $T \times T$ for every pair of points $x, x' \in X$). We impose the following two conditions on the above data:

(i) One has $K_0 = K|_{T \times T \times \{0\}} \simeq \mathcal{O}_{\Delta_T}$ where $\Delta_T \subset T \times T$ is the diagonal. Under this identification the isomorphisms $i_{0,x}$ and $i_{x,0}$ should become identities.

(ii) The following diagram should be commutative:

$$\begin{array}{ccc}
K_x * K_{x'} * K_{x''} & \xrightarrow{i_{x,x'}} & \mathcal{B}_{x,x'} \otimes K_{x+x'} * K_{x''} \\
\Big\downarrow{\scriptstyle i_{x',x''}} & & \Big\downarrow{\scriptstyle i_{x+x',x''}} \\
\mathcal{B}_{x',x''} \otimes K_x * K_{x'+x''} & \xrightarrow{i_{x,x'+x''}} & \mathcal{B}_{x,x'} \otimes \mathcal{B}_{x',x''} \otimes \mathcal{B}_{x,x''} \otimes K_{x+x'+x''}
\end{array} \tag{15.2.1}$$

As usual, one should think about this diagram as a symbolic notation of the corresponding diagram of sheaves and morphisms on $T \times T \times X \times X \times X$.

**Remarks.** 1. Instead of $\mathcal{B} = p_{14}^*\mathcal{P}$ we could have used any other biextension $\mathcal{B}$ of $X \times X$ such that $\mathcal{E} \simeq \mathcal{B} \otimes \sigma^*\mathcal{B}$, where $\sigma : X \times X \to X \times X$ is the permutation of factors. All the resulting Heisenberg groupoids are (noncanonically) equivalent.

2. The above definition gives a categorification of the notion of representation of the Heisenberg group on which the center $U(1)$ acts in the standard way.

3. Another definition of a representation of $\mathbf{H}$ on $D^b(T)$ can be given in terms of groupoids $\mathbf{H}(S)$ attached to every $k$-scheme $S$. Namely, such a representation is a collection of monoidal functors $\mathbf{H}(S) \to \mathcal{D}^b(T \times T \times S)$, compatible with natural functors $\mathbf{H}(S) \to \mathbf{H}(S')$ for every morphism $S' \to S$.

4. If the characteristic is zero, then by the theorem of Orlov [104] every exact autoequivalence $D^b(T) \to D^b(T)$ is represented by a kernel on $T \times T$, so we can rephrase the above definition in terms of exact autoequivalences instead of kernels.

By the construction we have a natural representation of $\mathbf{H}$ on $D^b(A)$. We urge the reader to try to write explicitly the sheaf on $A \times A \times X$ corresponding to this representation of $\mathbf{H}$ on $D^b(A)$ (before looking at the general answer given in Proposition 15.3).

More generally, we can construct a representation of $\mathbf{H}$ associated with an *isotropic* abelian subvariety $Y \subset X$. By definition, this means that the restriction $\mathcal{E}|_{Y \times Y}$ is trivial. Then the restriction $\mathcal{B}|_{Y \times Y}$ is a symmetric biextension. Hence, there exists a line bundle $\alpha$ on $Y$ such that

$$\mathcal{B}|_{Y \times Y} \simeq \Lambda(\alpha) = (p_1 + p_2)^* \alpha \otimes p_1^* \alpha^{-1} \otimes p_2^* \alpha^{-1}$$

(we trivialize all line bundles at 0 and require the isomorphisms to be trivial over 0). We fix such a line bundle $\alpha$ (this is analogous to fixing a lifting of an isotropic subgroup in $V$ to a subgroup in the Heisenberg group $\mathcal{H}(V)$). In addition, we assume that there exists a section $s : X/Y \to X$ of the projection $p : X \to X/Y$, so that $X \simeq Y \times X/Y$ (in the case of vector spaces this was automatic). Then we can define a representation of $\mathbf{H}$ on $D^b(X/Y)$. The idea is to mimic the construction of a representation $\mathcal{F}(L)$ of the Heisenberg group $\mathcal{H}(V)$. Namely, we identify $D^b(X/Y)$ with the full subcategory

$$D^b(X)_{Y,\alpha} \subset D^b(X)$$

consisting of sheaves that are "left-invariant" with respect to $(Y, \alpha)$. Informally, we can write the condition of left-invariance of $F \in D^b(X)$ with respect to $(Y, \alpha)$ as

$$\alpha_y \otimes \mathcal{B}_{y,x} \otimes F_{x+y} \simeq F_x$$

for every $y \in Y$. The reader will easily give a more formal definition involving an isomorphism on $X \times Y$ satisfying some cocycle condition on $X \times Y \times Y$.

Clearly, such $F$ is uniquely determined by its restriction to $s(X/Y) \subset X$, so we have an equivalence

$$D^b(X/Y) \simeq D^b(X)_{Y,\alpha}.$$

Now the action of $\mathbf{H}$ on such sheaves is given by the "right translations." In other words, the operator corresponding to $x_0 \in X$ sends $F$ as above to $F'$ such that

$$F'_x = \mathcal{B}_{x,x_0} \otimes F_{x+x_0}.$$

Rewriting this action in terms of sheaves on $X/Y$, we get the following result.

**Proposition 15.3.** *The following family of kernels $K_{Y,\alpha} \in D^b(X/Y \times X/Y \times X)$ gives a representation of $\mathbf{H}$ on $D^b(X/Y)$:*

$$K_{Y,\alpha} = p_3^* q^* \alpha^{-1} \otimes (qp_3, sp_1)^* \mathcal{B}^{-1} \otimes (sp_2, spp_3)^* \mathcal{B} \otimes f_* \mathcal{O}_{X/Y \times X},$$

*where $f : X/Y \times X \to X/Y \times X/Y \times X$ is the closed embedding given by $f(\overline{x}, x) = (\overline{x} + p(x), \overline{x}, x)$, $p : X \to X/Y$, is the natural projection, $q = \mathrm{id} - sp : X \to Y$ is the corresponding projection to $Y$.*

We can write the above kernel $K_{Y,\alpha}$ symbolically as follows:

$$(K_{Y,\alpha})_{\overline{x_1},\overline{x_2},x} = \alpha_{q(x)}^{-1} \otimes \mathcal{B}_{q(x),s(\overline{x_1})}^{-1} \otimes \mathcal{B}_{s(\overline{x_2}),s(p(x))} \otimes \delta_{f(X/Y \times X)}.$$

In the case $Y = \{0\} \times \hat{A} \subset X$ we can take trivial $\alpha$. The corresponding representation on $D^b(X/\hat{A}) = D^b(A)$ is the natural representation of $\mathbf{H}$ on $D^b(A)$.

## 15.3. Equivalences as Intertwining Functors

Guided by the analogy with the theory of Heisenberg groups, we want to construct equivalences between representations of $\mathbf{H}$ on $D^b(X/Y)$ for various $Y$. Of course, we have a chance of doing this only when $Y$ is *Lagrangian*, i.e., when the homomorphism $X \to \hat{X}$ induced by $\mathcal{E}$ restricts to an isomorphism $Y \to \widehat{X/Y}$. First, let us give a definition.

**Definition.** An *intertwining functor* between representations of $\mathbf{H}$ on $D^b(T)$ and $D^b(T')$ given by kernels $K \in D^b(T \times T \times X)$ and $K' \in D^b(T' \times T' \times X)$, respectively, is a functor of the form $\Phi_R : D^b(T) \to D^b(T')$, where $R \in D^b(T \times T')$ is an object satisfying

$$R * K'_x \simeq K_x * R,$$

for every $x \in X$. More precisely, this is a symbolic notation for an isomorphism

$$R\pi_{134,*}(L\pi_{12}^* R \otimes^L L\pi_{234}^* K) \widetilde{\rightarrow} Rp_{134,*}(Lp_{124}^* K \otimes^L Lp_{23}^* R)$$

of sheaves on $T \times T' \times X$, where $p_I$ (resp., $\pi_I$) are projections from $T \times T \times T' \times X$ (resp., $T \times T' \times T' \times X$) to products of factors. This isomorphism should be compatible with isomorphisms $i_{x,x'}$ from the definition of a representation of **H**. We leave to the reader to find the corresponding commutative diagram. It is easy to see that the composition of intertwining functors (given in terms of the convolution of kernels) is again an intertwining functor.

Now assume that we are given a pair of Lagrangian abelian subvarieties $Y_1, Y_2 \subset X$, equipped with additional data $\alpha_i \in \mathrm{Pic}(Y_i)$, $i = 1, 2$, such that $\Lambda(\alpha_i) \simeq \mathcal{B}|_{Y_i \times Y_i}$. Also we assume that projections $p_i : X \to X/Y_i$, $i = 1, 2$, admit splittings $s_i : X/Y_i \to X$. Then we claim that there exists an intertwining functor $D^b(X/Y_1) \to D^b(X/Y_2)$ between the corresponding **H**-representations, which is an equivalence of categories. We are going to construct it exlicitly in the case when the intersection $Y_1 \cap Y_2$ is finite. The existence in the general case will follow from Exercise 2.

The appearance of the finite group scheme $Y_1 \cap Y_2$ is a new feature for which there is no analogue in the theory of Chapter 4. This group scheme is equipped with an additional structure. Namely, we have a line bundle

$$G = \alpha_2|_{Y_1 \cap Y_2} \otimes \alpha_1^{-1}|_{Y_1 \cap Y_2}$$

on $Y_1 \cap Y_2$ equipped with a trivialization of $\Lambda(G)$. This means that the corresponding $\mathbb{G}_m$-torsor over $Y_1 \cap Y_2$ (which we still denote by $G$) has a structure of the central extension of $Y_1 \cap Y_2$ by $\mathbb{G}_m$. The following lemma implies that $G$ is a Heisenberg group scheme.

**Lemma 15.4.** *The commutator form of the central extension $G \to Y_1 \cap Y_2$ coincides with the canonical pairing associated with the biextension $\mathcal{E}|_{Y_1 \times Y_2}$, as defined in Section 10.4. In particular, it is nondegenerate.*

*Proof.* Recall that the structure of a central extension on $G$ is induced by the isomorphisms

$$\Lambda(\alpha_2)|_{Y_1 \cap Y_2} \simeq \mathcal{B}|_{(Y_1 \cap Y_2) \times (Y_1 \cap Y_2)} \simeq \Lambda(\alpha_1)|_{Y_1 \cap Y_2}.$$

Note that the isomorphism $\mathcal{B}|_{Y_i \times Y_i} \simeq \Lambda(\alpha_i)$ (where $i = 1, 2$) gives the symmetry isomorphism $\mathcal{B}_{y_i, y_i'} \simeq \mathcal{B}_{y_i', y_i}$, where $y_i, y_i' \in Y_i$. It is easy to see that the commutator form of $G$ measures the difference between two symmetry isomorphisms for $\mathcal{B}|_{(Y_1 \cap Y_2) \times (Y_1 \cap Y_2)}$ restricted from $Y_1 \times Y_1$ and from $Y_2 \times Y_2$.

In other words, it measures the difference between the restrictions to $Y_1 \cap Y_2$ of the trivializations of $\mathcal{E}$ on $Y_1 \times Y_1$ and $Y_2 \times Y_2$. It remains to note that the left and right kernels of the biextension $\mathcal{E}|_{Y_1 \times Y_2}$ coincide with $Y_1 \cap Y_2$ (embedded into $Y_1$ and $Y_2$, respectively). Indeed, the homomorphism $Y_1 \to \widehat{Y_2}$ associated with $\mathcal{E}|_{Y_1 \times Y_2}$ is the composition of the embedding $Y_1 \to X$ with the homomorphism $X \to \widehat{Y_2}$ induced by $\mathcal{E}|_{X \times Y_2}$. Since $Y_2$ is Lagrangian, the kernel of the latter homomorphism is exactly $Y_2$. Therefore, the kernel of the homomorphism $Y_1 \to \widehat{Y_2}$ is $Y_1 \cap Y_2$. Similarly, we check that the right kernel of $\mathcal{E}|_{Y_1 \times Y_2}$ coincides with $Y_1 \cap Y_2$. □

As in Section 15.2, we use the identification of the category $D^b(X/Y_i)$ (where $i = 1, 2$) with the category $D^b(X)_{Y_i,\alpha_i}$ of objects in $D^b(X)$, left-invariant with respect to $(Y_i, \alpha_i)$. As the first approximation, we can consider the intertwining functor

$$\widetilde{R} : D^b(X)_{Y_1,\alpha_1} \to D^b(X)_{Y_2,\alpha_2}$$

that performs the "averaging" with respect to the left action of $(Y_2, \alpha_2)$. Namely, $\widetilde{R}(F) = F'$, where

$$F'_x = \int_{y_2 \in Y_2} (\alpha_2)_{y_2} \otimes \mathcal{B}_{y_2,x} \otimes F_{y_2+x},$$

$x \in X$. It is easy to see that $F'$ is left-invariant with respect to $(Y_2, \alpha_2)$. One can check that $\widetilde{R}$ is given by some kernel on $X/Y_1 \times X/Y_2$ and that it is an intertwining functor from $D^b(X/Y_1)$ to $D^b(X/Y_2)$. The drawback of $\widetilde{R}$ is that it does not take into account the possible nontriviality of the intersection $Y_1 \cap Y_2$ and performs the integration over the entire $Y_2$. Recall that when defining intertwining operators for pairs of Lagrangian subgroups $L_1, L_2$ in a real Heisenberg group we integrated over $L_2/L_1 \cap L_2$. In our situation we cannot completely eliminate the integration over $Y_1 \cap Y_2$ because of the non-triviality of the central extension $G \to Y_1 \cap Y_2$ (which follows from Lemma 15.4). To see this let us denote

$$IF_{y_2,x} := (\alpha_2)_{y_2} \otimes \mathcal{B}_{y_2,x} \otimes F_{y_2+x} \in D^b(Y_2 \times X),$$

so that $F'_x = \int_{y_2} IF_{y_2,x}$. In order to eliminate the integration over some subgroup $I \subset Y_1 \cap Y_2$, we have to define a descent data on $IF$ for the morphism $Y_2 \times X \to Y_2/I \times X$. We have a canonical isomorphism

$$IF_{y_{12}+y_2,x} \simeq G_{y_{12}} \otimes IF_{y_2,x}, \tag{15.3.1}$$

where $y_{12} \in Y_1 \cap Y_2$ (as usual, this should be understood as an isomorphism on $(Y_1 \cap Y_2) \times Y_2 \times X$). Here we used the isomorphisms $(\alpha_2)_{y_{12}+y_2} \simeq (\alpha_2)_{y_{12}} \otimes$

$(\alpha_2)_{y_2} \otimes \mathcal{B}_{y_{12},y_2}$ and $F_{y_{12}+y_2+x} \simeq (\alpha_1)_{y_{12}}^{-1} \otimes \mathcal{B}_{y_{12},y_2+x}^{-1} \otimes F_{y_2+x}$. Thus, in order to define a descent data on $IF$ for the morphism $Y_2 \times X \to Y_2/I \times X$, we need to trivialize the central extension $G \to Y_1 \cap Y_2$ over $I$. According to Lemma 12.2, this is possible if and only if $I$ is isotropic with respect to the commutator pairing corresponding to $G$. We can write the corresponding modified functor $R_I : D^b(X)_{Y_1,\alpha_1} \to D^b(X)_{Y_2,\alpha_2}$ symbolically as follows:

$$R_I(F)_x = \int_{y_2 \in Y_2/I} (\alpha_2)_{y_2} \otimes \mathcal{B}_{y_2,x} \otimes F_{y_2+x},$$

where $x \in X$.

In order to eliminate "as much of excess integration as possible," let us pick a Lagrangian subgroup scheme $I \subset Y_1 \cap Y_2$ (recall that the existence of such a subgroup was established in Lemma 12.3) and choose a trivialization of $G|_I$ compatible with the central extension structure. We claim that the corresponding functor $R = R_I$ is an equivalence. More precisely, we are going to show that $R$ coincides with one of the equivalences constructed in Section 15.1.

**Remark.** Another way to pass from $\widetilde{R}$ to $R_I$ is to interpret the isomorphism (15.3.1) as an action of $G$ on the sheaf $IF$, compatible with the action of $Y_1 \cap Y_2$ on $Y_2 \times X$ by translations of the first argument. Then there is an induced action of $G$ on the derived push-forward of $IF$ to $X$, where $G$ acts on $X$ trivially. Now if $V$ is a Schrödinger representation for $G$ then we have

$$\int_{Y_2} IF \simeq V \otimes \mathrm{Hom}_G(V, IF),$$

where $\mathrm{Hom}_G(V, IF)$ is an object of $D^b(X)$ (see the remark after Corollary 12.5). If $V$ is the Schrödinger representation associated with a Lagrangian subgroup $I \subset Y_1 \cap Y_2$ then $\mathrm{Hom}_G(V, IF) \simeq R_I(F)$. In particular, we see that up to an isomorphism the functor $R_I$ does not depend on a choice of $I$ and that $\widetilde{R} \simeq V \otimes R$.

Our next observation is that the biextension $\mathcal{E}|_{Y_1 \times Y_2}$ on $Y_1 \times Y_2$ descends to a biextension $\overline{\mathcal{E}}$ on $Y_1/I \times Y_2/I$, which induces duality between the abelian varieties $Y_1/I$ and $Y_2/I$. Indeed, this follows immediately from Proposition 10.4 and from the identification of the commutator form of $G$ with the canonical pairing associated with $\mathcal{E}|_{Y_1 \times Y_2}$ (see Lemma 15.4). The fact that the descended biextension $\overline{\mathcal{E}}$ induces the duality between $Y_1/I$ and $Y_2/I$ follows from Exercise 8 of Chapter 10.

Now we claim that there is a natural equivalence of categories

$$\Psi_1 : D^b(X)_{Y_1,\alpha_1} \widetilde{\rightarrow} D^b(\mathrm{Coh}_{\overline{G},\phi}(Y_2/I)),$$

where $\overline{G}$ is the line bundle on $K = Y_1 \cap Y_2/I$ obtained from $G$ by the descent (the descent data is induced by the trivialization of $\Lambda(G)$), $\phi : K \to \widehat{Y_2/I}$ is the homomorphism sending $k \in K$ to $\overline{\mathcal{E}}^{-1}|_{k \times Y_2/I}$. Let us assume for simplicity that $I$ is reduced (in the nonreduced case the argument that follows has to be slightly modified). For a $(Y_1, \alpha_1)$-invariant sheaf $F$ on $X$ let us consider the sheaf $\widetilde{F} = F|_{Y_2} \otimes \alpha_2$ on $Y_2$. Then from $(Y_1, \alpha_1)$-invariance of $F$ we get a system of isomorphisms

$$i_z : \widetilde{F}_{y_2+z} \widetilde{\rightarrow} G_z \otimes \widetilde{F}_{y_2}, \tag{15.3.2}$$

where $y_2 \in Y_2$, $z \in Y_1 \cap Y_2$. Since we trivialized $G|_I$, from (15.3.2) we get a system of isomorphisms $\widetilde{F}_{y_2+z} \simeq \widetilde{F}_{y_2}$ for $z \in I$, which gives a descent data on $\widetilde{F}$ for the projection $Y_2 \to Y_2/I$. Let $\overline{F}$ be the descended object on $Y_2/I$. We would like to interpret the isomorphisms (15.3.2) as some additional data on $\overline{F}$. The difficulty is that, because of non-commutativity of the central extension $G \to Y_1 \cap Y_2$, the isomorphisms $i_z$ do not commute with each other, so we cannot descend them to $Y_2/I$. In order to kill the commutator form, we replace the data (15.3.2) by a system of isomorphisms

$$i'_z : \widetilde{F}_{y_2+z} \widetilde{\rightarrow} G_z \otimes \mathcal{E}^{-1}_{z,y_2} \otimes \widetilde{F}_{y_2}, \tag{15.3.3}$$

(where $y_2 \in Y_2$, $z \in Y_1 \cap Y_2$) induced by $i_z$ and by the trivialization of $\mathcal{E}|_{Y_2 \times Y_2}$. It is not difficult to check that the new isomorphisms $i'_z$ commute with $i_t$ for $t \in I$, so they induce isomorphisms

$$\overline{F}_{\overline{y_2}+k} \simeq \overline{G}_k \otimes \overline{\mathcal{E}}^{-1}_{k,\overline{y_2}} \otimes \overline{F}_{\overline{y_2}},$$

where $k \in K$, $\overline{y_2} \in Y_2/I$. In other words, $\overline{F}$ is equipped with a structure of an object of $\mathrm{Coh}_{\overline{G},\phi}(Y_2/I)$ that we take to be $\Psi_1(F)$. It is easy to see that $\Psi_1$ defines an equivalence of the category of $(Y_1, \alpha_1)$-invariant coherent sheaves on $X$ with $\mathrm{Coh}_{\overline{G},\phi}(Y_2/I)$, which extends to the derived categories in an obvious way. Similarly, we define an equivalence

$$\Psi_2 : D^b(X)_{Y_2,\alpha_2} \simeq D^b(\mathrm{Coh}_{\overline{G}^{-1},\phi'}(Y_1/I)),$$

where $K = Y_1 \cap Y_2/I$ is embedded into $Y_1/I$ in a natural way, and the homomorphism $\phi' : K \to \widehat{Y_1/I}$ sends $k$ to $\overline{\mathcal{E}}|_{Y_1/I \times k}$. By the definition, for a $(Y_2, \alpha_2)$-invariant sheaf $F$ on $X$ the sheaf $\Psi_2(F)$ on $Y_1/I$ is obtained by the descent of $F|_{Y_1} \otimes \alpha_1$ and is equipped with an additional structure in the same way as above.

Now we can rewrite the functor $R : D^b(X)_{Y_1,\alpha_1} \to D^b(X)_{Y_2,\alpha_2}$ using the equivalences $\Psi_1$ and $\Psi_2$. We have

$$\begin{aligned}\Psi_2(R(F))_{y_1} &\simeq (\alpha_1)_{y_1} \otimes \int_{y_2\in Y_2/I} (\alpha_2)_{y_2} \otimes \mathcal{B}_{y_2,y_1} \otimes F_{y_1+y_2} \\ &\simeq (\alpha_1)_{y_1} \otimes \int_{y_2\in Y_2/I} (\alpha_2)_{y_2} \otimes \mathcal{B}_{y_2,y_1} \otimes \left(\alpha_1^{-1}\right)_{y_1} \otimes \mathcal{B}^{-1}_{y_1,y_2} \otimes F_{y_2} \\ &\simeq \int_{y_2\in Y_2/I} \overline{\mathcal{E}}_{y_2,y_1} \otimes \Psi_1(F)_{y_2},\end{aligned}$$

where $y_1 \in Y_1/I$. In other words, this functor is compatible with the Fourier–Mukai transform $D^b(Y_2/I) \to D^b(Y_1/I)$, where the duality of $Y_1/I$ and $Y_2/I$ is given by the biextension $\overline{\mathcal{E}}$. This gives an informal proof of the fact that the functor $\Psi_2 \circ R \circ \Psi_1^{-1}$ coincides with the equivalence defined in Section 15.1. In order to give a real proof one has to rewrite the above argument in terms of kernels and descent data on them. The first step in doing this is to find line bundles $L_i$ on $Y_i/I$, where $i = 1, 2$, "untwisting" the categories $\mathrm{Coh}_{\overline{G},\phi}(Y_2/I)$ and $\mathrm{Coh}_{\overline{G}^{-1},\phi'}(Y_1/I)$. It is easy to see that the existence of a splitting of the projection $X \to X/Y_1$ is equivalent to the existence of a homomorphism $f_{21} : Y_2 \to Y_1$ such that $f_{21}|_{Y_1\cap Y_2} = \mathrm{id}_{Y_1\cap Y_2}$ (the splitting is induced by the map $Y_2 \to X : y_2 \mapsto y_2 - f_{21}(y_2)$). Now $L_2$ is obtained by descent from the line bundle $\alpha_2 \otimes (-f_{21})^*\alpha_1 \otimes (-f_{21}, \mathrm{id}_{Y_2})^*\mathcal{B}|_{Y_1\times Y_2}$ on $Y_1$. Indeed, the pull-back of the biextension $\Lambda(L_2)$ to $Y_2 \times Y_2$ can be written as

$$\begin{aligned}\Lambda(L_2)_{y_2,y_2'} &\simeq \mathcal{B}_{y_2-f_{21}(y_2),y_2'} \otimes \mathcal{B}_{-f_{21}(y_2'),y_2-f_{21}(y_2)} \\ &\simeq \mathcal{E}_{y_2',f_{21}(y_2)} \otimes \mathcal{B}_{y_2'-f_{21}(y_2'),y_2-f_{21}(y_2)}.\end{aligned}$$

Since the second factor descends to $Y_2/(Y_1\cap Y_2)\times Y_2/(Y_1\cap Y_2)$, it follows that the restriction of $\phi_{L_2}$ to $Y_1 \cap Y_2/I$ coincides with $\phi$. Similarly, $L_1$ is obtained by descent from the line bundle $\alpha_1 \otimes (-f_{12})^*\alpha_2 \otimes (-f_{12}, \mathrm{id}_{Y_1})^*\mathcal{B}|_{Y_2\times Y_1}$ where $f_{12} : Y_1 \to Y_2$ is a homomorphism such that $f_{12}|_{Y_1\cap Y_2} = \mathrm{id}_{Y_1\cap Y_2}$. We leave for the reader to write down the kernel on $X/Y_1 \times X/Y_2 = Y_2/Y_1\cap Y_2 \times Y_1/Y_1\cap Y_2$ corresponding to the functor $\Psi_2 \circ R \circ \Psi_1^{-1}$, and to check that it coincides with the kernel from Theorem 15.2.

One can also check that if $R' : D^b(X)_{Y_2,\alpha_2} \to D^b(X)_{Y_1,\alpha_1}$ is the equivalence constructed in the same way as above but with $Y_1$ and $Y_2$ switched, then $R' \circ R \simeq \mathrm{id}[-g]$.

The following theorem summarizes some of the theory developed above.

**Theorem 15.5.** *Let $A$ and $A'$ be abelian varieties, $(X, \mathcal{E})$ and $(X', \mathcal{E}')$ be the corresponding "symplectic" data defined in 15.2. Assume that there is an* *isomorphism of* $(X, \mathcal{E})$ *with* $(X', \mathcal{E}')$. *Then the categories* $D^b(A)$ *and* $D^b(A')$ *are equivalent.*

*Proof.* In this situation we can consider $\widehat{A'}$ as a Lagrangian subvariety in $X$. If $\hat{A}$ and $\widehat{A'}$ have finite intersection in $X$, the above construction gives the required equivalence. If not, we can find a Lagrangian subvariety $Y \subset X$ such that both intersections $Y \cap \hat{A}$ and $Y \cap \widehat{A'}$ are finite (see Exercise 2), apply the above construction to the pairs $(\hat{A}, Y)$ and $(Y, \widehat{A'})$ and then take the composition of the obtained equivalences. □

Examples of pairs of abelian varieties with equivalent derived categories can be found in Exercise 1.

Another application of the above techniques is to the construction of auto-equivalences of $D^b(A)$, which we only briefly sketch here. The main idea is to apply an automorphism $g$ of the symplectic data $(X, \mathcal{E})$ to the Lagrangian subvariety $\hat{A} \subset A$ and then consider the intertwining functor corresponding to the pair $(\hat{A}, g(\hat{A}))$. To be more precise, one has to make some extra choices, since for the construction of intertwining functors we need some additional structure on Lagrangian subvarieties (liftings to $\mathbf{H}$). The obtained functor will be an autoequivalence of $D^b(A)$. For example, in this way one can get the action of a central extension of $\mathrm{SL}_2(\mathbb{Z})$ on $D^b(A)$ considered in Section 11.5 (where $A$ is equipped with a nondegenerate line bundle $L$, such that $K(L) = 0$).

## 15.4. Analogue of the Maslov Index

Recall that in Section 4.3 we constructed the quadratic function $q_{L_1,L_2,L_3}$ associated with a triple of Lagrangian subgroups $(L_1, L_2, L_3)$ in a Heisenberg group. Now we are going to consider a categorification of this construction.

Let $Y_1, Y_2, Y_3$ be a triple of Lagrangian abelian subvarieties in $X$. As before we assume that every $Y_i$ is equipped with a line bundle $\alpha_i$ such that $\Lambda(\alpha_i) \simeq \mathcal{B}|_{Y_i \times Y_i}$. Then we can associate to $(Y_1, Y_2, Y_3)$ a line bundle on some group scheme. Namely, let $Z$ be the kernel of the homomorphism $Y_1 \times Y_2 \times Y_3 \to X : (y_1, y_2, y_3) \mapsto y_1 + y_2 + y_3$. Note that the connected component of zero $Z^0 \subset Z$ is an abelian variety, so $Z$ is an extension of the finite group scheme $Z/Z^0$ by the abelian variety $Z^0$. Now we define the line bundle $L(Y_1, Y_2, Y_3)$ on $Z$ which is an analogue of the quadratic form $q_{L_1,L_2,L_3}$ from Section 4.3:

$$L(Y_1, Y_2, Y_3) = \left([-1]_{Y_1}^* \alpha_1^{-1} \boxtimes \left(\alpha_2 \boxtimes \alpha_3 \otimes \mathcal{B}|_{Y_2 \times Y_3}\right)\right)\big|_Z.$$

**Proposition 15.6.** *(i) One has canonical isomorphisms*

$$L(Y_1, Y_2, Y_3) \simeq L(Y_2, Y_3, Y_1) \simeq [-1]^* L(Y_3, Y_2, Y_1)^{-1},$$
$$\Lambda(L(Y_1, Y_2, Y_3))_{(y_1,y_2,y_3),(y_1',y_2',y_3')} \simeq \mathcal{E}_{y_1',y_2} \simeq \mathcal{E}_{y_2',y_3},$$

*where* $(y_1, y_2, y_3), (y_1', y_2', y_3') \in Z$.

*(ii) If all intersections* $Y_i \cap Y_j$ *are finite, then* $L(Y_1, Y_2, Y_3)|_{Z^0}$ *is a non-degenerate line bundle.*

*(iii) Consider* $Y_1 \cap Y_2$ *as a subgroup in* $Z$ *via the embedding* $Y_1 \cap Y_2 \to Z : u \mapsto (-u, u, 0)$. *Then there is an action of the Heisenberg group* $G_{12} = \alpha_2|_{Y_1\cap Y_2} \otimes \alpha_1^{-1}|_{Y_1\cap Y_2}$ *on* $L(Y_1, Y_2, Y_3)$, *compatible with the action of* $Y_1 \cap Y_2$ *on* $Z$ *by translations and such that* $\mathbb{G}_m \subset G_{12}$ *acts in the standard way.*

*Proof.* (i) By the definition,

$$(L(Y_2, Y_3, Y_1) \otimes L(Y_1, Y_2, Y_3)^{-1})_{y_1,y_2,y_3} \simeq (\alpha_1)_{y_1} \otimes (\alpha_1)_{-y_1} \otimes \left(\alpha_2^{-1}\right)_{y_2} \otimes \left(\alpha_2^{-1}\right)_{-y_2} \otimes \mathcal{B}_{y_3,y_1} \otimes \mathcal{B}^{-1}_{y_2,y_3},$$

where $(y_1, y_2, y_3) \in Z$. Now we can use the isomorphisms $(\alpha_i)_{y_i} \otimes (\alpha_i)_{-y_i} \simeq \mathcal{B}^{-1}_{y_i,-y_i} \simeq \mathcal{B}_{y_i,y_i}$ and the equality $y_1 + y_2 + y_3 = 0$ to rewrite this as

$$\mathcal{B}_{y_1,y_1} \otimes \mathcal{B}_{y_3,y_1} \otimes \mathcal{B}^{-1}_{y_2,y_2} \otimes \mathcal{B}^{-1}_{y_2,y_3} \simeq \mathcal{B}_{-y_2,y_1} \otimes \mathcal{B}^{-1}_{y_2,-y_1} \simeq \mathcal{O}.$$

Similarly,

$$([-1]^* L(Y_3, Y_2, Y_1) \otimes L(Y_1, Y_2, Y_3))_{y_1,y_2,y_3}$$
$$\simeq (\alpha_2)_{y_2} \otimes (\alpha_2)_{-y_2} \otimes \mathcal{B}_{y_2,y_3} \otimes \mathcal{B}_{y_2,y_1} \simeq \mathcal{B}_{y_2,y_2} \otimes \mathcal{B}_{y_2,-y_2} \simeq \mathcal{O}.$$

Finally,

$$\begin{aligned}\Lambda(L(Y_1, Y_2, Y_3))_{(y_1,y_2,y_3),(y_1',y_2',y_3')} &\simeq \mathcal{B}^{-1}_{y_1,y_1'} \otimes \mathcal{B}_{y_2,y_2'} \otimes \mathcal{B}_{y_3,y_3'} \otimes \mathcal{B}_{y_2,y_3'} \otimes \mathcal{B}_{y_2',y_3} \\ &\simeq \mathcal{B}_{y_1,-y_1'} \otimes \mathcal{B}_{y_2,-y_1'} \otimes \mathcal{B}_{y_3,y_3'} \otimes \mathcal{B}_{y_2',y_3} \\ &\simeq \mathcal{B}_{y_3,y_1'} \otimes \mathcal{B}_{y_3,y_3'} \otimes \mathcal{B}_{y_2',y_3} \\ &\simeq \mathcal{B}^{-1}_{y_3,y_2'} \otimes \mathcal{B}_{y_2',y_3} = \mathcal{E}_{y_2',y_3}.\end{aligned}$$

(ii) As we have seen in (i), the symmetric homomorphism $Z^0 \to \widehat{Z^0}$ associated with $L(Y_1, Y_2, Y_3)|_{Z_0}$ is the composition

$$Z^0 \xrightarrow{p_1} Y_1 \xrightarrow{\phi} \widehat{Y_2} \xrightarrow{\hat{p_2}} \widehat{Z^0},$$

where $p_i : Z^0 \to Y_i$ are the natural projections, $\phi : Y_1 \to \widehat{Y_2}$ corresponds to the biextension $\mathcal{E}|_{Y_1\times Y_2}$. Now finiteness of $Y_2 \cap Y_3$ implies that $Y_2 + Y_3 = X$, hence, the morphism $p_1 : Z^0 \to Y_1$ is an isogeny. Similarly, finiteness of $Y_1 \cap Y_3$ implies that $p_2$ is an isogeny. Finally, finiteness of $Y_1 \cap Y_2$ implies that $\phi$ is an isogeny.

(iii) This follows from the fact that the restriction of the biextension $\Lambda(L(Y_1, Y_2, Y_3))$ to $(Y_1 \cap Y_2) \times Z$ is trivial and from the isomorphism

$$L(Y_1, Y_2, Y_3)_{(-u,u,0)} \simeq (\alpha_2)_u \otimes \left(\alpha_1^{-1}\right)_u. \qquad \square$$

**Remark.** Since $L(Y_1, Y_2, Y_3)$ is the restriction of a line bundle on an abelian variety $Y_1 \times Y_2 \times Y_3$ to the subgroup scheme $Z$, it has an additional structure coming from the theorem of the cube (it is called the *cube structure*). Essentially this structure is given by the biextension structure on the line bundle $\Lambda(L(Y_1, Y_2, Y_3))$ over $Z \times Z$. One can check that the isomorphisms of the above lemma are compatible with cube structures.

Now let us assume that all the projections $X \to X/Y_i$ admit a splitting and that all pairwise intersections $Y_i \cap Y_j$ for $i \neq j$ are finite. Let

$$R_{ij} : D^b(X)_{Y_i,\alpha_i} \to D^b(X)_{Y_j,\alpha_j}$$

be the equivalences constructed in Section 15.3. We want to compare the composition $R_{23} \circ R_{12}$ with $R_{13}$. Recall that in the picture with a triple of Lagrangian subgroups $(L_1, L_2, L_3)$ of a Heisenberg group, similar intertwining operators differ by a constant determined by the quadratic form $q_{L_1,L_2,L_3}$ (see Theorems 4.5 and 4.8). The following theorem provides an analogous result in the categorical setup.

**Theorem 15.7.** *Under the above assumptions one has $R_{23} \circ R_{12} \simeq R_{13}[m]$, where $m$ is the index of the nondegenerate line bundle $L(Y_1, Y_2, Y_3)|_{Z^0}$.*

*Proof.* It suffices to prove that $R_{23} \circ R_{12}$ is a direct summand in $V \otimes R_{13}[m]$ for some $k$-vector space $V$, or rather that the same statement holds for the kernels giving these functors. Indeed, since $R_{13}$ is an equivalence, the corresponding kernel is an indecomposable vector bundle. Hence, any indecomposable direct summand in $V \otimes R_{13}[m]$ should be isomorphic to $R_{13}[m]$.

Let us compute the composition

$$\widetilde{R}_{23} \circ \widetilde{R}_{12} : D^b(X)_{Y_1,\alpha_1} \to D^b(X)_{Y_3,\alpha_3}.$$

By the definition, we have

$$(\widetilde{R}_{23}\widetilde{R}_{12}F)_x = \int_{(y_2,y_3)\in Y_2\times Y_3} (\alpha_3)_{y_3} \otimes \mathcal{B}_{y_3,x} \otimes (\alpha_2)_{y_2} \otimes \mathcal{B}_{y_2,y_3+x} \otimes F_{y_2+y_3+x}$$
$$\simeq \int_{(y_2,y_3)\in Y_2\times Y_3} (\alpha_2)_{y_2} \otimes (\alpha_3)_{y_3} \otimes \mathcal{B}_{y_2,y_3} \otimes \mathcal{B}_{y_2+y_3,x} \otimes F_{y_2+y_3+x}.$$

This should be understood as the derived push-forward of an object in $D^b(Y_2 \times Y_3 \times X)$ under the projection $Y_2 \times Y_3 \times X \to X$. We can rewrite this expression as

$$(\widetilde{R}_{23}\widetilde{R}_{12}F)_x \simeq \int_{y_2\in Y_2} \mathcal{Q}_{y_2,x},$$

where

$$\mathcal{Q}_{y_2,x} := (\alpha_2)_{y_2} \otimes \int_{y_3\in Y_3} (\alpha_3)_{y_3} \otimes \mathcal{B}_{y_2,y_3} \otimes \mathcal{B}_{y_2+y_3,x} \otimes F_{y_2+y_3+x}.$$

First, we can calculate the pull-back of $\mathcal{Q}$ under the morphism $(\pi_2\times \mathrm{id}_X) : Z\times X \to Y_2\times X$ induced by the natural projection $\pi_2 : Z \to Y_2$. This corresponds to choosing elements $y_1' \in Y_1$ and $y_3' \in Y_3$ such that $y_2 = -y_1' - y_3'$. We have

$$\mathcal{Q}_{y_2,x} \simeq (\alpha_2)_{y_2} \otimes \int_{y_3\in Y_3} (\alpha_3)_{y_3} \otimes \mathcal{B}_{-y_3',y_3} \otimes \mathcal{B}_{-y_1',y_3+x}$$
$$\otimes \mathcal{B}_{-y_3'+y_3,x} \otimes F_{-y_1'-y_3'+y_3+x}.$$

From $(Y_1, \alpha_1)$-invariance of $F$ we get

$$F_{-y_1'-y_3'+y_3+x} \simeq \left(\alpha_1^{-1}\right)_{-y_1'} \otimes \mathcal{B}^{-1}_{-y_1',-y_3'+y_3+x} \otimes F_{-y_3'+y_3+x}.$$

On the other hand, we have an isomorphism $(\alpha_3)_{y_3} \otimes \mathcal{B}_{-y_3',y_3} \simeq (\alpha_3)_{-y_3'+y_3} \otimes (\alpha_3^{-1})_{-y_3'}$. Hence,

$$\mathcal{Q}_{y_2,x} \simeq (\alpha_2)_{y_2} \otimes \left(\alpha_1^{-1}\right)_{-y_1'} \otimes \left(\alpha_3^{-1}\right)_{-y_3'} \otimes \mathcal{B}^{-1}_{y_1',y_3'}$$
$$\otimes \int_{y_3\in Y_3} (\alpha_3)_{-y_3'+y_3} \otimes \mathcal{B}_{-y_3'+y_3,x} \otimes F_{-y_3'+y_3+x}$$
$$\simeq L(Y_2, Y_1, Y_3)^{-1}_{-y_2,-y_1',-y_3'} \otimes \int_{y_3\in Y_3} (\alpha_3)_{y_3} \otimes \mathcal{B}_{y_3,x} \otimes F_{y_3+x}.$$

Taking into account the isomorphism $[-1]^*L(Y_2, Y_1, Y_3)^{-1} \simeq L(Y_1, Y_2, Y_3)$ (see Proposition 15.6), we get a canonical isomorphism

$$(\pi_2 \times \mathrm{id}_X)^*\mathcal{Q} \simeq L(Y_1, Y_2, Y_3) \otimes p_2^*\widetilde{R}_{13}F, \qquad (15.4.1)$$

where $p_2 : Z \times X \to X$ is the natural projection. Recall that $\widetilde{R}_{13}F \simeq V_{13} \otimes R_{13}F$, where $V_{13}$ is the Schrödinger representation of the Heisenberg group $G_{13} = \alpha_3|_{Y_1 \cap Y_3} \otimes \alpha_1^{-1}|_{Y_1 \cap Y_3}$. We have an isomorphism $V_{13} \simeq V_{31}^*$, where $V_{31}$ is the Schrödinger representation of $G_{31} = G_{13}^{-1} \simeq G_{13}^{op}$. Thus, we can rewrite (15.4.1) as

$$(\pi_2 \times \mathrm{id}_X)^* Q \simeq L(Y_1, Y_2, Y_3) \otimes V_{31}^* \otimes p_2^* R_{13}F. \tag{15.4.2}$$

According to Proposition 15.6, there is a natural action of $G_{31}$ on $L(Y_1, Y_2, Y_3) \simeq L(Y_3, Y_1, Y_2)$ compatible with the action of $Y_1 \cap Y_3$ on $Z$ by translations. Hence, there is an induced action of $Y_1 \cap Y_3$ on $L(Y_1, Y_2, Y_3) \otimes V_{31}^*$. It is easy to check that this action corresponds to the descent data for the morphism $\pi_2 \times \mathrm{id}_X$ coming from the isomorphism (15.4.2) (note that the kernel of the homomorphism $\pi_2$ is precisely the subgroup $Y_1 \cap Y_3 \subset Z$). Now arguing similarly to Proposition 12.7, we obtain an isomorphism

$$Q \simeq \mathrm{Hom}_{G_{31}}(V_{31}, \pi_{2*} L(Y_1, Y_2, Y_3)) \boxtimes R_{13}F$$

on $Y_2 \times X$ (to be completely rigorous, instead of choosing particular $F$ one has to work with the kernels giving the corresponding functors). Tensoring this isomorphism with $V_{31}$ and taking the push-forward under the projection to $X$, we conclude that

$$V_{31} \otimes \widetilde{R}_{23} \widetilde{R}_{12} F \simeq R\Gamma(Z, L(Y_1, Y_2, Y_3)) \otimes R_{13}F.$$

Since $\widetilde{R}_{ij} \simeq V_{ij} \otimes R_{ij}$ for some vector spaces $V_{ij}$, it follows that

$$V \otimes R_{23} R_{12} F \simeq R\Gamma(Z, L(Y_1, Y_2, Y_3)) \otimes R_{13}F,$$

where $V$ is a vector space. Since $R_{ij}$ are equivalences, this implies that $R\Gamma(Z, L(Y_1, Y_2, Y_3))$ is concentrated in one degree. Since $Z/Z^0$ is affine, this degree coincides with the index of $L(Y_1, Y_2, Y_3)|_{Z^0}$. □

This theorem allows to derive certain equation for indexes of nondegenerate line bundles. As we have seen in Section 11.6, the index $i(L)$ depends only on $\phi_L$. Let us denote by $\mathrm{Hom}^{\mathrm{sym}}(A, \hat{A})$ the group of symmetric homomorphisms and by $U \subset \mathrm{Hom}^{\mathrm{sym}}(A, \hat{A})$ the subset of symmetric isogenies. Recall that by Theorem 13.7 every element of $U$ has form $\phi_L$ for some nondegenerate line bundle $L$. Let us also set $U_{\mathbb{Q}} = \mathbb{Q}^* U \subset \mathrm{Hom}^{\mathrm{sym}}(A, \hat{A}) \otimes \mathbb{Q}$. By Proposition 11.18, there is a well-defined function $i : U_{\mathbb{Q}} \to \mathbb{Z}$, such that $i(\phi_L) = i(L)$ and $i(r\phi) = i(\phi)$ for $r \in \mathbb{Q}_{>0}$. Note that for every $\phi \in U_{\mathbb{Q}}$ there is a well-defined element $\phi^{-1} \in \mathrm{Hom}^{\mathrm{sym}}(\hat{A}, A)_{\mathbb{Q}}$, such that $(r\phi)^{-1} = r^{-1}\phi^{-1}$ for $r \in \mathbb{Q}^*$. Indeed, it suffices to define $\phi^{-1}$ for $\phi \in U$. Let $\psi : \hat{A} \to A$ be an isogeny such that $\psi \circ \phi = [n]_A$ for some $n > 0$ (such $\psi$ exists by Exercise 4 of Chapter 9). Note that $\psi$ is necessarily symmetric. Then we set $\phi^{-1} = \psi/n$.

**Proposition 15.8.** *Let $\phi_1, \phi_2 \in U_{\mathbb{Q}}$ be such that $\phi_2 - \phi_1 \in U_{\mathbb{Q}}$. Then*

$$i\big(\phi_1 - \phi_1\phi_2^{-1}\phi_1\big) + i(\phi_2) = i(\phi_1) + i(\phi_2 - \phi_1) \tag{15.4.3}$$

*Proof.* Assume that we have four Lagrangian abelian subvarieties ($Y_i, i = 1, \dots, 4$) in $A \times \hat{A}$ as above, such that all pairwise intersections are finite. Then from Theorem 15.7 we get

$$i(L(Y_1, Y_2, Y_3)) + i(L(Y_1, Y_3, Y_4)) = i(L(Y_1, Y_2, Y_4)) + i(L(Y_2, Y_3, Y_4))$$

(this equation is analogous to (4.2.3)). Note that multiplying $\phi_1$ and $\phi_2$ by a sufficiently large positive integer, we can assume that $\phi_1, \phi_2$ and $\phi_1 - \phi_1\phi_2^{-1}\phi_1$ are actual homomorphisms from $A$ to $\hat{A}$. Now let us apply the above equality to the Lagrangians $Y_1 = A \times 0$, $Y_2 = \Gamma_{-\phi_1}$, $Y_3 = \Gamma_{-\phi_2}$ and $Y_4 = 0 \times \hat{A}$, where $\Gamma_{-\phi_i}$ is the graph of the isogeny $-\phi_i$. Each of these Lagrangians $Y_i$ is equipped with the natural line bundle $\alpha_i$ such that $\Lambda(\alpha_i) \simeq \mathcal{B}|_{Y_i \times Y_i}$. It remains to identify the four terms in the above equation with the four terms of (15.4.3) which amounts to calculating $\phi_{L(Y_i, Y_j, Y_k)}$. More precisely, we find that

$$\begin{aligned}
i(L(Y_1, Y_2, Y_3)) &= i\big(\phi_1 - \phi_1\phi_2^{-1}\phi_1\big),\\
i(L(Y_1, Y_3, Y_4)) &= i(\phi_2),\\
i(L(Y_1, Y_2, Y_4)) &= i(\phi_1),\\
i(L(Y_2, Y_3, Y_4)) &= i(\phi_2 - \phi_1).
\end{aligned} \tag{15.4.4}$$

We will only check the first equality (the proof of the other three is similar but simpler). Using the projection to $A$ we identify each of the varieties $Y_1$, $Y_2$ and $Y_3$ with $A$. Now we have an isogeny

$$A \to Z^0 \subset Y_1 \times Y_2 \times Y_3 \simeq A^3 : x \mapsto \big((n - n\phi_2^{-1}\phi_1)(x),\ -nx, n\phi_2^{-1}\phi_1(x)\big),$$

where $n$ is a large integer, so that $n\phi_2^{-1}\phi_1 \in \operatorname{Hom}(A, A)$. It suffices to compute the index of the pull-back $M$ of $L(Y_1, Y_2, Y_3)$ by this isogeny. According to Proposition 15.6, we have

$$\begin{aligned}
\Lambda(M)_{x,x'} &\simeq \mathcal{E}_{(nx' - n\phi_2^{-1}\phi_1(x'),0),(-nx, n\phi_1(x))} \simeq \mathcal{P}_{(n - n\phi_2^{-1}\phi_1)(x'), n\phi_1(x)}\\
&\simeq \mathcal{P}_{x, n^2(\phi_1 - \phi_1\phi_2^{-1}\phi_1)(x')}.
\end{aligned}$$

Hence, $\phi_M = n^2(\phi_1 - \phi_1\phi_2^{-1}\phi_1)$ and $i(M) = i(\phi_1 - \phi_1\phi_2^{-1}\phi_1)$. □

### Exercises

1. Let $f: A \to \hat{A}$ be a symmetric isogeny. Show that for every $n > 0$ the image of the homomorphism $A \to A \times \hat{A} : x \mapsto (nx, f(x))$ is Lagrangian with respect to the biextension $\mathcal{E}$. Assume in addition that $\ker(f) \subset A_{mn}$, for some $m > 0$ relatively prime to $n$. Set $f_n = f|_{A_n}$. Construct an equivalence of derived categories $D^b(A) \simeq D^b(A/\ker(f_n))$.
2. Let $Y, Y' \subset X$ be Lagrangian abelian subvarieties with respect to $\mathcal{E}$. Show that there exists a Lagrangian abelian subvariety $Z \subset X$, such that the intersections $Y \cap Z$ and $Y' \cap Z$ are finite. [This statement is the content of Lemma 4.1 in [107].]
3. Let $\phi_L : A \to \hat{A}$ be the symmetric involution associated with an ample line bundle $L$. Assume that there exists a number field $K$ contained in $\mathrm{Hom}(A, A)_{\mathbb{Q}}$ as a $\mathbb{Q}$-subalgebra and such that for every $x \in K$ one has $\hat{x} \circ \phi_L = \phi_L \circ x$. Note that for $x \in K^*$ we have $\phi_L \circ x \in U_{\mathbb{Q}}$. For $x \in K^*$ we denote $i(x) = i(\phi_L \circ x)$.
   (a) Show that $i(\lambda x) = i(x)$, where $x \in K^*$, $\lambda \in \mathbb{Q}_{>0}(K^*)^2$.
   (b) Prove that
   $$i((\lambda + \mu x)xy) = i(xy) + i(y) - i((\lambda + \mu x)y),$$
   where $x, y \in K^*$, $\lambda, \mu \in \mathbb{Q}_{>0}(K^*)^2$, $\lambda + \mu x \neq 0$.
   (c) Show that
   $$i\left(\left(\sum_{i=1}^{n} \lambda_i\right)x\right) = i(x),$$
   where $x \in K^*$, $\lambda_i \in \mathbb{Q}_{>0}(K^*)^2$. [*Hint*: Use induction by $n$. Apply (b) to $x = \sum_{i=1}^{n-1} \lambda_i$, $\mu = 1$.]
   (d) Show that $K$ is totally real.
   (e) Let $\sigma_i : K \to \mathbb{R}, i = 1, \ldots, d$ be all real embeddings of $K$. Prove that there exist some integers $m_i$, $i = 1, \ldots, d$, with $\sum_{i=1}^{d} m_i = \dim A$, such that
   $$i(x) = \sum_{i=1}^{d} m_i \frac{1 - \mathrm{sign}(\sigma_i(x))}{2}.$$
   [*Hint*: Use the fact that every totally positive element in $K^*$ is a sum of squares, cf. [80], Section XI.2.]

# Part III

## Jacobians

# 16

# Construction of the Jacobian

Let $C$ be a smooth projective (irreducible) curve of genus $g$ over an algebraically closed field $k$. By $\mathrm{Pic}(C)$ (resp. $\mathrm{Pic}^d(C)$) we denote the Picard group of $C$ (resp. the degree $d$ subset in it). In this chapter we introduce the structure of abelian variety on $\mathrm{Pic}^0(C)$. More precisely, we construct an abelian variety $J = J(C)$ called the *Jacobian* of $C$, such that the group of $k$-points of $J$ is isomorphic to $\mathrm{Pic}^0(C)$. The idea is to use the fact that every line bundle of degree $g$ on $C$ has a nonzero global section and that for generic line bundle $L$ of degree $g$ this section is unique (up to rescaling). Therefore, a big subset in $\mathrm{Pic}^g(C)$ can be described in terms of effective divisors on $C$. The set of effective divisors of degree $d$ on $C$ can be identified with the set of $k$-points of the symmetric power $\mathrm{Sym}^d C$ (the definition and the main properties of the varieties $\mathrm{Sym}^d C$ are given in Section 16.1). The subset in $\mathrm{Pic}^g(C)$ consisting of line bundles $L$ with $h^0(L) = 1$ corresponds to the set of $k$-points of an open subset in $\mathrm{Sym}^g C$. Translating this subset by various line bundles of degree $-g$ we obtain algebraic charts for $\mathrm{Pic}^0(C)$. We define the Jacobian variety $J$ by gluing these open charts. This construction also provides the Poincaré line bundle $\mathcal{P}_C$ on $C \times J(C)$, such that the restriction of $\mathcal{P}_C$ to $C \times \{\xi\}$ is the line bundle on $C$ corresponding to $\xi \in J(k)$. The fact that the pair $(J, \mathcal{P}_C)$ represents the functor of families of line bundles of degree 0 on $C$, is deduced from the fact that $\mathrm{Sym}^d C$ represents the functor of families of effective divisors of degree $d$ on $C$ (see Theorem 16.4 and Exercise 2).

## 16.1. Symmetric Powers of a Curve

For every $d > 0$ we denote by $\mathrm{Sym}^d C$ the $d$th symmetric power of a curve $C$. By the definition, $\mathrm{Sym}^d C$ is the quotient of $C^d$ by the action of the symmetric group $S_d$. In particular, we have a natural surjective morphism $\pi_d : C^d \to \mathrm{Sym}^d C$, and $\mathrm{Sym}^d C$ is a proper variety. Furthermore, we can identify $k$-points of $\mathrm{Sym}^d C$ with effective divisors of degree $d$ on $C$. On the other hand, if $A$ is the ring of functions on an affine open subset $U \subset C$, then we have $\mathcal{O}(\mathrm{Sym}^d U) = TS^d(A)$, where for every $k$-vector space $V$ we denote by $TS^d(V)$ the subspace of symmetric tensors in $V^{\otimes d}$ (since we work in arbitrary characteristic, we cannot identify $TS^d(V)$ with the usual symmetric power of $V$, which is a quotient of $V^{\otimes d}$)

For every $d_1, d_2 \geq 1$ we have a canonical morphism

$$s_{d_1,d_2} : \mathrm{Sym}^{d_1} C \times \mathrm{Sym}^{d_2} C \to \mathrm{Sym}^{d_1+d_2} C$$

induced by the isomorphism $C^{d_1} \times C^{d_2} \widetilde{\to} C^{d_1+d_2}$ and by the standard embedding of groups $S_{d_1} \times S_{d_2} \hookrightarrow S_{d_1+d_2}$. Locally it corresponds to the embedding of rings

$$TS^{d_1+d_2}(A) \hookrightarrow TS^{d_1}(A) \otimes TS^{d_2}(A).$$

We will use the following symbolic notation for this morphism: $s_{d_1,d_2}(D_1, D_2) = D_1 + D_2$.

Let us introduce some notation. Let $V$ be a $k$-vector space. For every pair of elements $x, y \in V$ and every subset $I \subset [1, d]$ we denote by $x_I y_{\overline{I}} \in V^{\otimes d}$ the decomposable tensor having $x$ at the factors corresponding to $i \in I$ and $y$ at all other factors (e.g., if $d = 3$ then $x_1 y_{\overline{1}} = x \otimes y \otimes y$). Using this notation, we can define the following sequence of elements in $TS^d(V)$ (analogues of elementary symmetric functions):

$$\sigma_n(x, y) = \sum_{I \subset [1,d]:|I|=n} x_I y_{\overline{I}},$$

where $n = 0, \dots, d$. Note that $\sigma_n(x, y) = \sigma_{d-n}(y, x)$.

**Definition.** We say that a map $f : V \to W$ between $k$-vector spaces is homogeneous of degree $d$ if $f(\lambda x) = \lambda^d f(x)$ and $\sum_{I \subset [0,d]} (-1)^{|I|} f(\sum_{i \in I} x_i) = 0$ for all $x, x_0, \dots, x_d \in V, \lambda \in k$.

**Lemma 16.1.** *(a) For every $k$-vector space $V$ the space $TS^d(V)$ is spanned by tensors of the form $x^{\otimes d}$, where $x \in V$.*

*(b) More generally, let $W \subset V$ be a subspace, and for every $n = 0, \dots, d$ let $TS^d_n(V, W) \subset TS^d(V)$ be the subspace consisting of symmetric tensors which can be written as a linear combination of tensors $x_1 \otimes \dots \otimes x_d$ with at least $n$ of $x_i$'s belonging to $W$. Then $TS^d_n(V, W)$ is spanned by the elements $\sigma_n(x, y)$ with $x \in W$, $y \in V$.*

*(c) For every $k$-vector space $W$ there is an isomorphism of* $\mathrm{Hom}_k(TS^d(V), W)$ *with the space of homogeneous maps $V \to W$ of degree $d$, such*

*that a linear map $\phi : TS^d(V) \to W$ and the corresponding degree $d$ map $f : V \to W$ are related by $f(x) = \phi(x^{\otimes d})$.*

*Proof.* (a) It suffices to prove this for finite-dimensional $V$. Then we can use induction in $\dim V$. Let $V = V_1 \oplus ke$. By induction assumption $TS^d(V)$ is spanned by the elements of the form $\sigma_n(x, e)$ with $x \in V_1$, $0 \leq n \leq d$. Now using the equations

$$(x + \lambda e)^{\otimes d} = \sum_{n=0}^{d} \lambda^{d-n} \sigma_n(x, e),$$

for $d$ distinct $\lambda \in k$, we can express $\sigma_n(x, e)$ with $n < d$ as linear combination of tensors of the form $y^{\otimes d}$.

(b) Let $T_n^d(V, W) \subset V^{\otimes d}$ be the subspace spanned by $x_1 \otimes \ldots \otimes x_d$ with at least $n$ of $x_i$'s belonging to $W$, so that $TS_n^d(V, W) = T_n^d(V, W) \cap TS^d(V)$. Then we have exact sequences of $S_d$-representations

$$0 \to T_{n-1}^d(V, W) \to T_n^d(V, W) \to \operatorname{Ind}_{S_n \times S_{n-d}}^{S_d} W^{\otimes n} \otimes (V/W)^{\otimes(d-n)},$$

where for a group $G$, a subgroup $H \subset G$ and a representation $U$ of $H$ we denote by $\operatorname{Ind}_H^G U$ the induced representation of $G$. Passing to $S_d$-invariants we get exact sequences

$$0 \to TS_{n-1}^d(V, W) \to TS_n^d(V, W) \xrightarrow{p} TS^n(W) \otimes TS^{d-n}(V/W).$$

Moreover, the map $p$ sends $\sigma_n(x, y)$ to $x^{\otimes n} \otimes \overline{y}^{\otimes(d-n)}$, where $x \in W$, $y \in V$, $\overline{y} = y \bmod W$. Now our assertion follows easily from part (a) by induction in $n$.

(c) Let $f : V \to W$ be a homogeneous map of degree $d$. We want to construct a linear map $\phi : TS^d(V) \to W$ such that $f(x) = \phi(x^{\otimes d})$. By part (a) such $\phi$ is unique (if exists), so it suffices to prove this assertion in the case when $V$ is finite-dimensional. Now we can use induction in $\dim V$. Let $V = V_1 \oplus ke$. Then $f$ should have form

$$f(x + \lambda e) = \sum_{n=0}^{d} \lambda^{d-n} f_n(x),$$

where $x \in V_1$, $f_n : V_1 \to W$ is a homogeneous map of degree $n$. On the other hand, we have an isomorphism

$$TS^d(V) \simeq \oplus_{n=0}^d TS^n(V_1),$$

such that the element $\sigma_n(x, e) \in TS^d(V)$, where $x \in V_1$, corresponds to $x^{\otimes n} \in TS^n(V_1)$. By induction assumption, we can define linear maps $\phi_n$ :

$TS^n(V_1) \to W$, where $n = 0, \dots, d$, such that $f_n(x) = \phi_n(x^{\otimes n})$. Now we define the linear map $\phi$ by the condition that $\phi|_{TS^n(V_1)} = \phi_n$. Then for every $x \in V_1$ and $\lambda \in k$ we have

$$\phi((x + \lambda e)^{\otimes d}) = \sum_{n=0}^{d} \lambda^{d-n}\phi_n(x^{\otimes n}) = \sum_{n=0}^{d} \lambda^{d-n} f_n(x) = f(x + \lambda e).$$

**Proposition 16.2.** *The variety* $\operatorname{Sym}^d C$ *is smooth.* □

*Proof.* Clearly, it suffices to prove the same statement with $C$ replaced by a smooth affine curve $\operatorname{Spec}(A)$. Moreover, using the maps $s_{d_1,d_2}$ one can reduce the problem to showing smoothness of $\operatorname{Sym}^d C$ near the most degenerate point $d \cdot p$, where $p$ is a point in $C$ (see Exercise 1). In other words, it suffices to prove a similar statement when $A$ is a local ring of $C$ at some point. Let $\mathfrak{m} \subset A$ be the maximal ideal of $A$, and let $\mathfrak{m}_d$ be the corresponding maximal ideal of $A^{\otimes d}$: $\mathfrak{m}_d$ is a linear span of tensors $x_1 \otimes \cdots \otimes x_d$ with at least one of $x_i$'s in $\mathfrak{m}$. Then $M = \mathfrak{m}_d \cap TS^d(A)$ is a maximal ideal in $TS^d(A)$ and we have to prove that it is generated by $d$ elements. It is easy to see (using Lemma 16.1 or directly) that $M$ is the linear span of elements of the form $\sigma_n(x, 1)$, where $x \in \mathfrak{m}$, $1 \le n \le d$. Let $t$ be a generator of $\mathfrak{m}$. Then we are going to show that $M$ is generated by the elements $(\sigma_n(t, 1), n = 1, \dots, d)$. Let $M' \subset M$ be the ideal generated by these elements. We are going to show by decreasing induction in $n = d, d-1, \dots, 1$ that $\sigma_n(x, 1) \in M'$ for all $x \in \mathfrak{m}$. The base of induction is clear: $\sigma_d(x, 1) = x \otimes \cdots \otimes x = \sigma_d(t, 1)\sigma_d(x/t, 1)$, where $x \in \mathfrak{m}$. Now assume that $n < d$ and that $\sigma_{n'}(x, 1) \in M'$ for all $x \in \mathfrak{m}$ and all $n' \ge n + 1$. From this we can derive the inclusion $TS^d_{n+1}(A, \mathfrak{m}) \subset M'$. Indeed, by Lemma 16.1, for every $n' \ge n+1$ the space $TS^d_{n'}(A, \mathfrak{m})$ is a linear span of $\sigma_{n'}(x, y)$ with $x \in \mathfrak{m}$, $y \in A$. Writing $y = \lambda + y'$ with $\lambda \in k$ and $y' \in \mathfrak{m}$ we see that $\sigma_{n'}(x, y) \equiv \lambda^{d-n'}\sigma_{n'}(x, 1) \bmod TS^d_{n'+1}(A, \mathfrak{m})$. Thus, using the descending induction in $n' = d, d-1, \dots, n+1$ we derive the desired inclusion. Now for every $x \in \mathfrak{m}$ the difference $\sigma_n(x, 1) - \sigma_n(t, 1)\sigma_n(x/t, 1)$ belongs to $TS^d_{n+1}(A, \mathfrak{m})$ so we are done. □

Let $\mathcal{D}_d \subset C \times \operatorname{Sym}^d C$ be the universal divisor consisting of $(p, D)$ such that $p \in \operatorname{supp}(D)$. More precisely, $\mathcal{D}_d$ is obtained by the descent of the $S_d$-invariant divisor $\sum_{i=1}^d \Delta_{0i}$ where $\Delta_{0i} \subset C \times C^d$ are the partial diagonals:

$$\Delta_{0i} = \{(x_0, x_1, \dots, x_d) | \; x_0 = x_i\}.$$

If $k = \mathbb{C}$ then we can describe local equation of $\mathcal{D}_d$ near a point $(p, p_1 + \cdots + p_d) \in \mathcal{D}_d$ as follows. Let $x$ (resp. $x_i$) be a local coordinate near $p$ (resp., $p_i$).

Then the local equation of $\mathcal{D}_d$ is $\prod_{i=1}^{d}(x - x_i)$ considered as a function of $x$ and elementary symmetric functions of $x_i$.

**Proposition 16.3.** *The morphism*

$$C \times \mathrm{Sym}^{d-1} C \to C \times \mathrm{Sym}^d C : (p, D) \mapsto (p, p + D) \qquad (16.1.1)$$

*is a closed embedding with the image $\mathcal{D}_d$.*

The proof is left to the reader.

The reason we called $\mathcal{D}_d$ the universal divisor is due to the following result.

**Theorem 16.4.** *For every (Noetherian) scheme $S$ and a relative (over $S$) effective Cartier divisor $D \subset C \times S$ such that for every $s \in S$ the degree of the corresponding divisor $D_s \subset C$ is equal to $d$, there exists a unique morphism $f_D : S \to \mathrm{Sym}^d C$ such that $D = (\mathrm{id} \times f_D)^{-1}(\mathcal{D}_d)$.*

*Proof.* The statement is local in $S$, so we will assume that $S$ is affine. Also, locally over $S$ we can replace $C$ by its open affine piece $U$ such that $D$ is contained in $U \times S$, so we can also assume that $C$ is affine. Let $C = \mathrm{Spec}(A)$, $S = \mathrm{Spec}(K)$, $D = \mathrm{Spec}(B)$. By definition, we have a surjective homomorphism of $K$-algebras $r : A \otimes_k K \to B$, and $B$ is a flat projective $K$-module of rank $d$. Localizing $S$ if necessary we can assume that $B$ is a free $K$-module. Then for every $K$-linear endomorphism $\phi : B \to B$ we can define its determinant $\det(\phi) \in K$. Applying this to the operators of multiplication by $a \in A$, we get a homogeneous map $A \to K$ of degree $d$. According to Lemma 16.1, it defines a $k$-linear map

$$\det_{B/K} : TS^d(A) \to K.$$

Using the multiplicativity of the determinant we immediately see that $\det_{B/K}$ is a homomorphism of $k$-algebras. We claim that the morphism $f_D : S \to \mathrm{Sym}^d C$ determined by $\det_{B/K}$ satisfies the required property. Indeed, let us consider the exact sequence of $B$-modules

$$0 \to J \to B \otimes_K B \xrightarrow{m} B \to 0, \qquad (16.1.2)$$

where $m(b_1 \otimes b_2) = b_1 b_2$. Then $J$ is a free $B$-module of rank $d - 1$. Hence, we can consider determinants of $B$-linear endomorphisms of $J$. In particular, for every $a \in A$ we can compute the determinant of the operator of multiplication by $a$ on $J$. This gives a homogeneous map of degree $d - 1$ from $A$ to $B$, hence a $k$-linear map

$$\det_{J/B} : TS^{d-1}(A) \to B.$$

From the exact sequence (16.1.2) we immediately get that for every $a \in A$ one has the following equality in $B$:

$$r(a)\det_{J/B}\left(a^{\otimes(d-1)}\right) = \det_{B/K}(a^{\otimes d})$$

(note that the RHS belongs to $K \subset B$). This implies that we have a commutative diagram

$$\begin{array}{ccc} D & \xrightarrow{\quad f' \quad} & C \times \operatorname{Sym}^{d-1} C \\ \Big\downarrow & & \Big\downarrow \alpha \\ C \times S & \xrightarrow{\ \mathrm{id} \times f_D\ } & C \times \operatorname{Sym}^d C, \end{array} \tag{16.1.3}$$

where $\alpha$ is the morphism (16.1.1), the first component of $f'$ is the natural map $D \to C$, and the second component of $f'$ corresponds to $\det_{J/B}$. Therefore, we get a morphism $D \to (\mathrm{id} \times f_D)^{-1}(\mathcal{D}_d)$ compatible with embeddings of both schemes into $C \times S$. This implies the equality of these subschemes of $C \times S$ (the corresponding homomorphism of algebras $B' \to B$ is surjective, while both $B$ and $B'$ are finitely generated projective modules of the same rank over $K$).

To prove uniqueness it suffices to show that for $D = \mathcal{D}_d$ (and $S = \operatorname{Sym}^d C$) the morphism $f_{\mathcal{D}_d} : \operatorname{Sym}^d C \to \operatorname{Sym}^d C$ constructed above is the identity. Since we have the canonical surjective projection $\pi = \pi_d : C^d \to \operatorname{Sym}^d C$, it suffices to prove that the morphism $f_{\pi^{-1}\mathcal{D}_d} : C^d \to \operatorname{Sym}^d C$ obtained by the above construction applied to $\pi^{-1}\mathcal{D}_d = \sum_{i=1}^d \Delta_{0i} \subset C \times C^d$, coincides with $\pi$. It remains to observe that the general construction $D \mapsto f_D$ is compatible with taking sums of relative Cartier divisors (over the same base) in the following sense:

$$f_{D_1+D_2} = s_{\deg D_1, \deg D_2} \circ (f_{D_1}, f_{D_2}).$$

Indeed, this follows easily by computing determinants of the action of $A$ in the exact sequence

$$0 \to \mathcal{O}_{D_2}(-D_1) \to \mathcal{O}_{D_1+D_2} \to \mathcal{O}_{D_1} \to 0.$$

Applying this property to the decomposition of $\pi^{-1}\mathcal{D}_d$ and using the compatibility of the construction of $f_D$ with the base change, we reduce the problem to the case $d = 1$ which is straightforward. □

## 16.2. Construction

We are going to glue the Jacobian from open pieces which are all isomorphic to the same open subset in $\mathrm{Sym}^g C$. We need the following technical result.

**Lemma 16.5.** *For every line bundle $L$ of degree $0$ on $C$ the subset $U_L \subset \mathrm{Sym}^g C$ consisting of the effective divisors $D$ of degree $g$ with $h^1(L(D)) = 0$, is Zariski open and nonempty. Furthermore, there is a natural surjective projective morphism $U_L \to U_{\mathcal{O}}$ mapping $D$ to the unique effective divisor $D'$ with $\mathcal{O}(D') \simeq L(D)$.*

*Proof.* The fact that $U_L$ is open follows from the semicontinuity theorem. To prove that it is nonempty we note that the condition $h^1(L(D)) = 0$ is equivalent to $h^0(\omega_C L^{-1}(-D)) = 0$. Now we are going to use the fact that for every line bundle $M$ with $h^0(M) \neq 0$ there exists a point $p \in C$ with $h^0(M(-p)) = h^0(M) - 1$. Applying this iteratively we find points $p_1, \dots, p_{g-1} \in C$ such that $h^0(\omega_C L^{-1}(-p_1 - \dots - p_{g-1})) = h^0(\omega_C L^{-1}) - g + 1 \leq 1$. Now we can choose a point $p_g$ such that $h^0(\omega_C L^{-1}(-p_1 - \dots - p_{g-1} - p_g)) = 0$. This proves that $U_L$ is nonempty. It remains to construct the morphism $U_L \to U_{\mathcal{O}}$. Let $\mathcal{D} = \mathcal{D}_g \cap (C \times U_L)$ be the universal family of effective divisors over $U_L$. Consider the line bundle $\mathcal{M} = p_1^* L(\mathcal{D})$ on $C \times U_L$, where $p_1 : C \times U_L \to C$ is the natural projection. Then the restriction of $\mathcal{M}$ to every fiber of $p_2 : C \times U_L \to U_L$ is a line bundle of degree $g$ with no $H^1$. It follows that $p_{2*}\mathcal{M}$ is a line bundle on $U_L$ and the canonical map $p_2^* p_{2*}\mathcal{M} \to \mathcal{M}$ vanishes on a relative divisor of degree $g$. Applying Theorem 16.4, we get a morphism $U_L \to \mathrm{Sym}^g C$. It is easy to see that the image of this morphism is $U_{\mathcal{O}}$. Here is another description of this morphism. Let us consider the closed subscheme $Z_L \subset \mathrm{Sym}^g C \times \mathrm{Sym}^g C$ consisting of $(D_1, D_2)$ such that $\mathcal{O}_C(D_1) \simeq L^{-1}(D_2)$ (we define the subscheme structure on $Z_L$ using Proposition 9.3). Let $p_1, p_2$ be projections of the product $\mathrm{Sym}^g C \times \mathrm{Sym}^g C$ to its factors. Then it is easy to see that $U := Z_L \cap p_1^{-1}(U_L) = Z_L \cap p_2^{-1}(U_{\mathcal{O}})$ and that the projection $p_1$ induces an isomorphism $U \xrightarrow{\sim} U_L$. Using this isomorphism our map from $U_L$ to $U_{\mathcal{O}}$ can be identified with the map $U \to U_{\mathcal{O}}$ induced by the second projection. Thus, our morphism is obtained by the base change from the map $p_2 : Z_L \to \mathrm{Sym}^g C$, hence it is projective. □

The open pieces from which we glue $J(C)$ are numbered by the isomorphism classes of line bundles of degree $g$. Namely, for every $M \in \mathrm{Pic}^g(C)$ we set $X_M = U_{\mathcal{O}}$, where $U_{\mathcal{O}}$ is the open subset in $\mathrm{Sym}^g C$ consisting of $D$ with $h^1(D) = 0$. We identify $k$-points of $X_M$ with the subset $U_{\mathcal{O}}(k) - [M]$ in $\mathrm{Pic}^0(C)$. Now for every pair of line bundles $M_1, M_2 \in \mathrm{Pic}^g(C)$ we set

$$X_{M_1,M_2} = U_{\mathcal{O}} \cap U_{M_2M_1^{-1}} \subset X_{M_1}.$$

The map $U_{M_2M_1^{-1}} \to U_{\mathcal{O}}$ constructed in Lemma 16.5 restricts to a natural isomorphism

$$i_{M_1,M_2} : X_{M_1,M_2} \to X_{M_2,M_1}$$

mapping $D \in U_{\mathcal{O}} \cap U_{M_2M_1^{-1}}$ to the unique effective divisor $D'$ such that $\mathcal{O}(D') \simeq M_2M_1^{-1}(D)$. It is easy to check that $i_{M_2,M_1}$ is inverse to $i_{M_1,M_2}$. Moreover, for every triple $M_1, M_2, M_3 \in \mathrm{Pic}^g(C)$ one has

$$i_{M_2,M_3} \circ i_{M_1,M_2} = i_{M_1,M_3}$$

on an open subset of $X_{M_1}$ where both morphisms are defined. Hence, we can glue the pieces $(X_M)$ along the identifications $(i_{M_1,M_2})$ into a scheme $X$ such that the natural embeddings $X_M \subset X$ are open. By construction, we have a natural identification of the set of $k$-points of $X$ with $\mathrm{Pic}^0(C)$ such that $X_M(k)$ corresponds to the subset of line bundles $\xi \in \mathrm{Pic}^0(C)$ with $h^1(\xi \otimes M) = 0$. Also, for every line bundle $M_0$ of degree $g$ we have a natural surjective proper morphism

$$\sigma_{M_0} : \mathrm{Sym}^g C \to X$$

such that $\sigma_{M_0}^{-1}(X_M) = U_{MM_0^{-1}}$ and $\sigma_{M_0}|_{U_{MM_0^{-1}}} : U_{MM_0^{-1}} \to U_{\mathcal{O}} = X_M$ is the morphism constructed in Lemma 16.5. Therefore, $X$ is a proper smooth irreducible variety. Next we want to introduce the structure of abelian variety on $X$. On the level of points the group law $m : X \times X \to X$ is given by the group structure on $\mathrm{Pic}^0(C)$. To prove that it is regular we can argue locally. Let $(L_1, L_2) \in \mathrm{Pic}^0(C) \times \mathrm{Pic}^0(C)$ be a point of $X \times X$. We can choose an open neighborhood of $(L_1, L_2)$ of the form $X_{M_1} \times X_{M_2}$. Let $\mathcal{D}_1$ (resp., $\mathcal{D}_2$) be the universal divisor on $C \times X_{M_1}$ (resp., $C \times X_{M_2}$). Let us also choose a line bundle $M$ of degree $g$ such that $h^1(L_1L_2M) = 0$. Then the line bundle

$$\mathcal{M} = p_1^*(M_1^{-1}M_2^{-1}M) \otimes p_{12}^*\mathcal{O}(\mathcal{D}_1) \otimes p_{13}^*\mathcal{O}(\mathcal{D}_2)$$

on $C \times X_{M_1} \times X_{M_2}$ has degree $g$ on every fiber of the projection $p_{23} : C \times X_{M_1} \times X_{M_2} \to X_{M_1} \times X_{M_2}$. Let $U \subset X_{M_1} \times X_{M_2}$ be an open neighborhood of $(L_1, L_2)$ such that $h^1(\mathcal{M}|_{C\times u}) = 0$ for $u \in U$. Then the natural morphism $p_{23}^*(p_{23})_*M|_{C\times U} \to M|_{C\times U}$ vanishes along a relative divisor of degree $g$, so we get a morphism $U \to U_{\mathcal{O}} = X_L \subset X$ which produces our group law on the level of points.

We call the abelian variety $(X, m)$ the *Jacobian* of $C$ and denote it by $J = J(C)$.

## 16.3. Poincaré Bundle

A *Poincaré line bundle* $\mathcal{P} = \mathcal{P}_C$ on $C \times J(C)$[6] is a universal family of line bundles of degree 0 on $C$ parametrized by $J(C)$. This means that the restriction of $\mathcal{P}$ to $C \times \{L\}$ is isomorphic to $L$ for every $L \in \mathrm{Pic}^0(C)$. Such a line bundle is not unique: we can tensor it with the pull-back of a line bundle on $J(C)$. Let us fix a point $p \in C$. A Poincaré line bundle is called *normalized at* $p$ if $\mathcal{P}|_{p\times J(C)}$ is trivial. Starting from any Poincaré line bundle $\mathcal{P}$ we can construct a normalized Poincaré line bundle by tensoring $\mathcal{P}$ with the pull-back of $\mathcal{P}^{-1}|_{p\times J(C)}$. We will show that for every $p \in C$ there exists a Poincaré line bundle normalized at $p$. The idea is to glue this line bundle from its restriction to open subsets $C \times X_M$ where $M \in \mathrm{Pic}^g(C)$. Note that we have a natural universal family on $C \times X_M$, namely,

$$\mathcal{P}'_M = p_1^* M^{-1}(\mathcal{D}),$$

where $\mathcal{D} \subset C \times X_M$ is the universal divisor of degree $g$. The problem is that although the restrictions of $\mathcal{P}'_{M_1}$ and $\mathcal{P}_{M_2}$ to $C \times (X_{M_1} \cap X_{M_2})$ are isomorphic there is no canonical choice of such an isomorphism. This is the reason for introducing normalizations at the point $p$. Namely, tensoring $\mathcal{P}'_M$ with the pull-back of $(\mathcal{P}'_M)^{-1}|_{p\times X_M}$ we obtain a universal family $\mathcal{P}_M$ on $C \times X_M$ together with a trivialization $\mathcal{O}_{X_M} \simeq \mathcal{P}_M|_{p\times X_M}$. Then we can find unique isomorphism between restrictions of $\mathcal{P}_{M_1}$ and $\mathcal{P}_{M_2}$ to $C \times (X_{M_1} \cap X_{M_2})$ which is compatible with trivializations at $p$. Indeed, two such isomorphism differ by an invertible function on $C \times (X_{M_1} \cap X_{M_2})$ which is equal to 1 over $p \times (X_{M_1} \cap X_{M_2})$. However, any such function is equal to 1 since $C$ is complete. These isomorphisms satisfy the cocycle condition on triple intersections, hence, we obtain the Poincaré bundle on $C \times J(C)$ normalized at $p$.

[6] We usually denote a Poincaré line bundle on $C \times J(C)$ simply by $\mathcal{P}$, which is the same notation as for the Poincaré line bundle on the product of dual abelian varieties. When we need to distinguish the two kinds of Poincaré bundles, we change the notation for the former one to $\mathcal{P}_C$.

Using Theorem 16.4 one can show that the pair $(J(C), \mathcal{P})$ represents certain functor called the Picard functor (see Exercise 2).

For every $d \in \mathbb{Z}$ we can introduce the variety $J^d = J^d(C)$ and the Poincaré line bundle $\mathcal{P}(d)$ on $C \times J^d$, normalized at $p \in C$, such that the map $\xi \mapsto \mathcal{P}(d)|_{C\times\{\xi\}}$ gives an isomorphism $J^d(k) \simeq \mathrm{Pic}^d(J)$. Namely, for every line bundle $L$ of degree $d$ on $C$, define a pair $(J^L, \mathcal{P}^L)$, where $\mathcal{P}^L$ is a line bundle on $C \times J^L$ by setting $J^L = J$, $\mathcal{P}^L = p_1^*L \otimes \mathcal{P}$, where $\mathcal{P}$ is the Poincaré bundle on $C \times J$ normalized at $p$. If $L'$ is another line bundle of degree $d$, then we have an isomorphism $J^{L'} \to J^L$ given by translation $t_{L'L^{-1}}$ on $J$. Under this isomorphism $\mathcal{P}^L$ corresponds to $\mathcal{P}^{L'}$. Thus, we have a transitive system of isomorphisms between data $(J^L, \mathcal{P}^L)$ and we define $(J^d, \mathcal{P}(d))$ to be the projective limit of this system.

### 16.4. Analytic Construction

Let $C$ be a complex projective curve. Then using the exponential sequence we see that the identity component in $H^1(C, \mathcal{O}^*)$ is identified with $J = H^1(C, \mathcal{O})/H^1(C, \mathbb{Z})$. It is easy to see that the natural map $H^1(C, \mathbb{R}) \to H^1(C, \mathcal{O})$ is an isomorphism, so $J$ is a complex torus. To see the algebraic structure on $J$ we have to find a polarization on it. The cup product on $H^1(C, Z)$ gives a symplectic form $E : H^1(C, \mathbb{Z}) \times H^1(C, \mathbb{Z}) \to \mathbb{Z}$. On the other hand, we can identify $H^1(C, \mathcal{O})$ with the space of closed (0,1)-forms on $C$. There is a natural negative hermitian form $H$ on this space given by

$$H(\alpha, \beta) = 2i \cdot \int \alpha \wedge \overline{\beta}.$$

Now if $\alpha + \overline{\alpha}$ and $\beta + \overline{\beta}$ represent integer cohomology classes then we have

$$E(\alpha + \overline{\alpha}, \beta + \overline{\beta}) = \int (\alpha + \overline{\alpha}) \wedge (\beta + \overline{\beta}) = \mathrm{Im}\, H(\alpha, \beta).$$

Hence, $-E$ gives a polarization on $J$. In the next chapter we will show how to algebraically construct this polarization.

### Exercises

1. Let $D_1$ and $D_2$ be a pair of disjoint effective divisors on $C$ of degrees $d_1$ and $d_2$ respectively. Show that the morphism $s_{d_1,d_2} : \mathrm{Sym}^{d_1} C \times \mathrm{Sym}^{d_2} C \to \mathrm{Sym}^{d_1+d_2} C$ is étale near $(D_1, D_2)$.
2. Let us consider the functor on the category of $k$-schemes which associates to $S$ the set $\mathrm{Pic}(C \times S)/\mathrm{Pic}(S)$, where the embedding

Pic($S$) $\to$ Pic($C \times S$) is given by the pull-back. Show that this functor is represented by $J(C)$, so that Poincaré bundles on $C \times J(C)$ (which form a coset for the subgroup Pic($J(C)$) $\subset$ Pic($C \times J(C)$)) correspond to the identity morphism $J(C) \to J(C)$.

3. Show that the tangent space to $J(C)$ at any point is canonically isomorphic to $H^1(C, \mathcal{O}_C)$. [*Hint*: Use the previous Exercise and Cech description of $H^1$.]
4. Assume that the genus of $C$ is 2.
   (a) Prove that $\mathrm{Sym}^2 C$ is isomorphic to the blow-up of $J(C)$ at one point.
   (b) Let $\mathbb{P}^1 \subset \mathrm{Sym}^2 C$ be the complete linear series of canonical divisors on $C$. Prove that $\mathbb{P}^1$ is the canonical divisor on $\mathrm{Sym}^2 C$.

# 17

# Determinant Bundles and the Principal Polarization of the Jacobian

In this chapter we discuss the natural principal polarization of the Jacobian $J$ of a curve $C$, i.e., a class of ample line bundles $\mathcal{L}$ on $J$ (that differ by translations), such that $\phi_{\mathcal{L}} : J \to \hat{J}$ is an isomorphism. These line bundles are called *theta line bundles* and their sections are called *theta functions of degree* 1 on $J$. Let $\mathcal{F}$ be a flat family of coherent sheaves on a relative (smooth projective) curve $\pi : \mathcal{C} \to S$, such for each member of this family one has $\chi(\mathcal{C}_s, \mathcal{F}_s) = 0$. In Section 17.1 we describe a general determinantal construction that associates to such $\mathcal{F}$ a line bundle $\det^{-1} R\pi_*(\mathcal{F})$ (up to an isomorphism) equipped with a section $\theta_{\mathcal{F}}$. We define the theta line bundle on $J$ associated with a line bundle $L$ of degree $g-1$ on $C$ by applying this construction to the family $p_1^*L \otimes \mathcal{P}$ on $C \times J$. Zeroes of the corresponding theta function $\theta_L$ constitute the *theta divisor* $\Theta_L = \{\xi \in J : h^0(L(\xi)) > 0\}$. To prove that in this way we get a principal polarization of $J$, we note that isomorphism classes of theta line bundles have form $\det^{-1} \mathcal{S}(L)$, where $\mathcal{S}(L)$ is the Fourier–Mukai transform of a line bundle $L$ of degree $g-1$ considered as a coherent sheaf on $\hat{J}$ supported on $C$ (the embedding $C \hookrightarrow \hat{J}$ is defined using the Poincaré bundle $\mathcal{P}$ on $C \times J$), $\det \mathcal{S}(L)$ is an element of $\operatorname{Pic}(J)$ defined using a locally free resolution of $\mathcal{S}(L)$ on $J$. Using the properties of the Fourier–Mukai transform, we obtain an explicit description of the morphism $\phi_{\mathcal{L}} : J \to \hat{J}$ associated with a theta line bundle, from which it is easy to see that $\phi_{\mathcal{L}}$ is an isomorphism.

In Section 17.3 we show that the restriction of the theta line bundle $\det^{-1} \mathcal{S}(L)$ to the curve $C$ embedded into its Jacobian, is isomorphic to the tensor product of $L^{-1}$ with a fixed line bundle of degree $2g-1$. This implies Riemann's theorem stating that for appropriate line bundle $L$ of degree $g-1$, the rational map $\xi \mapsto C \cap (\Theta_L + \xi)$ from $J$ to $\operatorname{Sym}^g C$ is inverse to the map $\operatorname{Sym}^g C \to J : D \mapsto \mathcal{O}_C(D - gp_0)$ (where $p_0$ is a fixed point on $C$).

Similarly, in Section 17.4 we compute the pull-back of the biextension $\mathcal{B}$ on $J^2$ associated with the principal polarization to $C^4 = (C \times C)^2$, under the map $C \times C \to J : (x, y) \mapsto x - y$ (the obtained bundle on $C^4$ is trivial outside all diagonals). The result of this computation can be interpreted as certain identity for theta functions on $J$, which is a particular case of Fay's trisecant identity considered in the next chapter.

Symmetric theta line bundles on $J$ correspond to line bundles $L$ on $C$ such that $L^2 \simeq \omega_C$. Such line bundles $L$ are called *theta characteristics* of $C$. In 17.6 using the results of Chapter 13 we show that the map $L \mapsto (-1)^{h^0(L)}$ is a quadratic function on the set of theta characteristics, such that the associated bilinear form is the Weil pairing on points of order 2 in $J$. We also prove that (for char$(k) \neq 2$) the number of theta characteristics with even (resp., odd) $h^0(L)$ is equal to $2^{g-1}(2^g + 1)$ (resp., $2^{g-1}(2^g - 1)$).

In Section 17.5, we prove that in the case $k = \mathbb{C}$ our definition of the principal polarization of $J$ agrees with the analytic definition that uses the intersection pairing on $H^1(C, \mathbb{Z})$.

## 17.1. Determinants

Let $V$ be a vector bundle on a variety $X$. We denote by $\det(V)$ the top-degree exterior power of $V$, which is a line bundle on $X$. This definition can be extended to bounded complexes of vector bundles on $X$ by setting

$$\det(V_\bullet) = \bigotimes_i \det(V_i)^{\otimes(-1)^i}.$$

It is well known that for every exact triple of vector bundles

$$0 \to V' \to V \to V'' \to 0$$

one has an isomoprhism $\det(V) \simeq \det(V') \otimes \det(V'')$. This fact can be generalized as follows.

**Proposition 17.1.** *Let $V_\bullet \to V'_\bullet$ be a quasiisomorphism of bounded complexes of vector bundles. Then* $\det(V_\bullet) \simeq \det(V'_\bullet)$.

*Proof.* Let $W_\bullet$ be the cone of our morphism of complexes. Then clearly $\det(W_\bullet) \simeq \det(V_\bullet) \otimes \det^{-1}(V'_\bullet)$. On the other hand, $W_\bullet$ is an exact sequence of vector bundles, so it can be divided into a number of exact triples. This implies the triviality if $\det(W_\bullet)$. □

**Remark.** One can define the isomorphism in the above proposition canonically up to a sign. To get rid of this sign ambiguity one should consider the pair

$$\left(\det(V_\bullet), \operatorname{rk}(V_\bullet) = \sum_i (-1)^i \operatorname{rk} V_i\right)$$

as an object of the category of graded line bundles and use the twisted commutativity constraint in this category (cf. Section 22.1, [31], Section 4). In this chapter we will be considering the isomorphism class of the line bundle $\det(V_\bullet)$, so this sign problem is irrelevant for us.

If $X$ is smooth and projective, then every object in the bounded derived category of coherent sheaves on $X$ can be represented by a finite complex of vector bundles, so for every $A \in D^b(X)$ we can define an element $\det(A) \in \operatorname{Pic}(X)$ (the above proposition shows that this element is well defined). Furthermore, det defines a homomorphism from the Grothendieck group of $D^b(X)$ to $\operatorname{Pic}(X)$.

The following construction will be used in Section 17.3 to define theta divisors in the Jacobian of a curve. Let $\mathcal{F}$ be a flat family of coherent sheaves on a relative (smooth projective) curve $\pi : C \to S$ such that $\chi(C_s, \mathcal{F}_s) = 0$ for all $s \in S$. Then one can define a line bundle $\det(R\pi_*(\mathcal{F}))$ on $S$ and a section

$$\theta_{\mathcal{F}} \in H^0(S, \det^{-1}(R\pi_*(\mathcal{F})))$$

(canonically up to an isomorphism). Namely, let us embed $\mathcal{F}$ into a coherent sheaf $\mathcal{E}_0$, flat over $S$, such that $R^1\pi_*(\mathcal{E}_0) = 0$ and the quotient-sheaf $\mathcal{E}_1 = \mathcal{E}_0/\mathcal{F}$ is flat over $S$ (e.g., we can take $\mathcal{E}_0 = \mathcal{F}(D)$, where $D$ is a relative effective Cartier divisor of sufficiently large degree). From the exact sequence of sheaves

$$0 \to \mathcal{F} \to \mathcal{E}_0 \xrightarrow{\alpha} \mathcal{E}_1 \to 0$$

we derive that $R^1\pi_*(\mathcal{E}_1) = 0$. Hence, both sheaves $\pi_*(\mathcal{E}_0)$ and $\pi_*(\mathcal{E}_1)$ are vector bundles. Then $R\pi_*(\mathcal{F})$ is represented by the complex of vector bundles

$$V_\bullet = [\pi_*(\mathcal{E}_0) \xrightarrow{\pi_*(\alpha)} \pi_*(\mathcal{E}_1)]$$

concentrated in degrees 0 and 1, and we set

$$\det R\pi_*(\mathcal{F}) = \det V_\bullet = \det(\pi_*(\mathcal{E}_0)) \otimes \det(\pi_*(\mathcal{E}_1))^{-1}.$$

Since the rank of $R\pi_*(\mathcal{F})$ is zero, the ranks of $\pi_*(\mathcal{E}_0)$ and $\pi_*(\mathcal{E}_1)$ are equal. Thus, we can define $\theta_{\mathcal{F}}$ to be the determinant of the morphism $\pi_*(\alpha)$, so that $\theta_{\mathcal{F}}$ is a section of $\det(R\pi_*(\mathcal{F}))^{-1}$. We claim that when we change the resolution $\mathcal{E}_0 \to \mathcal{E}_1$ for $\mathcal{F}$, the pair $(\det(R\pi_*(\mathcal{F})), \theta_{\mathcal{F}})$ gets replaced by an isomorphic one. Indeed, in the case when there is a morphism from one resolution to another this follows from Exercise 1 below applied to the induced morphism of resolutions of $R\pi_*(\mathcal{F})$. On the other hand, we claim that every resolution $\mathcal{E}_0 \to \mathcal{E}_1$ maps to the resolution $\mathcal{F}(ND) \to \mathcal{F}(ND)|_{ND}$ for ample divisor $D$ and for $N$ large enough. Indeed, in this case $\mathrm{Ext}^1(\mathcal{E}_1, \mathcal{F}(ND))$ vanishes, so we can use the following lemma.

**Lemma 17.2.** *Let $V_\bullet = [V_0 \to V_1]$ and $V'_\bullet = [V'_0 \to V'_1]$ be a pair of complexes in some abelian category $\mathcal{A}$ (placed in degrees 0 and 1). Assume that $\mathrm{Ext}^1_{\mathcal{A}}(V_1, V'_0) = 0$. Then every morphism $C \to C'$ in the derived category $D^b(\mathcal{A}) \to D^b(\mathcal{A})$ can be represented by a chain map between these complexes.*

*Proof.* Let us consider the exact triangles

$$V_0[-1] \to V_1[-1] \xrightarrow{\alpha} C \xrightarrow{\delta} V_0,$$

$$V'_0[-1] \to V'_1[-1] \xrightarrow{\alpha'} C' \xrightarrow{\delta'} V'_0$$

in $D^b(\mathcal{A})$. Now given $f \in \mathrm{Hom}_{D^b(\mathcal{A})}(C, C')$ let us consider the composition $\delta' \circ f \circ \alpha \in \mathrm{Hom}_{D^b(\mathcal{A})}(V_1[-1], V'_0)$. By our assumption, it is equal to zero. Hence, $f \circ \alpha$ factors as a composition of some morphism $f_1 : V_1[-1] \to V'_1[-1]$ with $\alpha'$. Therefore, we obtain a commutative square

$$\begin{array}{ccc} V_1[-1] & \xrightarrow{\alpha} & C \\ \downarrow{\scriptstyle f_1} & & \downarrow{\scriptstyle f} \\ V'_1[-1] & \xrightarrow{\alpha'} & C'. \end{array} \qquad (17.1.1)$$

Since $D^b(\mathcal{A})$ is triangulated, this commutative square extends to the morphism between the above exact triangles. In particular, we obtain a chain map $f_\bullet$ between our complexes. Let $\widetilde{f} \in \mathrm{Hom}_{D^b(\mathcal{A})}(C, C')$ be the map determined by $f_\bullet$. By construction $\widetilde{f} \circ \alpha = f \circ \alpha$. Therefore, the map $g = \widetilde{f} - f$ factors as a composition of $\delta : C \to V_0$ with some map $h : V_0 \to C'$ in $D^b(\mathcal{A})$. Now we observe that every such map $h$ should factor as a composition of some map $h_0 : V_0 \to H^0(C')$ with the canonical map $i : H^0(C') \to C'$. Since $i$ can be represented by a chain map, it follows that $h$ can be represented by a chain map from $V_0$ (considered as a complex concentrated in degree 0) to $C'$. Thus, both $h$ and $\delta$ are represented by a chain map, hence the same is true for $g$, and therefore for $f$. □

## 17.2. A Curve Mapping to an Abelian Variety

Let $a : C \to A$ be a nonconstant morphism from a curve to an abelian variety. Then for a coherent sheaf $\mathcal{F}$ on $C$ (considered as a sheaf on $A$ via push-forward by $a$) we can consider the corresponding Fourier transform which is a sheaf on the dual abelian variety $\hat{A}$:

$$\mathcal{S}(\mathcal{F}) = Rp_{2*}(p_1^* a_* \mathcal{F} \otimes \mathcal{P})$$

where $\mathcal{P}$ is a Poincaré line bundle on $A \times \hat{A}$. Using the notation of Section 11.1 we can rewrite this as

$$\mathcal{S}(\mathcal{F}) = \Phi_{(a \times \mathrm{id}_{\hat{A}})^* \mathcal{P}, C \to \hat{A}}(\mathcal{F})$$

where $(a \times \mathrm{id}_{\hat{A}})^* \mathcal{P}$ is a line bundle on $C \times \hat{A}$. Since $\mathcal{S}(\mathcal{F})$ is an object of the bounded derived category of coherent sheaves on $\hat{A}$, we can consider its determinant $\det(\mathcal{S}(\mathcal{F})) \in \mathrm{Pic}(\hat{A})$. Note that if $\mathcal{F} = \mathcal{O}_p$, the skyscraper sheaf at point $p \in C$, then $\mathcal{S}(\mathcal{F}) = \mathcal{P}_{a(p)} := \mathcal{P}|_{a(p) \times \hat{A}}$. Now for every line bundle $L$ on $C$ we can consider the corresponding determinant bundle $\det(\mathcal{S}(L))$. From the exact sequence

$$0 \to L \to L(p) \to L(p)|_p \to 0$$

we derive an isomorphism

$$\det(\mathcal{S}(L(p))) \simeq \det(\mathcal{S}(L)) \otimes \mathcal{P}_{a(p)}.$$

Iterating, we get that for every divisor $D = \sum n_i p_i$ one has

$$\det(\mathcal{S}(L(D))) \simeq \det(\mathcal{S}(L)) \otimes \mathcal{P}_{\sum n_i a(p_i)}. \qquad (17.2.1)$$

The LHS depends only on the rational equivalence class of $D$, hence the map $D \mapsto \sum n_i a(p_i)$ factors through a homomorphism $a_* : \mathrm{Pic}(C) \to A(k)$. It follows also from (17.2.1) that the algebraic equivalence class of $\det(\mathcal{S}(L))$ does not depend on $L$.

**Proposition 17.3.** *Let $\phi_a = \phi_{\det(\mathcal{S}(L))} : \hat{A}(k) \to A(k)$ be the homomorphism associated with the line bundle* $\det(\mathcal{S}(L))$ *on $\hat{A}$ (see Chapter 8), where $L$ is an arbitrary line bundle on $C$. Then $\phi_a$ coincides with the composition*

$$\mathrm{Pic}^0(A) \xrightarrow{a^*} \mathrm{Pic}^0(C) \xrightarrow{a_*} A(k)$$

*Proof.* By the definition, $\phi_a(x)$ corresponds to the line bundle $t_x^* \det(\mathcal{S}(L)) \otimes \det(\mathcal{S}(L))^{-1}$ on $\hat{A}$. Now we have an isomorphism

$$t_x^* \det \mathcal{S}(L) \simeq \det(\mathcal{S}(L \otimes a^*\mathcal{P}_x)) \simeq \det(\mathcal{S}(L)) \otimes \mathcal{P}_{a_*(a^*(x))}$$

which implies our claim. □

**Remark.** It is clear that the map $a^* : \mathrm{Pic}^0(A) \to \mathrm{Pic}^0(C)$ is algebraic (with respect to the structures of algebraic varieties on the source and the target). We will show that the map $a_*$ is algebraic in the case when $a$ is the standard map from $C$ to $\widehat{J(C)}$. The general case is considered in Exercise 3 at the end of this chapter.

Note that if $f : A \to B$ is a homomorphism of abelian varieties then we have $(f \circ a)_* = f \circ a_*$, while $(f \circ a)^* = a^* \circ \hat{f}$. Hence, applying the above theorem, we obtain

$$\phi_{f \circ a} = f \circ \phi_a \circ \hat{f}. \tag{17.2.2}$$

## 17.3. Principal Polarization of the Jacobian and Theta Divisors

Let $C$ be a curve, $p_0 \in C$ be a fixed point. We denote by $J$ the Jacobian of $C$ and by $\mathcal{P}_C$ the Poincaré line bundle on $C \times J$ normalized at $p_0$. Considering $\mathcal{P}_C$ as a family of line bundles on $J$ we get a morphism $a : C \to \hat{J}$ such that

$$\mathcal{P}_C \simeq (a \times \mathrm{id})^* \mathcal{P}, \tag{17.3.1}$$

where $\mathcal{P}$ is the Poincaré line bundle on $\hat{J} \times J$. The restriction morphism $a^* : J = \mathrm{Pic}^0(\hat{J}) \to \mathrm{Pic}^0(C) = J$ is the identity (indeed, this statement is a tautology which follows immediately from (17.3.1)). Thus, by Proposition 17.3 the morphism $a_*|_J : J(k) \to \hat{J}(k)$ coincides with the symmetric morphism $\phi_a$ (in particular, it is algebraic). To prove that $a_*|_J$ is an isomorphism, we are going to use the morphism $i : C \to J : p \mapsto \mathcal{O}(p - p_0)$. More precisely, $i$ corresponds to the family of line bundles on $C$ trivialized at $p_0$, given by $\mathcal{O}_{C \times C}(\Delta) \otimes p_1^* \mathcal{O}_C(-p_0) \otimes p_2^* \mathcal{O}_C(-p_0)$, where $p_1, p_2 : C \times C \to C$ are the projections. It is clear from the definition that the morphism $i_*|_{\mathrm{Pic}^0(C)} : J \to J$ is the identity. On the other hand, we have $i^* \circ a = i$ (this follows from the fact that the above line bundle on $C \times C$ is symmetric). Hence, we have $i^* \circ a_* = (i^* \circ a)_* = i_*$. Therefore, the map $\phi_a = a_*|_J : J \to \hat{J}$ is an isomorphism with the inverse $i^* : \hat{J} \to J$.

Now let $L$ be a bundle of degree $g-1$ on $C$. Then we claim that the line bundle $\det(\mathcal{S}(L))^{-1}$ on $J$ has a canonical (up to a nonzero scalar) global section $\theta_L$. Indeed, $\mathcal{S}(L)$ is the derived push-forward of a line bundle on $C \times J$ to $J$ and $\operatorname{rk}(\mathcal{S}(L)) = \chi(L) = 0$, hence we can apply the construction of Section 17.1. On the other hand, the line bundle $\det \mathcal{S}(L)^{-1}$ is nondegenerate, as we have seen above. Therefore, by Theorem 8.11 it is ample, so $-\phi_a$ is the principal polarization of $J$.

The zeros of $\theta_L \in H^0(J, \det(\mathcal{S}(L))^{-1})$ constitute a divisor $\Theta_L \subset J$ supported on the set of points $\xi \in J$ such that $L(\xi)$ has a nonzero global section. We call $\theta_L$ (resp., $\Theta_L$; resp., $\mathcal{O}_J(\Theta_L)$) a *theta function of degree* 1 (resp., a *theta divisor*; resp., a *theta line bundle*) on $J$. Since $-\phi_a$ is the principal polarization, all the divisors $\Theta_L$ (and the corresponding theta functions $\theta_L$) are translations of each other. This is not surprising, since by the definition the divisor $\Theta_L \subset J$ is the translation by $L^{-1}$ of the divisor $\Theta \subset J^{g-1}$ supported on line bundles of degree $g-1$ that have a section. The latter divisor $\Theta$ coincides the image of the natural morphism $\sigma^{g-1} : \operatorname{Sym}^{g-1} C \to J^{g-1}$. Set-theoretically this is clear, while on the scheme-theoretical level this is the content of Exercise 6.

Now we are going to look at intersections of the divisors $\Theta_L \subset J$ with the curve $C$ embedded into the Jacobian by the morphism $i : p \mapsto \mathcal{O}_C(p - p_0)$. The next theorem implies the beautiful property of the theta divisor discovered by Riemann: for appropriate line bundle $L$ of degree $g-1$ (depending on $p_0$), the rational map $\xi \mapsto C \cap (\Theta_L + \xi) = C \cap \Theta_{L(-\xi)}$ from $J$ to $\operatorname{Sym}^g C$ is the inverse of the birational map $\operatorname{Sym}^g C \to J : D \mapsto \mathcal{O}_C(D - gp_0)$.

**Theorem 17.4.** *For every line bundle L of degree d on C one has*

$$\det \mathcal{S}(L)|_C \simeq \omega_C^{-1} \otimes L((g-d-2)p_0).$$

*In particular, if* $\deg L = g-1$ *then*

$$\mathcal{O}_J(\Theta_L)|_C \simeq \omega_C(p_0) \otimes L^{-1}.$$

*Proof.* The pull-back of the Poincare bundle on $\hat{J} \times J$ by the map $a \times i : C \times C \to \hat{J} \times J$ is isomorphic to $\mathcal{O}_{C\times C}(\Delta) \otimes p_1^*\mathcal{O}_C(-p_0) \otimes p_2^*\mathcal{O}_C(-p_0)$, where $p_1, p_2 : C \times C \to C$ are the projections. Hence, by the base change formula of a flat morphism (see Appendix C) we get

$$\mathcal{S}(L)|_C \simeq Rp_{2*}(p_1^*(L(-p_0))(\Delta))(-p_0),$$

where in the LHS we consider the derived functor of the restriction to $C$. Let us denote $L_1 = L(-p_0)$. Note that $\operatorname{rk} \mathcal{S}(L) = \chi(L) = d - g + 1$, so we have

$$\det \mathcal{S}(L)|_C \simeq \det(Rp_{2*}p_1^*L_1(\Delta))(-(d-g+1)p_0).$$

Now from the exact sequence of sheaves on $C \times C$

$$0 \to p_1^*L_1 \to p_1^*L_1(\Delta) \to \Delta_*\big(L_1 \otimes \omega_C^{-1}\big) \to 0$$

we get

$$\det Rp_{2*}(p_1^*L_1(\Delta)) \simeq L_1 \otimes \omega_C^{-1},$$

which implies our formula for $\det \mathcal{S}(L)|_C$. □

**Remark.** The statement of the Theorem 17.4 becomes more natural if instead of the embedding $i$ one considers the canonical embedding $i_1 : C \to J^1 : p \mapsto \mathcal{O}_C(p)$. Then the theorem implies that for every line bundle $M$ of degree $g-2$ one has

$$i_1^*\mathcal{O}_{J^1}(\Theta_M) \simeq \omega_C \otimes M^{-1},$$

where $\Theta_M \subset J^1$ is the translation of the theta divisor $\Theta \subset J^{g-1}$ by $M^{-1}$.

For a divisor $D$ of degree $g-1$ on $C$ we set $\Theta_D = \Theta_{\mathcal{O}_C(D)}, \theta_D = \theta_{\mathcal{O}_C(D)}$.

**Corollary 17.5.** *Let $D \subset C$ be an effective divisor of degree $g-1$ such that $h^0(D) = 1$ and $p_0$ does not belong to the support of $D$. Then we have an equality of subschemes*

$$\Theta_D \cap C = D' + p_0,$$

*where $D'$ is the unique effective divisor on $C$ such that $\mathcal{O}_C(D+D') \simeq \omega_C$.*

*Proof.* Set-theoretically the intersection $\Theta_D \cap C$ consists of points $x \in C$ such that $h^0(D+x-p_0) \neq 0$, or equivalently $h^0(D'+p_0-x) \neq 0$. Our assumptions on $D$ and $p_0$ imply that $h^0(D-p_0) = 0$, hence $h^0(D'+p_0) = 1$. It follows that set-theoretically $\Theta_D \cap C$ coincides with the support of $D'+p_0$. In particular, $\theta_D|_C \neq 0$. By Theorem 17.4, we have $\mathcal{O}_J(\Theta_D)|_C \simeq \mathcal{O}_C(D'+p_0)$. Since this line bundle has 1-dimensional space of global sections, the divisor of zeros of $\theta_D|_C$ is precisely $D'+p_0$. □

A useful way to reformulate the previous corollary is the following. Let $D$ be an effective divisor of degree $g-1$, such that $h^0(D) = 1$, $x \in C$ be a point, not contained in the support of $D$. Then the divisor of zeroes of $y \mapsto \theta_D(y-x)$ is precisely $D'+x$, where $\mathcal{O}_C(D+D') \simeq \omega_C$.

## 17.4. Some Canonical Isomorphisms and an Identity for Theta Functions

Let us denote by $\mathcal{B}$ the biextension on $J \times J$ corresponding to the isomorphism $\phi_a = a_*|_J : J \widetilde{\to} \hat{J}$, so that we have $\mathcal{B} = \Lambda(\det \mathcal{S}(L)) = (\phi_a \times \mathrm{id}_J)^* \mathcal{P}$, where $\mathcal{P}$ is the Poincaré line bundle on $\hat{J} \times J$.

**Proposition 17.6.** *Consider the morphism $d : C \times C \to J : (x, y) \mapsto (y - x)$. Then one has a canonical isomorphism*

$$(d \times \mathrm{id}_J)^* \mathcal{B} \simeq p_{23}^* \mathcal{P}_C \otimes p_{13}^* \mathcal{P}_C^{-1}$$

*on $C \times C \times J$, where $p_{13}$ and $p_{23}$ are projections to $C \times J$, $\mathcal{P}_C$ is a Poincaré line bundle on $C \times J$.*

*Proof.* Note that both sides are trivial over $\Delta_C \times J$ where $\Delta_C \subset C \times C$ is the diagonal. Thus, if an isomorphism exists it can be chosen canonically (by rigidifying everything along $\Delta_C \times J$). Let $\widetilde{d} : C \times C \to \hat{J}$ be the map obtained as the composition of $a \times a : C \times C \to \hat{J} \times \hat{J}$ and the difference map $\hat{J} \times \hat{J} \to \hat{J}$. Then our assertion is equivalent to the equality $\widetilde{d} = a_* \circ d$. We have seen above that $i = i^* \circ a$. Composing both sides with $a_*$ and using the fact that $i^*$ and $a_*|_J$ are inverse to each other, we get that $a_* \circ i = a$. This immediately implies that $a_* \circ d = \widetilde{d}$. □

Let us denote by $\langle \xi, \xi' \rangle$ the fiber of $\mathcal{B}$ at the point $(\xi, \xi') \in J \times J$. The isomorphism of the previous proposition can be written as follows:

$$\langle y - x, \xi \rangle \simeq \xi_y \otimes \xi_x^{-1}, \tag{17.4.1}$$

where $x, y \in C$, $\xi \in J$. Note that for fixed $\xi$ similar isomorphism can be derived from Theorem 17.4. Indeed, applying this theorem to $L$ and $L \otimes \xi$ we get

$$\xi \simeq \det \mathcal{S}(L \otimes \xi) \otimes \det \mathcal{S}(L)^{-1}|_C$$

Since, $\det \mathcal{S}(L \otimes \xi) \simeq t_\xi^* \det \mathcal{S}(L)$ by (11.3.2), we derive that

$$\xi_y \otimes \xi_x^{-1} \simeq \langle i(y), \xi \rangle \otimes \langle i(x), \xi \rangle^{-1} \simeq \langle y - x, \xi \rangle.$$

This isomorphism is the same as (17.4.1), since both sides are trivial for $y = x$. In particular, taking $\xi$ to be $\mathcal{O}_C(z - t)$, where $z$ and $t$ are distinct points on $C$, we obtain from (17.4.1) a canonical trivialization of $\langle y - x, z - t \rangle$ for $x, y$ disjoint from $z, t$. On the other hand, applying (17.4.1) to $\xi = \mathcal{O}_C(z - t)$ and

$y = z$, we get a trivialization of $\omega_z \otimes \langle z - x, z - t \rangle$, where the points $x$, $z$, and $t$ are distinct, $\omega_z = \omega_C|_z$.

Now we want to interpret these trivializations in terms of theta functions on $J$. The following proposition expresses $\langle y - x, z - t \rangle$ in terms of a theta line bundle on $J$.

**Proposition 17.7.** *Let $\mathcal{L}$ be a theta line bundle on $J$. Then there is a canonical isomorphism*

$$\langle y - x, z - t \rangle \simeq \mathcal{L}_{x-t} \otimes \mathcal{L}_{y-z} \otimes \mathcal{L}_{x-z}^{-1} \otimes \mathcal{L}_{y-t}^{-1}$$

*of line bundles on $C \times C \times C \times C$, restricting to the identity for $y = x$ or $z = t$.*

*Proof.* Since $x - t = (x - z) + (z - t)$ we have

$$\mathcal{L}_0 \otimes \mathcal{L}_{x-t} \simeq \mathcal{L}_{x-z} \otimes \mathcal{L}_{z-t} \otimes \langle x - z, z - t \rangle^{-1}.$$

Replacing in this isomorphism $x$ by $y$ and dividing the obtained isomorphism by the original one, we get the result. □

Let us fix points $z, t \in C$ and consider the line $\langle y - x, z - t \rangle$ for varying $x, y$, as a line bundle on $C \times C$. Then (17.4.1) implies that there exists a unique rational section $\phi(x, y)$ of this line bundle with poles of order 1 at $x = z$ and $y = t$, zeros of order 1 at $x = t$ and $y = z$ (and no other poles or zeros), such that $\phi(x, x) = 1$. We claim that under the isomorphism of the previous proposition applied to $\mathcal{L} = \mathcal{O}_J(\Theta_D)$, where $D$ is a generic effective divisor of degree $g - 1$, we have

$$\phi(x, y) = \frac{\theta_D(x - t)\theta_D(y - z)}{\theta_D(x - z)\theta_D(y - t)}. \tag{17.4.2}$$

Indeed, the RHS reduces to 1 for $x = y$ and has correct zeros and poles as follows from Corollary 17.5. Varying $D$, we get an identity for theta functions.

**Corollary 17.8.** *Let $\theta$ be a theta function of degree* 1 *on $J$ such that $\theta(0) = 0$, and let $\xi \in J$ be such that $\theta(\xi) = 0$. Then for $x, y, z, t \in C$ we have*

$$\begin{aligned}&\theta(\xi + x - z)\theta(\xi + y - t)\theta(x - t)\theta(y - z)\\&= \theta(\xi + x - t)\theta(\xi + y - z)\theta(x - z)\theta(y - t).\end{aligned}$$

*Proof.* Since the theta divisor $\Theta \subset J^{g-1}$ is irreducible (being the image of $\mathrm{Sym}^{g-1} C$), it suffices to take $\theta = \theta_D, \xi = D' - D$, where $D$ and $D'$ are generic effective divisors of degree $g-1$. Then our identity follows immediately from (17.4.2) applied to $D$ and $D'$. □

The condition $\theta(\xi)=0$ suggests that this identity is a special case of some more general identity with an additional term divisible by $\theta(\xi)$. We will derive this more general identity in Chapter 18.

## 17.5. Albanese Variety

Let us assume that the ground field is $\mathbb{C}$. The Albanese variety of a complex projective variety $W$ is defined as follows:

$$\mathrm{Alb}(W)=H^0\big(W,\Omega_W^1\big)^\vee/H_1(W,\mathbb{Z}),$$

where a cycle $\gamma\in H_1(W,\mathbb{Z})$ determines the functional on $H^0(W,\Omega_W^1)$ via integration. Because the spaces $H^{0,1}=H^1(W,\mathcal{O}_W)$ and $H^{1,0}=H^0(W,\Omega_W^1)$ are conjugate to each other, the dual abelian variety is identified with the Jacobian variety

$$J(W)=H^1(W,\mathcal{O}_W)/H^1(W,\mathbb{Z}).$$

Recall that for an abelian variety $A=V/\Gamma$ the Hodge decomposition $H^1(A,\mathbb{C})=H^{0,1}\oplus H^{1,0}$ coincides with the natural decomposition $\mathrm{Hom}(\Gamma,\mathbb{C})\simeq V^\vee\oplus\overline{V}^\vee$. In the case $A=\mathrm{Alb}(W)$ we have $H^1(\mathrm{Alb}(W),\mathbb{C})=H^1(W,\mathbb{C})$ and the Hodge decompositions for $\mathrm{Alb}(W)$ and $W$ coincide.

When $W=C$ is a curve, the Serre duality gives an isomorphism between $\mathrm{Alb}(C)$ and $J(C)$. We want to compare this isomorphism with the one given by the principal polarization on $J=J(C)$ constructed in Section 17.3. First, we claim that the morphism $a:C\to\hat{J}$ (considered in Section 17.3) coincides with the natural embedding

$$a':C\to\mathrm{Alb}(C):p\mapsto\left(\omega\mapsto\int_{p_0}^{p}\omega\right).$$

Indeed, since $a(p_0)=a'(p_0)=0$, according to Exercise 4 of Chapter 11, it suffices to check that the induced morphism $a'^*:J\to J$ is the identity. This can be easily checked using the exponential sequence and the fact that the map on $H_1(\cdot,\mathbb{Z})$ induced by $a'$ is the identity. Thus, we have $a'=a$.

Now let us compute the tangent map to the induced map $a_*:\mathrm{Pic}^0(C)\to\mathrm{Alb}(C)$. Fix a sufficiently large $d$ and let

$$\sigma_d:C^d\to\mathrm{Pic}^0(C)$$

be the map sending $(p_1, \ldots, p_d)$ to $\mathcal{O}_C(p_1 + \ldots p_d - dp_0)$. Since $\sigma_d$ is a submersion for $d > 2g - 2$, it is enough to compute the tangent maps to $\sigma_d$ and to $a_d = a_* \circ \sigma_d$. The latter map sends $p_1, \ldots, p_d$ to the functional $\omega \mapsto \sum_{i=1}^{d} \int_{p_0}^{p_i} \omega$. Hence, the tangent map to $a_d$ is the natural map

$$\bigoplus_{i=1}^{d} \omega_{C,p_i}^{-1} \to H^0(C, \omega_C)^\vee$$

with components dual to the evaluation map.

**Lemma 17.9.** *The tangent map to $\sigma_d$ is the map*

$$\bigoplus_{i=1}^{d} \omega_{C,p_i}^{-1} \to H^1(C, \mathcal{O}_C)$$

*that is given by the natural map $H^0(C, \mathcal{O}_C(p_i)|_{p_i}) \to H^1(C, \mathcal{O}_C)$ under the identification $\omega_{C,p_i}^{-1} \simeq \mathcal{O}_C(p_i)|_{p_i}$.*

*Proof.* It suffices to prove the statement in the case $d = 1$. The tangent map to $\sigma_1$ at $p$ corresponds to the family of line bundles $L(\epsilon)$ on $C$ parametrized by $\operatorname{Spec} k[\epsilon]/\epsilon^2$ defined as follows. Let $U$ be a small neighborhood of $p$, $x$ be a generator of the maximal ideal of $p$ in $\mathcal{O}(U)$. By the definition, $L(\epsilon)$ is trivialized on the covering of $C$ consisting of $U$ and $C \setminus p$. The transition function defining $L(\epsilon)$ is $\frac{x+\epsilon}{x} = 1 + \frac{\epsilon}{x}$. Let $v$ be the tangent vector to $C$ at $p$, such that $dx(v) = 1$. Then the tangent map to $\sigma_1$ at the point $p$ sends $v$ to the class of $\frac{1}{x}$ considered as a Cech 1-cocycle with values in $\mathcal{O}$ with respect to the covering $(U, C \setminus p)$. This is equivalent to the assertion of the lemma. □

It is easy to check that for every point $p \in C$ the following diagram is commutative:

$$\begin{array}{ccc} H^0(C, \mathcal{O}_C(p)|_p) & \longrightarrow & \omega_{C,p}^{-1} \\ \downarrow & & \downarrow \\ H^1(C, \mathcal{O}_C) & \longrightarrow & H^0(C, \omega_C)^\vee \end{array} \qquad (17.5.1)$$

where the bottom arrow is Serre duality. Together with the above lemma this implies that the tangent map to $a_* : J \to \mathrm{Alb}(C)$ coincides with the Serre duality morphism $H^1(C, \mathcal{O}_C) \to H^0(C, \omega_C)^\vee$. In particular, the polarization on $J$ defined in Section 17.3, coincides with the one given by the cup-product on $H^1(C, \mathbb{Z})$ (see Section 16.4).

## 17.6. Theta Characteristics

Among theta line bundles on the Jacobian of special interest are the symmetric ones: corresponding theta functions play a prominent role in the classical theory. In the following proposition we determine which of the line bundles $\det \mathcal{S}(L)$ are symmetric.

**Proposition 17.10.** *Let L be the line bundle of degree $g-1$. Then*

$$[-1]^* \det \mathcal{S}(L) \simeq \det \mathcal{S}(\omega_C \otimes L^{-1}).$$

*In particular,* $\det \mathcal{S}(L)$ *is symmetric if and only if* $L^2 \simeq \omega_C$.

*Proof.* Using relative duality (see [61]) for the morphism $C \times J \to J$ we get

$$\mathcal{S}(L)^\vee \simeq [-1]^* \mathcal{S}(\omega_C \otimes L^{-1})[1].$$

Taking the determinants of both parts we obtain the assertion. □

**Definition.** A line bundle $L$ on $C$ with $L^2 \simeq \omega_C$ is called a *theta characteristic* (or *spin structure*) on $C$.

Let us assume until the end of this section that char($k$) $\neq 2$. Some important properties of theta characteristics are formulated in terms of the sign function

$$\epsilon(L) = (-1)^{h_0(L)},$$

where $L$ is a theta characteristic.

**Definition.** A theta characteristic $L$ is called *even* (resp., *odd*) if $h^0(L)$ is even (resp., $h^0(L)$ is odd).

The Riemann–Kempf singularity theorem that will be proven later implies (see Corollary 20.10) that

$$\text{mult}_L(\Theta) = h^0(L),$$

where $\Theta \subset J^{g-1}$ is the theta divisor. Combining this fact with the theory developed in Chapter 13 we derive the following result.

**Theorem 17.11.** *Assume that* char$(k) \neq 2$. *For a fixed theta characteristic L the map*

$$J_2 \to \pm 1 : \xi \mapsto \epsilon(L \otimes \xi)\epsilon(L)$$

*is a quadratic form on $J_2$ whose associated bilinear form is the Weil pairing $e_2$, where $J$ is identified with $\hat{J}$ by means of its principal polarization. The Arf-invariant of this form is equal to $(-1)^{h^0(L)}$.*

*Proof.* Let us use $L$ to identify $J^{g-1}$ with $J$, so that the theta divisor $\Theta \subset J^{g-1}$ gets identified with the symmetric divisor $\Theta_L \subset J$. By Proposition 13.3 the map $\xi \mapsto \epsilon(L \otimes \xi)$ coincides with the function $\epsilon_{\Theta_L}$ defined in Section 13.2. Now the first assertion follows from the equality (13.2.1) and Proposition 13.1. The statement about the Arf-invariant follows from Corollary 13.6. □

**Corollary 17.12.** *Assume that* char $(k) \neq 2$. *Then the number of even (resp. odd) theta characteristics is equal to* $2^{g-1}(2^g + 1)$ *(resp.,* $2^{g-1}(2^g - 1)$*).*

*Proof.* Let $L$ be a theta characteristic. According to Exercise 1(a) of Chapter 5, the number of $\xi \in J_2(k)$ with $\epsilon(L \otimes \xi)\epsilon(L) = 1$ is equal to $2^{g-1}(2^g + 1)$ if this quadratic form is even, and to $2^{g-1}(2^g - 1)$ otherwise. Since the Arf-invariant of this form is equal to $\epsilon(L)$, the assertion follows. □

## Exercises

1. Let $[V_0 \stackrel{d}{\to} V_1] \to [V_0' \stackrel{d'}{\to} V_1']$ be a quasi-isomorphism of complexes of vector bundles concentrated in degrees 0 and 1. Assume that rk $V_\bullet =$ rk $V_\bullet' = 0$. Prove that there is an isomorphism $\det^{-1} V_\bullet \widetilde{\to} \det^{-1} V_\bullet'$ mapping the section $\det(d)$ to $\det(d')$.
2. Let $a : C \to A$ be a nonconstant morphism from a curve to an abelian variety, $x \in A(k)$ be a point, $t_x : A \to A$ be a translation by $x$.
   (a) Check that $(t_x \circ a)_*(D) = a_*(D) + \deg(D) \cdot x$ for any $D \in \mathrm{Pic}(C)$.
   (b) Prove that one has $\phi_{t_x \circ a} = \phi_a$.
   (c) Show that $\phi_{-a} = \phi_a$, where $-a = [-1]_A \circ a$.
3. Let $a : C \to A$ be a nonconstant morphism from a curve to an abelian variety, let $J$ be the Jacobian of $C$, and let $a_0 : C \to \hat{J}$ be the standard morphism (denoted by $a$ in Section 17.3).
   (a) Let $\widehat{a^*} : \hat{J} \to A$ be the dual morphism to $a^* : \hat{A} \to J$. Show that $\widehat{a^*} \circ a_0 = a$. Thus, every morphism from $C$ to an abelian variety factors through $a_0$.

(b) Prove that $\widehat{a^*}$ induces the homomorphism $a_*|_J : J(k) \to A(k)$ introduced in Section 17.2 (upon the standard identification of $\hat{J}$ with $J$). Thus, the map $a_*$ is algebraic.

4. Let $f : C \to E$ be a morphism from a curve to an elliptic curve. Using the standard identification of $\hat{E}$ with $E$ we can consider the map $f^* : E \simeq \hat{E} \to J(C)$. Show that the composition of $f^*$ with $f_* : J(C) \to E$ coincides with the morphism $[\deg(f)]_E : E \to E$.
5. Prove that the natural morphism

$$\sigma^d : \mathrm{Sym}^d C \to J^d : D \mapsto \mathcal{O}_C(D)$$

for $d \le g$ is birational onto its image.

6. Show that the image of the morphism $\sigma^{g-1} : \mathrm{Sym}^{g-1} C \to J^{g-1}$ coincides with the theta divisor $\Theta \subset J^{g-1}$. [*Hint*: This image is reduced, so it suffices to prove that $\Theta$ is reduced. Now use the fact that $\Theta$ defines a principal polarization.]
7. Let $C$ be a curve of genus 2, $\tau : C \to C$ be a hyperelliptic involution, $p_1, \ldots, p_6 \in C$ be stable points of $\tau$. Assume that char$(k) \neq 2$.
   (a) Show that $\tau$ acts on $\mathrm{Pic}^0(C)$ as $L \mapsto L^{-1}$.
   (b) Show that $h^0(p_i + p_j) = 1$ for $i \neq j$.
   (c) Prove that the line bundles $(\mathcal{O}_C(p_i - p_j), 1 \le i < j \le 6)$ are precisely 15 non-trivial line bundles of order 2 on $C$.
   (d) Show that odd theta characteristics on $C$ are $(\mathcal{O}_C(p_i), i = 1, \ldots, 6)$.
   (e) Prove that even theta characteristics on $C$ are $(\mathcal{O}_C(p_i + p_j - p_6), 1 \le i < j \le 5)$.
   (f) Show that the complete linear system $|p_i + p_j + p_k|$ is base point free provided that the indices $i, j, k$ are distinct.

# 18

# Fay's Trisecant Identity

In this chapter we prove Fay's trisecant identity for theta functions of degree 1 on Jacobian $J$ of a curve $C$ (Theorem 18.6). A special case of this identity was considered in Section 17.4 as an illustration of some canonical isomorphisms related to the canonical biextension on $J \times J$. To prove the general identity, we study *Cauchy-Szegö kernels* $S(L, x, y)$ associated with a line bundle $L$ of degree $g - 1$ such that $h^0(L) = 0$ and a pair of distinct points $x, y \in C$. It is defined as the ratio $s(y)/\operatorname{Res}_x s$, where $s$ a nonzero rational section of $L$ with the only pole of order 1 at $x$. Using the residue theorem we derive a three-term identity for such kernels. On the other hand, using the results of the previous chapter, we express them in terms of theta functions on $J$. Then the above identity turns out to be equivalent to Fay's trisecant identity.

## 18.1. Cauchy-Szegö Kernel

Let $L$ be a line bundle of degree $g - 1$ on a curve $C$, such that $h^0(L) = 0$, and let $(x, y)$ be a pair of distinct points of $C$. Let $s$ be the unique (up to rescaling) nonzero section of the line bundle $L(x)$. The Cauchy-Szegö kernel is defined by the formula

$$S(L, x, y) = \frac{s(y)}{\operatorname{Res}_x s},$$

where $\operatorname{Res}_x s$ is an element of $L(x)|_x \simeq L_x \omega_x^{-1}$. Note that $S(L, x, y)$ is a well-defined element of the one-dimensional space $(\omega \otimes L^{-1})_x \otimes L_y$. One can easily globalize this construction to get a morphism

$$p_1^* L \to p_1^* \omega_C \otimes p_2^* L(\Delta_C)$$

of line bundles on $C \times C$ (where $p_1, p_2 : C \times C \to C$ are projections, $\Delta_C \subset C \times C$ is the diagonal) whose residue on the diagonal is equal to the identity.

Note that $S(L, x, y) = 0$ if and only if $h^0(L(x - y)) \neq 0$.

It is convenient to generalize this construction to some triples $(L, x, y)$ with $L$ not necessarily of degree $g-1$. Recall that a point $p$ is called a *base point* of a line bundle $M$ if the evaluation map $H^0(C, M) \to M|_p$ is zero.

**Definition.** We call a triple $(L, x, y)$ *good* if the following three conditions are satisfied:

1. $x \neq y$;
2. $x$ is a base point of $\omega_C L^{-1}$;
3. $y$ is a base point of $L$.

For a good triple $(L, x, y)$ one can still define $S(L, x, y) \in (\omega \otimes L^{-1})_x \otimes L_y$ by the same formula taking as $s$ any element of $H^0(C, L(x)) \setminus H^0(C, L)$. For example, one has the following easy result that follows immediately from the definition.

**Lemma 18.1.** *Let $x, y, z \in C$ be points such that $x \neq z$ and $x \neq y$. Assume that $g \geq 1$. Then the triple $(\mathcal{O}_C(z-x), x, y)$ is good and*

$$S(\mathcal{O}_C(z-x), x, y) = \begin{cases} 1, & y \neq z; \\ 0, & y = z. \end{cases}$$

## 18.2. Identity for Kernels

**Theorem 18.2.** *Let $x_1, \ldots, x_n$ be a collection of points, $L_1, \ldots, L_n$ be a collection of line bundles on $C$ such that $L_1 \otimes \ldots \otimes L_n \simeq \omega_C$. Let $\alpha \in H^0(C, \omega_C L_1^{-1} \ldots L_n^{-1})$ be a trivialization. Then one has*

$$\sum_{i=1}^{n} \alpha(x_i) \cdot \prod_{j \neq i} S(L_j, x_j, x_i) = 0$$

*provided that all triples $(L_j, x_j, x_i)$ for $j \neq i$ are good.*

*Proof.* For every $j = 1, \ldots, n$, let us choose $s_j \in H^0(C, L_j(x_j)) \setminus H^0(C, L_j)$. Then we can consider the product $s_1 \ldots s_n$ as a rational 1-form with poles of order 1 at $x_1, \ldots, x_n$ (note that all these points are distinct). Applying the residue theorem, we obtain

$$\sum_{i=1}^{n} \operatorname{Res}_{x_i}(s_i) \prod_{j \neq i} s_j(x_i) = 0.$$

Dividing by $\prod_i \operatorname{Res}_{x_i}(s_i)$ we get the result. □

We will use only the cases $n = 2$ and $n = 3$ of the above theorem. First, assume that $n = 2$. Then we obtain that for a pair of line bundles $L_1, L_2$, such that $L_1 \otimes L_2 \simeq \omega_C$, one has

$$\alpha(y) \cdot S(L_1, x, y) = -\alpha(x) \cdot S(L_2, y, x), \tag{18.2.1}$$

where $\alpha \in H^0(C, \omega_C L_1^{-1} L_2^{-1})$ is the corresponding trivialization. Note that by the definition, the triple $(L_1, x, y)$ is good if and only if $(L_2, y, x)$ is good.

Next, let us consider the case $n = 3$. Then Theorem 18.2 implies the following result.

**Corollary 18.3.** *Assume that $g \geq 1$. Let $L_1$ and $L_2$ be line bundles of degree $g - 1$ with $H^0(C, L_1) = H^0(C, L_2) = 0$, $z, t$ be points on $C$, such that $L_1 \otimes L_2 \simeq \omega_C(z-t)$. Let us fix a trivialization $\alpha \in H^0(C, \omega_C L_1^{-1} L_2^{-1}(z-t))$. Then for a pair of points $x, y \in C$ one has*

$$\begin{aligned} &\operatorname{Res}_z(\alpha) \cdot S(L_1, x, z) S(L_2, y, z) + \alpha(x) \cdot S(L_2, y, x) \\ &\quad + \alpha(y) \cdot S(L_1, x, y) = 0. \end{aligned}$$

*provided that the points $(x, y, z, t)$ are distinct.*

*Proof.* Apply Theorem 18.2 to the line bundles $L_1$, $L_2$ and $L_3 = \mathcal{O}_C(t - z)$ and the points $x, y, z$, and then use the equality $S(\mathcal{O}_C(t - z), z, y) = 1$ for $y \neq t$ (see Lemma 18.1). □

## 18.3. Expression in Terms of Theta-Functions

In this section we will express the Cauchy-Szegö kernels in terms of theta functions on the Jacobian $J = J(C)$. Then the identity of Corollary 18.3 will give us the trisecant identity.

Let $D_1$ and $D_2$ be a pair of effective divisors on $C$, such that $\omega_C \simeq \mathcal{O}_C(D_1 + D_2)$ and $h^0(D_1) = h^0(D_2) = 1$. Then there exists a unique (up to rescaling) non-zero global 1-form $\eta$ on $C$ with zeroes at $D_1 + D_2$. We want to calculate $S(\xi(D_2), x, y)$ where $\xi \in J$ is such that $h^0(\xi(D_2)) = 0$ and $x \neq y$ are points on $C$, in terms of the theta function $\theta_{D_1}$. We will do this under the additional assumption that $x$ is not contained in the support of $D_1 + D_2$ and $y$ is not contained in the support of $D_2$. In order to apply the definition, we have to construct a nonzero rational section $s$ of $\xi$ with $D_2 + x$ as the divisor of poles. Note that since $y$ is not contained in the support of $D_2$, we have $h^0(D_1 + y) = 1$. Therefore, $h^0(D_1 + y - x) = 0$, so $\theta_{D_1}(y - x) \neq 0$. Thus, we can set

$$s(t) = \frac{\theta_{D_1}(t - x - \xi)}{\theta_{D_1}(t - x)}.$$

We claim that $s$ is a rational section of a line bundle isomorphic to $\xi$ with $D_2 + x$ as the divisor of poles. Indeed, embedding $C$ into $J$ by $p \mapsto \mathcal{O}_C(p-x)$ and using Theorem 17.4 we get

$$\mathcal{O}_J(\Theta_{D_1-\xi})|_C \simeq \omega_C(x - D_1 + \xi),$$
$$\mathcal{O}_J(\Theta_{D_1})|_C \simeq \omega_C(x - D_1) \simeq \mathcal{O}_C(x + D_2).$$

It remains to note that $h^0(D_2 + x) = 1$ (this follows from the vanishing of $h^0(D_2 + x - y)$), hence the function $t \mapsto \theta_{D_1}(t - x)$ vanishes precisely on $x + D_2$. Now we have

$$S(\xi(D_2), x, y) = \frac{s(y)}{\operatorname{Res}_x(s)} = \frac{\theta_{D_1}(y - x - \xi)}{\theta_{D_1}(y - x)\theta_{D_1}(-\xi)\operatorname{Res}_{t=x}\frac{1}{\theta_{D_1}(t-x)}}.$$

Note that the residue appearing here is nonzero, since $x$ is a simple zero of $\theta_{D_1}(t - x)$ as $x \notin D_2$ (see Corollary 17.5). Using the natural identification of the cotangent space to $J$ at 0 with $H^0(C, \omega_C)$, we can consider the derivative $\theta'_{D_1}(0)$ as a global 1-form on $C$ (more precisely, it is a 1-form with values in the 1-dimensional vector space $\mathcal{O}_J(\Theta_{D_1})|_0$). Furthermore, the natural map $\Omega^1_J|_0 \to \omega_x$ induced by the embedding $C \to J : p \mapsto p - x$ coincides with the evaluation map $H^0(C, \omega_C) \to \omega_x$. It follows that $\theta'_{D_1}(0)(x) \neq 0$ and

$$\operatorname{Res}_{t=x}\frac{1}{\theta_{D_1}(t - x)} = \frac{1}{\theta'_{D_1}(0)(x)}.$$

Substituting this expression in the above formula for $S(\xi(D_2), x, y)$, we obtain the following result.

**Lemma 18.4.** *Under the canonical isomorphism*

$$\omega_x\xi_x^{-1}\xi_y \simeq \omega_x\langle y - x, -\xi\rangle^{-1}$$

*one has*

$$S(\xi(D_2), x, y) = \frac{\theta_{D_1}(y - x - \xi)\theta'_{D_1}(0)(x)}{\theta_{D_1}(y - x)\theta_{D_1}(-\xi)} \tag{18.3.1}$$

We need one more ingredient for the proof of the Fay's trisecant identity.

**Lemma 18.5.** *The global* 1*-form* $\theta'_{D_1}(0)$ *vanishes exactly on* $D_1 + D_2$.

*Proof.* Since $\Theta_{D_1} = -\Theta_{D_2}$, we can normalize $\theta_{D_2}$ in such a way that $\theta_{D_2}(\xi) = \theta_{D_1}(-\xi)$. Now the skew-symmetry condition (18.2.1)

$$\eta(y)S(\xi(D_2), x, y) = -\eta(x)S(\xi^{-1}(D_1), y, x)$$

combined with Lemma 18.4 gives

$$\eta(y)\theta'_{D_1}(0)(x) = -\eta(y)\theta'_{D_2}(0)(y) = \eta(y)\theta'_{D_1}(0)(y).$$

Therefore, $\theta'_{D_1}(0)$ is proportional to $\eta$. On the other hand, we have seen above that $\theta'_{D_1}(0)(x) \neq 0$ for generic $x$. □

Now we are ready to prove the Fay's trisecant identity.

**Theorem 18.6.** *Let* $\theta$ *be a theta function of degree* 1 *on* $J$ *such that* $\theta(0) = 0$. *Then one has*

$$\frac{\theta(x-t)\theta(y-z)}{\theta(x-z)\theta(y-t)} \cdot \theta(\xi)\theta(\xi + y - x + z - t) + \frac{\theta(z-t)\theta(y-x)}{\theta(z-x)\theta(y-t)}$$
$$\times\, \theta(\xi + z - x)\theta(\xi + y - t) = \theta(\xi + z - t)\theta(\xi + y - x),$$

*where* $x, y, z, t \in C$, $\xi \in J$.

*Proof.* With the above notation let us apply Corollary 18.3 to $L_1 = \xi^{-1}(D_1)$, $L_2 = \xi(D_2 + z - t)$, $x$ and $y$, where $z$ and $t$ are generic points on $C$. Then we have $\omega_C L_1^{-1} L_2^{-1}(z - t) \simeq \omega_C(-D_1 - D_2)$, so we get

$$\begin{aligned}&\eta(z)S(\xi^{-1}(D_1), x, z)S(\xi(D_2 + z - t), y, z)\\ &\quad+\eta(x)S(\xi(D_2 + z - t), y, x) + \eta(y)S(\xi^{-1}(D_1), x, y) = 0,\end{aligned} \tag{18.3.2}$$

where $\eta$ is a nonzero global 1-form vanishing on $D_1 + D_2$. According to Lemma 18.5, we can take $\eta = \theta'_{D_2}(0)$ (trivializing the fiber of $\mathcal{O}_J(D_2)$ at 0 we consider $\theta'_{D_2}(0)$ as a global 1-form on $C$). Now we apply Lemma 18.4 to express all terms of this equality via $\theta_{D_1}$ and $\theta_{D_2}$. Furthermore, using the equality $\theta_{D_1} = [-1]^*\theta_{D_2}$ we can express everything in terms of $\theta = \theta_{D_2}$. Thus, denoting the three terms in the LHS of equation (18.3.2) by $T_1$, $T_2$ and

$T_3$, we have:

$$T_1 = -\theta'(0)(z)\cdot\frac{\theta(z-x+\xi)\theta'(0)(x)}{\theta(z-x)\theta(\xi)}\cdot\frac{\theta(y-t+\xi)\theta'(0)(y)}{\theta(y-z)\theta(z-t+\xi)}$$
$$\in \omega_z\otimes\omega_x\otimes\omega_y\otimes\langle z-x,\xi\rangle^{-1}\otimes\langle y-z,z-t+\xi\rangle^{-1},$$
$$T_2 = -\theta'(0)(x)\cdot\frac{\theta(y-x+z-t+\xi)\theta'(0)(y)}{\theta(y-x)\theta(z-t+\xi)}$$
$$\in \omega_x\otimes\omega_y\otimes\langle y-x,z-t+\xi\rangle^{-1},$$
$$T_3 = \theta'(0)(y)\cdot\frac{\theta(y-x+\xi)\theta'(0)(x)}{\theta(y-x)\theta(\xi)}\in\omega_y\otimes\omega_x\otimes\langle y-x,\xi\rangle^{-1}.$$

We see that in order to get three elements in the same space, we have to multiply the first term by the canonical section of $\omega_z^{-1}\otimes\langle y-z,z-t\rangle$ and the second term by the canonical section of $\langle y-x,z-t\rangle$. As we have seen in Section 17.4, for $x$, $y$ distinct from $z$, $t$, the section

$$\phi(x,y) = \frac{\theta(x-t)\theta(y-z)}{\theta(x-z)\theta(y-t)}$$

gives the canonical trivialization of $\langle y-x,z-t\rangle$. On the other hand, the residue of $\phi$ with respect to $x$ at $x=z$ (for $y\neq t$) gives the canonical trivialization of $\omega_z^{-1}\langle y-z,z-t\rangle$ and is equal to

$$\frac{\theta(z-t)\theta(y-z)}{\theta'(0)(z)\theta(y-t)}.$$

Thus, we obtain the equality

$$\theta'(0)(z)\cdot\frac{\theta(z-x+\xi)\theta'(0)(x)}{\theta(z-x)\theta(\xi)}\cdot\frac{\theta(y-t+\xi)\theta'(0)(y)}{\theta(y-z)\theta(z-t+\xi)}\cdot\frac{\theta(z-t)\theta(y-z)}{\theta'(0)(z)\theta(y-t)}$$
$$+\,\theta'(0)(x)\cdot\frac{\theta(y-x+z-t+\xi)\theta'(0)(y)}{\theta(y-x)\theta(z-t+\xi)}\cdot\frac{\theta(x-t)\theta(y-z)}{\theta(x-z)\theta(y-t)}$$
$$-\,\theta'(0)(y)\cdot\frac{\theta(y-x+\xi)\theta'(0)(x)}{\theta(y-x)\theta(\xi)}=0.$$

Making the obvious cancellations and multiplying by $\theta(\xi)\theta(z-t+\xi)\theta(y-x)$ we get the result. □

**Remarks.** 1. In our formulation the three terms of the trisecant identity are sections of three line bundles that are canonically isomorphic. More precisely, the isomorphism between these line bundles is a direct consequence of the theorem of the cube. When the ground field is $\mathbb{C}$, we can choose an isomorphism of $\mathcal{O}_J(\Theta)$ with the line bundle of the form $L(H,\alpha)$ (see Chapter 1), where

$H$ is the Hermitian form on $V = H^1(C, \mathcal{O}_C)$ corresponding to the principal polarization of $J$. Then $\theta$ will correspond to a theta function on $V$. Since the natural trivialization of the pull-back of $L(H, \alpha)$ to $V$ respects the isomorphism of the theorem of the cube, it follows that the identity of Theorem 18.6 holds for theta functions on $V$ (where $\xi$, $i(x)$, $i(y)$, $i(z)$ and $i(t)$ are lifted to points of $V$).

2. The reason for the name "trisecant identity" comes from the relation to trisecants on the Kummer variety of $J$ (=quotient of $J$ by the involution $[-1]_J$) embedded into the projective space by second-order theta functions (see, e.g., [39]).

## Exercises

1. Check that in the case of elliptic curve the trisecant identity is equivalent to the identity of Exercise 6(c) in Chapter 12.
2. Prove that for every theta function $\theta$ of degree 1 on $J$ such that $\theta(0) = 0$, one has

$$\theta(x - y)\theta(y - z)\theta(z - x) = -\theta(y - x)\theta(z - y)\theta(x - z),$$

where $x, y, z \in C$ (this identity appears from the change of variables $\xi' = \xi + z - x$ in the trisecant identity). [*Hint*: Study the limit of the identity of Corollary 17.8 as $z \to y$ for an odd theta function $\theta$.]

# 19

# More on Symmetric Powers of a Curve

In this chapter we prove various results about the geometry of the varieties $\mathrm{Sym}^d C$, where $C$ is a curve. Some of them will be used in the proof of Torelli theorem in Chapter 21 and some are of independent interest. One approach to studying the varieties $\mathrm{Sym}^d C$ is to use the fact that for sufficiently large $d$ the morphism $\mathrm{Sym}^d C \to J = J(C)$ sending $D$ to $\mathcal{O}_C(D - dp)$, where $p \in C$ is a fixed point, is a projective bundle. On the other hand, for every $d$ we can consider $\mathrm{Sym}^{d-1} C$ as a divisor in $\mathrm{Sym}^d$ via the map $D \mapsto D + p$. This approach allows us to prove that for $d \geq 2$ the Picard group of $\mathrm{Sym}^d C$ is isomorphic (noncanonically) to $\mathrm{Pic}(J) \oplus \mathbb{Z}$. Indeed, for sufficiently large $d$ this is clear, while the relation between $\mathrm{Pic}(\mathrm{Sym}^{d-1} C)$ and $\mathrm{Pic}(\mathrm{Sym}^d C)$ can be studied using the Lefschetz theorem for Picard groups. In order to apply this theorem one has to prove that some cohomology groups of natural line bundles on $\mathrm{Sym}^d C$ vanish. We establish a general result of this kind stating that $H^i(\mathrm{Sym}^d C, V^{(d)}) = 0$ for $i > 0$ (resp., $i < d$) where $V^{(d)}$ is the symmetric power of a vector bundle $V$ on $C$ such that $H^1(C, V) = 0$ (resp., $H^0(C, V) = 0$).

In Section 19.5 we express the Chern classes of the vector bundle $E_d$ on $J$, such that $\mathbb{P}E_d$ is isomorphic to $\mathrm{Sym}^d C$ (for sufficiently large $d$) in terms of some natural cycles on $J$. This leads to an interesting relation in the Chow group of $J$ due to Mattuck (see equation (19.5.3)).

Throughout this chapter we fix a point $p \in C$.

### 19.1. Some Natural Divisors on $\mathrm{Sym}^d C$

**Proposition 19.1.** *The normal bundle to the closed embedding*

$$C \times \mathrm{Sym}^{d-1} \simeq \mathcal{D}_d \hookrightarrow C \times \mathrm{Sym}^d C$$

*is isomorphic to* $p_1^* \omega_C^{-1}(\mathcal{D}_{d-1})$, *where* $p_1 : C \times \mathrm{Sym}^{d-1} \to C$ *is the projection.*

*Proof.* We start by considering the closed embedding

$$f : C \times C \times \mathrm{Sym}^{d-1} C \to C \times C \times \mathrm{Sym}^d C : (p_1, p_2, D) \\ \mapsto (p_1, p_2, p_1 + D).$$

Let us denote by $p_{ij}$ the projections to products of 2 factors from the triple Cartesian products. Let $\mathcal{D}_d^{23} \subset C \times C \times \mathrm{Sym}^d C$ be the pull-back of the universal divisor under the projection $p_{23}$. Then it is easy to see that

$$f^*\mathcal{O}(\mathcal{D}_d^{23}) \simeq \mathcal{O}(\mathcal{D}_{d-1}^{23} + \Delta_{12}),$$

where $\Delta_{12} \subset C \times C \times \mathrm{Sym}^{d-1} C$ is the pull-back of the diagonal under the projection $p_{12}$. Making the base change $\Delta \times \mathrm{id} : C \times \mathrm{Sym}^d C \to C \times C \times \mathrm{Sym}^d C$, induced by the diagonal embedding of $C$, we immediately get the required formula. □

Making the base change of the closed embedding considered in the above proposition by the natural morphism $\{p\} \times \mathrm{Sym}^d C \hookrightarrow C \times \mathrm{Sym}^d C$, we get a closed embedding

$$s_p = s_p^d : \mathrm{Sym}^{d-1} C \to \mathrm{Sym}^d C : D \mapsto p + D$$

with the image $R_p^d := \mathcal{D}_d \cap p \times \mathrm{Sym}^d C$ (the intersection is taken inside $C \times \mathrm{Sym}^d C$). Proposition 19.1 implies that the normal bundle to this embedding is isomorphic to $\mathcal{O}_{\mathrm{Sym}^{d-1} C}(R_p^{d-1})$. Thus, we have

$$s_p^*\mathcal{O}(R_p^d) \simeq \mathcal{O}(R_p^{d-1}). \tag{19.1.1}$$

## 19.2. Morphisms to the Jacobian

For every $d \geq 1$ we have a natural morphism $\sigma^d : \mathrm{Sym}^d C \to J^d$ sending $D$ to the isomorphism class of $\mathcal{O}_C(D)$. Let us identify $J^d$ with $J$ using the line bundle $\mathcal{O}_C(dp)$. Then we can consider $\sigma^d$ as a morphism $\mathrm{Sym}^d C \to J$ sending $D$ to $\mathcal{O}_C(D - dp)$. More precisely, $\sigma^d$ corresponds to the following family of line bundles on $C$ trivialized at $p$, parametrized by $\mathrm{Sym}^d C$:

$$\mathcal{O}_{C \times \mathrm{Sym}^d C}(\mathcal{D}_d - d(p \times \mathrm{Sym}^d C) - C \times R_p^d).$$

The fiber of $\sigma^d$ over $L \in J^d$ is the variety of effective divisors $D$ such that $\mathcal{O}_C(D) \simeq L$. Thus, $(\sigma^d)^{-1}(L)$ can be identified with the projective space $\mathbb{P}H^0(C, L)$. In the following proposition we compute the tangent map to $\sigma^d$ at the point of $\mathrm{Sym}^d C$ corresponding to an effective divisor $D \subset C$.

**Proposition 19.2.** *The tangent space to* $\mathrm{Sym}^d C$ *at a point corresponding to an effective divisor* $D \subset C$ *is canonically isomorphic to* $H^0(C, \mathcal{O}_D(D))$. *The tangent map to the morphism* $\sigma^d : \mathrm{Sym}^d C \to J$ *is given by the coboundary homomorphism* $H^0(C, \mathcal{O}_D(D)) \to H^1(C, \mathcal{O}_C)$ *coming from the exact sequence* $0 \to \mathcal{O}_C \to \mathcal{O}_C(D) \to \mathcal{O}_D(D) \to 0$.

*Proof.* Set $C[\epsilon] = C \times \mathrm{Spec}(k[\epsilon]/\epsilon^2)$. According to Theorem 16.4, the tangent space $T_D \mathrm{Sym}^d C$ coincides with the set of effective Cartier divisors $\widetilde{D} \subset C[\epsilon]$ such that $\widetilde{D} \cap C = D$. Let $f_\alpha + g_\alpha \epsilon$ be local equations of $\widetilde{D}$ on $U_\alpha[\epsilon]$, where $(U_\alpha)$ is some open affine covering of $C$. Then $c_\alpha := g_\alpha / f_\alpha$ can be considered as a section of $\mathcal{O}_C(D)$ on $U_\alpha$. If we change local equation by an invertible function, $c_\alpha$ will change to $c_\alpha + u_\alpha$ for some regular function $u_\alpha$ on $U_\alpha$. Hence, $c_\alpha \bmod \mathcal{O}$ glue into a global section of $\mathcal{O}_C(D)/\mathcal{O}_C$. This gives the required isomorphism. The second assertion is an immediate consequence of this construction. □

**Remark.** In the case when $D$ consists of $d$ distinct points the above form of the tangent map to $\sigma^d$ at $D$ can be also deduced from Lemma 17.9.

**Corollary 19.3.** *Assume that* $d \leq g$. *Then the morphism* $\mathrm{Sym}^d C \to \sigma^d(\mathrm{Sym}^d C)$ *is an isomorphism over the (nonempty) open subset of* $\sigma^d(\mathrm{Sym}^d C)$ *consisting of* $L$ *with* $h^0(L) = 1$.

*Proof.* If $h^0(D) = 1$ then from the above proposition we get injectivity of the tangent map to $\sigma^d$ at $D$. □

On the other hand, using Proposition 19.2, one can easily check that for $d > 2g - 2$ the morphism $\sigma^d$ is a projective bundle over $J$. We are going to identify the corresponding vector bundle on $J$. Let $\mathcal{P}_C$ be the Poincaré bundle on $C \times J$ normalized at $p$. For $d > 2g - 2$ we denote by $E_d$ the following bundle on $J$:

$$E_d = p_{2*}(\mathcal{P}_C(dp \times J)).$$

In other words, $E_d = \mathcal{S}(\mathcal{O}_C(dp))$. By the base change formula, we have

$$\sigma^{d*} E_d \simeq p_{2*}(\mathcal{O}_{C \times \mathrm{Sym}^d C}(\mathcal{D}_d))(-R_p^d).$$

Thus, we have a natural inclusion

$$\mathcal{O}_{\mathrm{Sym}^d C}\big(-R_p^d\big) \to \sigma^{d*} E_d$$

as a subbundle. Hence, there is a morphism $v_d : \mathrm{Sym}^d C \to \mathbb{P}(E_d)$ of $J$-schemes, such that $v_d^* \mathcal{O}(1) \simeq \mathcal{O}_{\mathrm{Sym}^d C}(R_p^d)$. It is easy to see that $v_d$ is in fact an isomorphism.

**Remark.** One can rephrase the above construction in more invariant terms. Namely, let $L$ be a line bundle of degree $d > 2g-2$ on $C$. Then the morphism $\sigma_L : \mathrm{Sym}^d C \to J$ sending $D$ to $\mathcal{O}_C(D) \otimes L^{-1}$ can be identified with the projective bundle associated with $\mathcal{S}(L)$, the Fourier transform of $L$.

The natural embedding $\mathcal{O}_C((d-1)p) \to \mathcal{O}_C(dp)$ induces a morphism $k_d : E_{d-1} \to E_d$, which identifies $E_{d-1}$ with a subbundle in $E_d$. Furthermore, the following diagram is commutative:

$$\begin{array}{ccc} \mathrm{Sym}^{d-1} C & \xrightarrow{\ s_p\ } & \mathrm{Sym}^d C \\ \Big\downarrow{\scriptstyle v_{d-1}} & & \Big\downarrow{\scriptstyle v_d} \\ \mathbb{P}(E_{d-1}) & \xrightarrow{\ k_d\ } & \mathbb{P}(E_d) \end{array} \qquad (19.2.1)$$

The relation $k_d^* \mathcal{O}(1) \simeq \mathcal{O}(1)$ is equivalent to the isomorphism $s_p^* \mathcal{O}(R_p^d) \simeq \mathcal{O}(R_p^{d-1})$ proved in Section 19.1.

## 19.3. Symmetric Powers of Vector Bundles

Let $\mathcal{F}$ be a coherent sheaf on $C$. Then the coherent sheaf $\mathcal{F} \boxtimes \cdots \boxtimes \mathcal{F}$ ($d$ times) on $C^d$ has a natural action of $S_d$. Hence, it descends to a coherent sheaf on $\mathrm{Sym}^d C$ which we denote $\mathcal{F}^{(d)}$. The following theorem gives some information about cohomology of this sheaf in the case when $\mathcal{F}$ is a vector bundle.

**Theorem 19.4.** *Let $V$ be a vector bundle on $C$.*
*1. If $H^1(C, V) = 0$ then $H^i(\mathrm{Sym}^d C, V^{(d)}) = 0$ for all $i > 0$.*
*2. If $H^0(C, V) = 0$ then $H^i(\mathrm{Sym}^d C, V^{(d)}) = 0$ for all $i < d$.*

We want to reduce the proof to the case of $C = \mathbb{P}^1$. For this we need two preparatory lemmas. Recall that for every $d_1, d_2 \geq 0$ we have a natural map

$$s_{d_1,d_2} : \mathrm{Sym}^{d_1} C \times \mathrm{Sym}^{d_2} C \to \mathrm{Sym}^{d_1+d_2} C.$$

**Lemma 19.5.** *For every pair of coherent sheaves $\mathcal{F}_1$ and $\mathcal{F}_2$ on $C$ one has a natural isomorphism of vector bundles on* $\mathrm{Sym}^d C$*:*

$$(\mathcal{F}_1 \oplus \mathcal{F}_2)^{(d)} \widetilde{\rightarrow} \oplus_{d_1+d_2=d}\, s_{d_1,d_2,*}\big(\mathcal{F}_1^{(d_1)} \boxtimes \mathcal{F}_2^{(d_2)}\big). \tag{19.3.1}$$

*Proof.* The sheaf $(\mathcal{F}_1 \oplus \mathcal{F}_2)^{(d)}$ is obtained by descent from the sheaf $(\mathcal{F}_1 \oplus \mathcal{F}_2)^{\boxtimes d}$ on $C^d$. Now we have an $S_d$-equivariant decomposition

$$(\mathcal{F}_1 \oplus \mathcal{F}_2)^{\boxtimes d} \simeq \oplus_{d_1+d_2=d}\mathcal{E}_{d_1,d_2},$$

where $\mathcal{E}_{d_1,d_2}$ is the direct sum of all the products $\mathcal{F}_{i_1} \boxtimes \cdots \boxtimes \mathcal{F}_{i_d}$ on $C^d$ with $d_1$ factors $\mathcal{F}_1$ and $d_2$ factors $\mathcal{F}_2$. It remains to check that the pull-back of $s_{d_1,d_2,*}(\mathcal{F}_1^{(d_1)} \boxtimes \mathcal{F}_2^{(d_2)})$ to $C^d$ is isomorphic to $\mathcal{E}_{d_1,d_2}$ as a sheaf with $S_d$-action. To compute this pull-back one can use the flat base change ([61], Proposition 5.6) and the fact that the fibered product of $\mathrm{Sym}^{d_1} C \times \mathrm{Sym}^{d_2} C$ with $C^d$ over $\mathrm{Sym}^d C$ is the disjoint union of $S_d/(S_{d_1} \times S_{d_2})$ copies of $C^d$ (see Exercise 1 of Chapter 9). This leads to the required isomorphism. □

**Lemma 19.6.** *Let $f: C \to C'$ be a finite morphism of curves, $f^{(d)}$:* $\mathrm{Sym}^d C \to \mathrm{Sym}^d C'$ *be the induced morphism of symmetric powers. Then for every vector bundle $V$ on $C$ there is a natural isomorphism on* $\mathrm{Sym}^d C'$*:*

$$(f_*V)^{(d)} \widetilde{\rightarrow} f_*^{(d)}\big(V^{(d)}\big). \tag{19.3.2}$$

*Proof.* It is easy to see that the construction $V \to V^{(d)}$ commutes with pull-backs under finite morphisms. Therefore, we have

$$\big(f^{(d)}\big)^*(f_*V)^{(d)} \simeq (f^*f_*V)^{(d)}.$$

The natural map $f^*f_*V \to V$ induces the map $(f^*f_*V)^{(d)} \to V^{(d)}$, hence, by adjunction we get the map (19.3.2). The fact that it is an isomorphism can be proven locally, so we can assume that $C$ and $C'$ are affine, and that $f_*\mathcal{O}_C$ and $V$ are trivial bundles. Let $A$ and $A'$ be the rings of functions on $C$ and $C'$. Then $V$ corresponds to a free $A$-module of finite rank $M$. The ring of functions on $\mathrm{Sym}^d C$ (resp. $\mathrm{Sym}^d C'$) is $TS^d(A)$ (resp., $TS^d(A')$). The global sections of $V^{(d)}$ is the $A$-module $TS^d(M)$, same as the global sections of $(f_*V)^{(d)}$. □

*Proof of Theorem 19.4.* Let $f : C \to \mathbb{P}^1$ be a finite morphism. Then applying isomorphism (19.3.2) we get

$$H^i\big(\operatorname{Sym}^d C, V^{(d)}\big) \simeq H^i\big(\operatorname{Sym}^d \mathbb{P}^1, (f_* V)^{(d)}\big).$$

Therefore, it suffices to prove the theorem in the case $C = \mathbb{P}^1$. Since every vector bundle on $\mathbb{P}^1$ is a direct sum of line bundles, using (19.3.1) we can reduce the proof to the case of a line bundle on $\mathbb{P}^1$. It remains to use the fact that under the natural isomorphism $\operatorname{Sym}^d \mathbb{P}^1 \simeq \mathbb{P}^d$, the line bundle $(\mathcal{O}_{\mathbb{P}^1}(n))^{(d)}$ corresponds to $\mathcal{O}_{\mathbb{P}^d}(n)$ (see Exercise 3). □

## 19.4. Picard Groups

Now we can compute Picard groups of the symmetric powers of $C$.

**Theorem 19.7.** *For $d \geq 2$ there is an exact sequence of abelian groups*

$$0 \longrightarrow \operatorname{Pic}(J) \xrightarrow{\sigma^{d*}} \operatorname{Pic}(\operatorname{Sym}^d C) \xrightarrow{\deg} \mathbb{Z} \longrightarrow 0,$$

*where the homomorphism* deg *is normalized by the condition that* $\deg(\mathcal{O}(R^d_x)) = 1$ *for every $x \in C$. If $\mathbb{P} \subset \operatorname{Sym}^d C$ is a complete linear system of positive dimension, then* $\deg(L) = \deg(L|_{\mathbb{P}})$.

*Proof.* For $d$ sufficiently large, the statement is true since $\operatorname{Sym}^d C$ is a projective bundle over $J$. Let $\pi_d : C^d \to \operatorname{Sym}^d C$ be the canonical projection. Then we have $\pi_d^* \mathcal{O}(R^d_p) \simeq \mathcal{O}_C(p) \boxtimes \cdots \boxtimes \mathcal{O}_C(p)$. It follows that $\mathcal{O}(R^d_p)$ is ample. On the other hand, it is easy to see that

$$\mathcal{O}_{\operatorname{Sym}^d C}\big(R^d_p\big) \simeq (\mathcal{O}_C(p))^{(d)}.$$

Hence, applying Theorem 19.4 we get $H^i(\operatorname{Sym}^d C, \mathcal{O}(-nR^d_p)) = 0$ for $n > 0$, provided that $i \leq 1$, $d \geq 2$ or $i \leq 2$, $d \geq 3$. Therefore, we can apply the Lefschetz theorem for Picard groups (see Appendix C) to conclude that the map $s_p^{d*} : \operatorname{Pic}(\operatorname{Sym}^d C) \to \operatorname{Pic}(\operatorname{Sym}^{d-1} C)$ is an isomorphism for $d > 3$ and is an embedding for $d = 3$. It remains to check that $s_p^{3*}$ is surjective. Let us denote by $K \subset \operatorname{Pic}(C \times C)$ the subgroup of line bundles $L$, such that the restrictions $L|_{p \times C}$ and $L|_{C \times p}$ are trivial. Then there is an isomorphism $u : \operatorname{End}(J) \to K : \phi \mapsto (i_p \times \phi i_p)^* \mathcal{P}$, where $i_p : C \to J$ is the embedding corresponding to $p$, $\mathcal{P}$ is the Poincaré line bundle on $J \times J$. Let $\operatorname{Pic}^+(C \times C)$ be the subgroup of line bundles stable under the involution $(p_1, p_2) \mapsto (p_2, p_1)$, and let $K^+ = K \cap \operatorname{Pic}^+(C \times C)$. Then $u^{-1}$ identifies $K^+$ with the subgroup $\operatorname{End}^+(J) \subset \operatorname{End}(J)$ consisting of all symmetric endomorphisms of $J$. Let $r : \operatorname{Pic}(C \times C) \to K$

be the homomorphism given by $r(F) = F \otimes [(F^{-1}|_{C\times p}) \boxtimes (F^{-1}|_{p\times C})]$. Then $r(\mathrm{Pic}^+(C \times C)) = K^+$ and we have the commutative diagram

$$\begin{array}{ccccc} \mathrm{Pic}(J) & \xrightarrow{\sigma^{2*}} & \mathrm{Pic}(\mathrm{Sym}^2 C) & \xrightarrow{\pi_2^*} & \mathrm{Pic}(C \times C)^+ \\ \downarrow s & & & & \downarrow r \\ \mathrm{End}^+(J) & & \xrightarrow{u} & & K^+ \end{array} \tag{19.4.1}$$

where $s : \mathrm{Pic}(J) \to \mathrm{End}^+(J)$ is the homomorphism $L \mapsto \phi_L$ (recall that $\phi_L(a) = t_a^* L \otimes L^{-1}$). Since $s$ is surjective by Theorem 13.7, it follows that the composition $r \circ \pi_2^* \circ \sigma_2^*$ is surjective. Thus, in proving that some line bundle $L \in \mathrm{Pic}(\mathrm{Sym}^2 C)$ comes from $\mathrm{Pic}(\mathrm{Sym}^3 C)$, we are reduced to the case when $\pi_2^* L$ belongs to the kernel of $r$. In other words, we can assume that $\pi_2^* L \simeq L_1 \boxtimes L_1$ for some line bundle $L_1$ on $C$. The $S_2$-action on $L_1 \boxtimes L_1$ either coincides with the standard one, or differs from it by $-1$. In accordance with this dichotomy we equip $\widetilde{L} = L_1 \boxtimes L_1 \boxtimes L_1$ either with the standard $S_3$-action or with the standard action twisted by the sign character. Then if we consider $\widetilde{L}$ as a line bundle on $\mathrm{Sym}^3 C$, we will have $s_x^{3*} \widetilde{L} = L$. □

**Remark.** There is a conjecture that the restriction by $s_p^d$ induces an isomorphism of Chow groups of codimension $i$, provided that $d \geq 2i + 1$ (see [29]). The above theorem deals with the case $i = 1$.

## 19.5. Chern Classes

For every smooth projective variety $X$ we denote by $\mathrm{CH}^*(X)$ the Chow ring of $X$ (graded by codimension). Iterative application of (19.1.1) shows that upon the identification of $\mathbb{P}(E_d)$ with $\mathrm{Sym}^d C$ (where $d$ is sufficiently large), the class $c_1(\mathcal{O}(1))^i \in \mathrm{CH}^i(\mathrm{Sym}^d C)$ is represented by the cycle $\mathrm{Sym}^{d-i} C \subset \mathrm{Sym}^d C$. Let us consider the cycles $w_i = \sigma_*^{g-i}[\mathrm{Sym}^{g-i} C] \subset J$, $0 \leq i \leq g$. Note that $w_0 = 1$, while for $i > 0$, $w_i$ is the class of the subvariety $\sigma^{g-i}(\mathrm{Sym}^{g-i} C) \subset J$ (this follows from Exercise 5 of Chapter 17). Then we can express Chern classes of $E_d$ in terms of $w_i$'s. Indeed, for every vector bundle $E$ of rank $r + 1$ over a variety $X$ one has the following formula for Segre classes[7] of $E$:

$$s_i(E) = \pi_*(c_1(\mathcal{O}(1))^{r+i}),$$

[7] Recall that the Segre polynomial $s(E) = \sum s_i(E) t^i$ is the inverse of the Chern polynomial $c(E) = \sum c_i(E) t^i$.

where $\pi : \mathbb{P}(E) \to X$ is the associated projective bundle. In our situation $r = d - g$, so we deduce that $s_i(E_d) = w_i$. In particular, Chern classes of $E_d$ do not depend on $d$ (which is always assumed to be sufficiently large).

Let $\delta : J \to J$ be the involution sending $L$ to $\omega_C(-(2g-2)p) \otimes L^{-1}$. We claim that for every $d > 2g - 2$ there is an exact sequence

$$0 \to E_d \to F \to \delta^* E_d^\vee \to 0, \tag{19.5.1}$$

where $F$ is a successive extension of trivial line bundles. Indeed, we have

$$\delta^* E_d \simeq p_{2*}\big(\mathcal{P}_C^{-1} \otimes p_1^* \omega_C((d - 2g + 2)p)\big).$$

Therefore, by relative duality (see [61]) in the fibers of $p_2 : C \times J \to J$ we get

$$\delta^* E_d^\vee \simeq R^1 p_{2*}((\mathcal{P}_C(2g - 2 - d)p)).$$

Now applying the functor $Rp_{2*}$ to the exact sequence

$$0 \to \mathcal{P}_C((2g - 2 - d)p) \to \mathcal{P}_C(dp) \to \mathcal{P}_C(dp)|_{(2d-2g+2)p} \to 0$$

we get the sequence (19.5.1) with $F = p_{2*}(\mathcal{P}_C \otimes p_1^*(\mathcal{O}_C(dp)|_{(2d-2g+2)p}))$. This bundle is a successive extension of trivial line bundles, since $\mathcal{P}_C|_{p \times J}$ is trivial. Because $c(F) = 1$ we deduce from (19.5.1) the following relation between characteristic classes:

$$c(E_d) \cdot c\big(\delta^* E_d^\vee\big) = 1. \tag{19.5.2}$$

In other words, $c(\delta^* E_d^\vee) = s(E_d)$. It follows that $(-1)^i c_i(\delta^* E_d) = s_i(E_d) = w_i$. Hence,

$$c_i(E_d) = (-1)^i \delta^* w_i.$$

Note also that (19.5.2) gives the following quadratic identity in the Chow ring of $J$:

$$\left(\sum_{i \geq 0} (-1)^i \delta^* w_i\right) \left(\sum_{i \geq 0} w_i\right) = 1, \tag{19.5.3}$$

which implies that the subring of $\mathrm{CH}^*(J)$ generated by all the cycles $w_i$, is stable under the involution $\delta^*$. For example, we have

$$\delta^* w_1 = w_1, \quad \delta^* w_2 = w_1^2 - w_2, \quad \delta^* w_3 = w_1^3 - 2w_1 w_2 + w_3, \ldots$$

## Exercises

1. (a) Prove that there is a canonical isomorphism

$$s_{1,d-1}^{*}\omega_{\mathrm{Sym}^d C} \simeq \omega_{C\times \mathrm{Sym}^{d-1} C}(-\mathcal{D}_{d-1}).$$

(b) Let $\pi_d : C^d \to \mathrm{Sym}^d C$ be the natural projection. Show that

$$\pi_d^{*}\omega_{\mathrm{Sym}^d C} \simeq \omega_{C^d}\left(-\sum_{i<j}\Delta_{ij}\right),$$

where $\Delta_{ij}$ is the pull-back of the diagonal in $C \times C$ under the projection $p_{ij} : C^d \to C^2$.

(c) Prove that for every two points $p, p' \in C$ one has

$$s_{p'}^{*}\mathcal{O}(R_p^d) \simeq \mathcal{O}(R_p^{d-1}). \tag{19.5.4}$$

2. (a) Let $R_1, \ldots, R_d$ be a collection of mutually disjoint finite subsets of $C$. Show that the complement to the open subset

$$\cup_{i=1}^{d} \mathrm{Sym}^d(C - R_i)$$

in $\mathrm{Sym}^d C$ coincides with the image of $R_1 \times \cdots \times R_d$ under the natural projection $C^d \to \mathrm{Sym}^d C$. In particular, it consists of a finite number of points.

(b) Prove that $\mathrm{Sym}^d C$ can be covered by $d+1$ open affine subsets of the form $\mathrm{Sym}^d(C - S)$, where $S \subset C$ is finite.

3. Prove that $\mathrm{Sym}^d \mathbb{P}^1$ is isomorphic to $\mathbb{P}^d$. Show that under this isomorphism the line bundle $(\mathcal{O}_{\mathbb{P}^1}(1))^{(d)}$ on $\mathrm{Sym}^d \mathbb{P}^1$ corresponds to $\mathcal{O}_{\mathbb{P}^d}(1)$.

4. Let $E$ be an elliptic curve. Consider the addition morphism

$$\mathrm{Sym}^d E \to E : (p_1, \ldots, p_d) \mapsto p_1 + \cdots + p_d.$$

Prove that $\mathrm{Sym}^d E$ is isomorphic over $E$ to the projectivization of the vector bundle $\mathcal{S}(\mathcal{O}_E(de))$ on $E$, where $e \in E$ is the neutral element, $\mathcal{S}$ is the Fourier–Mukai transform. Deduce that $\mathrm{Sym}^d E \simeq \mathbb{P}(V)$ where $V$ is the unique (up to tensoring with line bundles) stable bundle on $E$ of rank $d$ and degree $-1$.

5. This exercise gives an alternative proof of the cohomology vanishing used in the proof of Theorem 19.7. Let $X$ be a quasi-projective variety equipped with an action of the finite group $G$, $Y = X/G$ be the quotient, $\pi : X \to Y$ be the natural projection. Let $L$ be a line bundle on $Y$.

(a) Construct a complex $V_\bullet = (V_0 \to V_1 \to \cdots)$ of $G$-modules computing the cohomology of $\pi^* L$ on $X$, such that the complex of $G$-invariants $(V_\bullet)^G$ computes the cohomology of $L$. [*Hint*: Use the

Cech cohomology; start with an open affine covering $(U_\alpha)$ of $Y$, such that for every $\alpha$ the restriction of $L$ to $U_\alpha$ is trivial.]

(b) Show that $H^0(Y, L) \simeq H^0(X, \pi^* L)^G$.

(c) Using two spectral sequences computing group cohomology of the complex $V_\bullet$, show that if $H^0(X, \pi^* L) = H^1(X, \pi^* L) = 0$, then $H^1(Y, L) = 0$.

(d) Assume that $X = C^d$, $G = S_d$, $Y = \mathrm{Sym}^d C$, $L$ is of the form $M^{(d)}$ for some line bundle $M$ on $C$. Show that if the characteristic of the ground field is different from 2 and $H^i(C^d, M^{\boxtimes d}) = 0$ for $i \leq 2$, then $H^i(\mathrm{Sym}^d C, M^{(d)}) = 0$ for $i \leq 2$. [*Hint*: Use the open affine covering as above with all $U_\alpha$ of the form $\mathrm{Sym}^d(C - S)$ such that $M|_{C-S}$ is trivial; then use the spectral sequence and the fact that for any vector space $V$ over a field of characteristic $\neq 2$, the cohomology $H^1(S_d, V^{\otimes d})$ vanishes.]

6. Prove that $\deg(\mathcal{O}_{\mathrm{Sym}^2 C}(\Delta)) = 2g + 2$, where $\Delta \subset \mathrm{Sym}^2 C$ is the diagonal, $\deg : \mathrm{Pic}(\mathrm{Sym}^2 C) \to \mathbb{Z}$ is the homomorphism introduced in Theorem 19.7. [*Hint*: Use the isomorphism

$$\mathcal{O}_{C \times C}(\Delta - p \times C - C \times p) \simeq (i_p \times i_p)^* \mathcal{B}^{-1},$$

where $i_p : C \to J$ is the embedding sending $x$ to $\mathcal{O}_C(x - p)$, $\mathcal{B}$ is the biextension on $J \times J$ defined in Section 17.4.]

7. Using the relation $E_d \simeq \mathcal{S}(\mathcal{O}_C(dp))$, the formula for the Chern classes of $E_d$ from Section 19.5, and Proposition 11.20, derive the Poincaré formula:

$$w_n = \frac{w_1^n}{n!}$$

modulo homological equivalence.

# 20
# Varieties of Special Divisors

In this chapter we present some results about varieties of special divisors $W_d \subset J^d$ defined as loci in $J^d$ where $h^0$ jumps ($d \leq g - 1$). In particular, we prove that the multiplicity of the theta divisor $\Theta = W_{g-1} \subset J^{g-1}$ at a point $L \in J^{g-1}$ is equal to $h^0(L)$. Also, we prove that the singular locus of $\Theta$ has codimension $\geq 2$ in $\Theta$ (codimension $\geq 3$ if $C$ is non-hyperelliptic). This fact will be used in the proof of Torelli theorem in Chapter 21. The natural framework for these results involves more general varieties of special divisors $W_d^r \subset J^d$ consisting of line bundles of degree $d$ with $h^0 \geq r + 1$ (we have $W_d = W_d^0$). The computation of multiplicities of $\Theta$ is the consequence of the following Riemann-Kempf singularity theorem (Theorem 20.8): The tangent cone to $W_d^r$ at $L$ coincides with the reduced subvariety of $H^1(\mathcal{O}_C)$ swept by the kernels of the natural maps $H^1(\mathcal{O}_C) \to H^1(\mathcal{O}_C(D))$, where $D$ runs through the linear system $\mathbb{P}H^0(L)$. An important role in the local study of $W_d^r$ near a point $L$ is played by the cup-product map $\mu : H^0(L) \otimes H^0(\omega_C L^{-1}) \to H^0(\omega_C) \simeq H^1(\mathcal{O}_C)^*$. For example, the above description of the tangent cone depends only on $\mu$. On the other hand, if $L \in W_d^r \setminus W_d^{r+1}$ then the Zariski tangent space to $W_d^r$ at $L$ can be identified with the orthogonal complement to the image of $\mu$. This leads to the estimate on the dimension of $W_d^r$. The singular locus of theta divisor $\Theta = W_{g-1}$ coincides with the subvariety $W_{g-1}^1 \subset \Theta$, so we get an estimate on its dimension.

### 20.1. Definitions

Let $C$ be a curve. The closed subset $W_d^r \subset J^d = J^d(C)$ consists of all line bundles $L$ of degree $d$ such that $h^0(L) > r$. One has the canonical scheme structure on $W_d^r$, since it can be described as the degeneration locus of some morphism of vector bundles on $J^d$. More precisely, if we fix an effective divisor $D$ of sufficiently large degree then the exact sequence

$$0 \to H^0(L) \to H^0(L(D)) \xrightarrow{\alpha_L} H^0(L(D)|_D),$$

where $L \in J^d$, tells us that $h^0(L) > r$ if and only if $\operatorname{rk} \alpha_L < h^0(L(D)) - r$. Let us define the vector bundles $\mathcal{V}_0$ and $\mathcal{V}_1$ on $J^d$ by $\mathcal{V}_0 = p_{2*}(\mathcal{P}(d)(p_1^{-1}D))$, $\mathcal{V}_1 = p_{2*}(\mathcal{P}(d)(p_1^{-1}D)|_{D\times J^d})$, where $\mathcal{P}(d)$ is a Poincaré line bundle on $C \times J^d$, and $p_1$ and $p_2$ are the projections from the product $C \times J^d$ to its factors. Then we have a natural morphism $\alpha : \mathcal{V}_0 \to \mathcal{V}_1$ and $W_d^r$ is the locus of points in $J^d$, where the rank of $\alpha$ is $< k = d + \deg D - g + 1 - r$. Locally, we can trivialize our vector bundles and represent $\alpha$ by a matrix of functions. Then the ideal sheaf of $W_d^r$ is generated by $k \times k$ minors of $\alpha$. We omit the proof of the fact that this ideal sheaf does not depend on a choice of $D$ (essentially, this follows from the fact that all complexes $\mathcal{V}_0 \to \mathcal{V}_1$ obtained in this way, are quasiisomorphic). Note that $W_{g-1}^0$ is exactly the theta divisor $\Theta \subset J^{g-1}$.

To study $W_d^r$, it is convenient to introduce an auxiliary scheme $G_d^r$ parametrizing pairs $(L, V)$, where $L \in J^d$, $V \subset H^0(L)$ is the subspace of dimension $r + 1$. It can be defined as follows. Pick a divisor $D$ as above. To choose a subspace $V \subset H^0(L)$ is the same as to choose a subspace $V \subset H^0(L(D))$ such that $\alpha_L(V) = 0$. Thus, we can define $G_d^r$ as a closed subscheme of the relative Grassmanian $G_{r+1}(\mathcal{V}_1)$ associated with $\mathcal{V}_1$, given by the equation $\alpha(V) = 0$.

**Lemma 20.1.** *The natural proper morphism* $f : G_d^r \to W_d^r : (L, V) \mapsto L$, *is an isomorphism over* $W_d^r \setminus W_d^{r+1}$.

*Proof.* By definition, over $W_d^r$ the rank of $\alpha : \mathcal{V}_0 \to \mathcal{V}_1$ is $<k = d + \deg D - g + 1 - r$. On the other hand, locally near every point of $W_d^r \setminus W_d^{r+1}$ some $(k-1) \times (k-1)$ minor of $\alpha$ is invertible. It follows that over $W_d^r \setminus W_d^{r+1}$ the image of $\alpha$ is a subbundle of $\mathcal{V}_1$. This easily implies the assertion of the lemma. □

## 20.2. Tangent Spaces

In the following proposition we compute the Zariski tangent spaces to all points $L \in W_d^r$.

**Proposition 20.2.** *(a) Assume that* $L \in W_d^r \setminus W_d^{r+1}$. *Then* $T_L W_d^r \simeq \operatorname{Im}(\mu)^\perp \subset H^1(\mathcal{O}_C)$ *where*

$$\mu : H^0(L) \otimes H^0(\omega_C L^{-1}) \to H^0(\omega_C) \simeq H^1(\mathcal{O}_C)^* \qquad (20.2.1)$$

*is the cup-product homomorphism.*

*(b) If* $L \in W_d^{r+1}$, *then* $T_L W_d^r = H^1(\mathcal{O})$.

*Proof.* (a) Since $L \notin W_d^{r+1}$, by Lemma 20.1 the tangent space $T_L W_d^r$ can identified with the tangent space $T_{(L,V)} G_d^r$, where $V = H^0(L)$. Let $(L(\epsilon), V(\epsilon))$ be the first order deformation of $(L, V)$ in $G_d^r$. In other words, we consider a morphism $\mathrm{Spec}(k[\epsilon]/\epsilon^2) \to G_d^r$ passing through $(L, V)$. Thus, $L(\epsilon)$ is a line bundle over $C \times \mathrm{Spec}(k[\epsilon]/\epsilon^2)$ and $V(\epsilon)$ is a vector bundle of rank $r+1$ over $\mathrm{Spec}(k[\epsilon]/\epsilon^2)$ equipped with an embedding $V(\epsilon) \to p_{2*}(L(\epsilon))$, where $p_2 : C \times \mathrm{Spec}(k[\epsilon]/\epsilon^2) \to \mathrm{Spec}(k[\epsilon]/\epsilon^2)$ is the projection. Let $g_{ij} \in \mathcal{O}^*(U_{ij})$ be the transition functions of $L$ with respect to some open covering $U_i$ of $C$. Then the transition functions of $L(\epsilon)$ have form $\widetilde{g}_{ij} = g_{ij}(1 + \epsilon\phi_{ij})$ where $\phi_{ij}$ is a Cech 1-cocycle with values in $\mathcal{O}_C$. Now $V(\epsilon)$ is a vector bundle over $\mathrm{Spec}(k[\epsilon]/\epsilon^2)$ and $V(\epsilon)|_{\mathrm{Spec}(k)} = V$, hence we can identify $V(\epsilon)$ with $V \otimes k[\epsilon]/\epsilon^2$. Thus, for every element $s \in V = H^0(L)$ we have a section $\widetilde{s}$ in $L(\epsilon)$ extending $s$. Let $\widetilde{s}$ be given by a collection of functions $\widetilde{s}_i \in \mathcal{O}(U_i) \otimes k[\epsilon]/\epsilon^2$ such that $\widetilde{s}_i = \widetilde{g}_{ij}\widetilde{s}_j$ over $U_{ij}$. We can write $\widetilde{s}_i = s_i + \epsilon s_i'$, where $s_i = s|_{U_i}$. Then we have

$$\phi_{ij} s_i = s_i' - g_{ij} s_j'$$

in $U_{ij}$. Therefore, the Cech 1-cocycle $(\phi_{ij} s_i)$ with values in $L$ is a coboundary. But this cocycle represents the cup-product of classes $(\phi_{ij}) \in H^1(C, \mathcal{O}_C)$ and $(s_i) \in H^0(C, L)$. So we obtain that the tangent space to $G_d^r$ at $(L, V)$ can be identified with the space of classes $\phi \in H^1(C, \mathcal{O}_C)$, such that $s \cdot \phi = 0$ in $H^1(C, L)$. It remains to use Serre duality.

(b) Locally there exists a morphism from $J^d$ to the variety $M(n, m)$ of $n \times m$ matrices, such that $W_d^r$ is the pull-back of the variety of matrices $M_k(n, m)$ of rank $\leq k$, while $W_d^{r+1}$ is the pull-back of $M_{k-1}(n, m)$. Hence, the result follows from the analogous statement for the tangent space to $M_k(n, m)$ at the point of $M_{k-1}(n, m)$, which can be checked by explicit computation in coordinates. □

**Corollary 20.3.** *$W_d = W_d^0$ is singular at a point $L$ if and only if $L \in W_d^1$.*

*Proof.* Since $G_d^0$ is just $\mathrm{Sym}^d C$, we have $\dim W_d^0 = d$. On the other hand, for $L \in W_d^0 \setminus W_d^1$ we have $h^0(L) = 1$, hence for such $L$ the map (20.2.1) is injective. It follows that the tangent space to $W_d^0$ at $L$ has dimension $d$. □

In particular, we obtain that the singular locus of $\Theta = W_{g-1} \subset J^{g-1}$, coincides with $W_{g-1}^1$.

## 20.3. Dimension Estimates

Recall that Clifford's theorem asserts that if $D$ is a special divisor on a non-hyperelliptic curve $C$ (i.e., $h^0(D) > 0$ and $h^1(D) > 0$) then the inequality $2\dim|D| \geq \deg D$ implies either $D = 0$ or $\mathcal{O}_C(D) \simeq \omega_C$ (see [62],IV,5.4). We will use this theorem to prove the following result due to Martens.

**Theorem 20.4.** *Assume that $C$ is not hyperelliptic, and that $d$ and $r$ satisfy $2 \leq d \leq g-1$, $0 < 2r \leq d$. Then* $\dim W_d^r \leq d - 2r - 1$.

*Proof.* Assume that the assertion is wrong. Then we can choose minimal $d$ such that $\dim W_d^r \geq d - 2r$. Note that $d - 2r > 0$ by Clifford's theorem (since $C$ is not hyperelliptic). Let $L \in W_d^r$ be a point in the component of maximal dimension. We may assume that $h^0(L) = r + 1$ (otherwise we have $\dim W_d^{r+1} \geq d - 2r \geq d - 2(r+1)$, so we can replace $r$ by $r+1$). We also can assume that the linear system $|L|$ has no base points. Indeed, by the assumption of minimality of $d$ we have $\dim W_{d-1}^r \leq d-2r-2$. Hence, the points of $W_d^r$ of the form $L'(p)$ where $L' \in W_{d-1}^r$, $p \in C$ form a closed subset of dimension $\leq d - 2r - 1$, and we can choose $L \in W_d^r$ outside this subset.

According to Proposition 20.2, the dimension of the tangent space to $W_d^r$ at $L$ is equal to $g - \mathrm{rk}(\mu) = g - (r+1)h^1(L) + \dim(\ker(\mu))$. By our assumption this should be $\geq d - 2r$, so we get the inequality

$$\dim(\ker(\mu)) \geq d - 2r - g + (r+1)h^1(L).$$

Let us choose two global sections of $L$, $s_1$ and $s_2$, without common zeros. Let $W \subset H^0(L)$ be the (2-dimensional) subspace spanned by $s_1$ and $s_2$. Consider the restriction of $\mu$ to $W \otimes H^0(\omega_C L^{-1})$:

$$\mu_W : W \otimes H^0(\omega_C L^{-1}) \to H^0(\omega_C).$$

Then the above inequality implies that

$$\dim(\ker(\mu_W)) \geq d - 2r - g + 2h^1(L) = g - d.$$

On the other hand, from the exact sequence

$$0 \to \omega_C L^{-2} \to W \otimes \omega_C L^{-1} \to \omega_C \to 0$$

we deduce that $\dim(\ker(\mu_W)) = h^0(\omega_C L^{-2})$. Therefore, $h^0(L^2) \geq d + 1$. If $d \leq g-2$, or $d = g-1$ and $L^2 \not\simeq \omega_C$, then we get a contradiction with Clifford's theorem. Note that since dimension of $W_d^r$ is positive we can always choose $L$ in such a way that $L^2 \not\simeq \omega_C$. The obtained contradiction proves the theorem. □

**Corollary 20.5.** *The dimension of the singular locus of* $\Theta \subset J^{g-1}$ *is* $\leq g-3$. *If* $C$ *is nonhyperelliptic, then this dimension is* $\leq g-4$.

*Proof.* According to Corollary 20.3, the singular locus of $\Theta = W^0_{g-1}$ is $W^1_{g-1}$. If $C$ is nonhyperelliptic, we can apply the above theorem to get the required estimate. In the case when $C$ is hyperelliptic we can use the equality $\dim W^r_d = d - 2r$, which we leave for the reader to check. □

## 20.4. Tangent Cones

Now we will determine the tangent cones to $W_d := W^0_d$ considered as subschemes of $J$. The idea is to consider the (total space of the) normal bundle $N$ to the projective space $\mathbb{P} = \mathbb{P}(H^0(L))$ sitting inside $\mathrm{Sym}^d C$, where $L \in W_d$. Then the map $\sigma = \sigma^d : \mathrm{Sym}^d C \to J^d$ induces a morphism $\tau : N \to T$, where $T = H^1(\mathcal{O}_C)$ is the tangent space to $J^d$ at the point $L$ (since $\mathbb{P} = \sigma^{-1}(L)$). The image of $\tau$ is contained in the tangent cone to $W_d$ at $L$ and set-theoretically coincides with it. On the other hand, it is easy to see that $N$ is described inside $\mathbb{P} \times T$ by explicit equations. This will give us enough information to conclude that the tangent cone is reduced and coincides with $\tau(N) \subset T$.

To describe equations of $N$ in $\mathbb{P} \times T$ let us consider the cup-product tensor (20.2.1) as a map from $H^0(\omega_C L^{-1})$ to the space of sections of $\mathcal{O}_{\mathbb{P}}(1) \otimes \mathcal{O}_T$. Let $s_1, \ldots, s_{h^1}$ be the sections of $\mathcal{O}_{\mathbb{P}}(1)$ obtained via this map from a basis in $H^0(\omega_C L^{-1})$, where $h^1 = h^1(L)$.

**Proposition 20.6.** *The vanishing locus of* $s_1, \ldots, s_{h^1}$ *coincides with* $N \subset \mathbb{P} \times T$.

*Proof.* Recall that the tangent space to $\mathrm{Sym}^d C$ at $D$ is canonically isomorphic to $H^0(\mathcal{O}_D(D))$, so that the tangent map to $\sigma : \mathrm{Sym}^d C \to J^d(C)$ at $D$ is just the boundary homomorphism

$$\delta_D : H^0(\mathcal{O}_D(D)) \to H^1(\mathcal{O}_C)$$

coming from the exact sequence $0 \to \mathcal{O}_C \to \mathcal{O}_C(D) \to \mathcal{O}_D(D) \to 0$ (see Proposition 19.2). Therefore, the fiber of the normal bundle $N$ at $D \in \mathbb{P}$ is $\mathrm{im}(\delta_D) = \ker(H^1(\mathcal{O}_C) \to H^1(\mathcal{O}_C(D)))$. Equivalently, this is a linear subspace in $T$ defined by the subspace of linear forms $H^0(\omega_C(-D)) \subset H^0(\omega_C) = T^*$. Let $N' \subset \mathbb{P} \times T$ be the vanishing locus of $s_1, \ldots, s_{h^1}$. Then the above argument shows that $N' \cap (D \times T)$ coincides with $N \cap (D \times T)$ for every $D$.

In particular, $\dim N' = \dim N = \dim(\mathbb{P} \times T) - h^1$. Hence, $N'$ is a complete intersection in $\mathbb{P} \times T$ and $N = N'$. $\square$

Another way to look at the tensor $\mu$ is to consider it as a matrix of linear functions on $T$ of the size $h^0 \times h^1$, where $h^i = h^i(L)$. Let $M(h^0, h^1)$ be the variety of matrices of this size. Then $\mu$ gives a morphism $m : T \to M(h^0, h^1)$ and the above proposition can be interpreted as saying that $\tau : N \to T$ is the pull-back of the standard family of projective spaces $\varphi : PS \to M(h^0, h^1)$ under $m$. Namely, $PS$ is the closed subset in $\mathbb{P}^{h^0-1} \times M$ given by the equations $\sum_k x_k m_{kl} = 0$ where $(m_{kl})$ are the coordinates in $M$, $(x_k)$ are homogeneous coordinates in $\mathbb{P}^{h^0-1}$. In particular, set-theoretically $\tau(N)$ is the pull-back of $\varphi(PS)$ under $m$. Therefore, the codimension of $\tau(N)$ in $T$ is $\leq \operatorname{codim}_M \varphi(PS) = h^1 - h^0 + 1 = g - d$. Hence, $\dim \tau(N) \geq d = \dim N$. It follows that $\dim \tau(N) = d$. The next ingredient we need is the following statement.

**Lemma 20.7.** *The canonical morphism $\mathcal{O}_T \to \tau_* \mathcal{O}_N$ is surjective.*

*Proof.* Since $N$ is the complete intersection of $h^1$ sections of $\mathcal{O}_{\mathbb{P}}(1) \otimes \mathcal{O}_T$ in $\mathbb{P} \times T$, we have the Koszul resolution for $\mathcal{O}_N$ as $\mathcal{O}_{\mathbb{P}\times T}$-module:

$$0 \to \mathcal{O}_{\mathbb{P}}(-h^1) \otimes \mathcal{O}_T = K_{-h^1} \to \cdots \to K_{-2} \to K_{-1} \to K_0$$
$$= \mathcal{O}_{\mathbb{P}\times T} \to \mathcal{O}_N \to 0,$$

where $K_{-j} = \mathcal{O}_{\mathbb{P}}(-j) \times \mathcal{O}_T^{\oplus \binom{h^1}{j}}$. Computing $\tau_* \mathcal{O}_N$ with the help of this resolution, we immediately obtain the result. $\square$

The above lemma implies that $\tau$ is birational onto its image and that $\tau(N)$ is normal. The last technical step is to show that $\tau(N)$ coincides with the tangent cone $X_L$ of $W_d$ at $L$.

**Theorem 20.8.** *One has $X_L = \tau(N)$ as subschemes in $T$.*

*Proof.* Consider the surjection $\mathcal{O}_{X_L} \to \mathcal{O}_{\tau(N)} = \tau_*(\mathcal{O}_N)$. We want to show that it is an injection, i.e., the map $H^0(\mathcal{O}_{X_L}) \to H^0(\mathcal{O}_N)$ is an injection. In other words, we want to prove that the surjective maps $\alpha_i : \mathfrak{m}^i/\mathfrak{m}^{i+1} \to H^0(I^i/I^{i+1})$ are injective for all $i \geq 0$ (here $\mathfrak{m}$ is the ideal of the point $L$ in $W_d$, $I$ is the ideal of $\mathbb{P}$ in $\operatorname{Sym}^d C$). Let $A$ denote the local ring of $W_d$ at $L$. Consider the canonical maps $\beta_i : A/\mathfrak{m}^i \to H^0(\mathcal{O}_{\operatorname{Sym}^d C}/I^i)$. Then the surjectivity of $\alpha_i$ implies that the following sequence is exact for every $i$:

$$0 \to \ker(\alpha_i) \to \ker(\beta_{i+1}) \to \ker(\beta_i) \to 0.$$

In particular, every element of $\ker(\beta_i)$ can be lifted to an element of the completion $A\hat{} = \text{proj.}\lim_n A/\mathfrak{m}^n$, that lies in the kernel of the canonical morphism

$$A\hat{} \to \text{proj.}\lim_n H^0(\mathcal{O}_{\mathrm{Sym}^d}/I^n). \tag{20.4.1}$$

It remains to prove that this morphism is injective. Indeed, this would imply that $\ker(\beta_i) = 0$ for all $i$, hence $\ker(\alpha_i) = 0$. Let us apply the theorem on formal functions (see [62], III, Section 11.1) to the morphism $\sigma : \mathrm{Sym}^d C \to J$ and the point $L \in J$. We obtain that the natural morphism

$$(\sigma_* \mathcal{O}_{\mathrm{Sym}^d C})_L\hat{} \to \text{proj.}\lim_n H^0(\mathcal{O}_{\mathrm{Sym}^d}/I^n)$$

is an isomorphism. Since $A$ (considered as $(\mathcal{O}_J)_L$-module) is a submodule of $(\sigma_* \mathcal{O}_{\mathrm{Sym}^d C})_L$, this implies injectivity of (20.4.1). □

Thus, the geometric description of $X_L$ is the following: $X_L$ is the reduced subvariety of $T$ swept by linear subspaces $\ker(H^1(\mathcal{O}_C) \to H^1(\mathcal{O}_C(D)))$, where $D$ runs through the linear system $\mathbb{P}H^0(L)$.

**Corollary 20.9.** *The schemes $W_d$ are reduced.*

**Corollary 20.10.** *The degree of the projectivized tangent cone $\mathbb{P}X \subset \mathbb{P}T$ is equal to the binomial coefficient $\binom{h^1(L)}{h^0(L)-1}$.*

*Proof.* The rational equivalence class of the subvariety $\mathbb{P}N \subset \mathbb{P} \times \mathbb{P}T$ is $(h + h')^{h^1(L)}$ where $h = c_1(\mathcal{O}(0, 1))$, $h' = c_1(\mathcal{O}(1, 0))$. Therefore, the degree of $\mathbb{P}X$ is equal to $h^{g-d} \cdot (h + h')^{h^1(L)} = \binom{h^1(L)}{h^0(L)-1}$. □

In particular, for $d = g - 1$ we have $h^0(L) = h^1(L)$, so the multiplicity of the theta divisor $\Theta = W_{g-1}$ at $L$ is equal to $h^0(L)$.

# 21

# Torelli Theorem

In this chapter we prove the Torelli theorem asserting that a curve $C$ can be recovered from its Jacobian $J$ considered as a principally polarized abelian variety. The proof is based on the observation that the Fourier–Mukai transform of a line bundle of degree $g - 1$ on $C$, is a coherent sheaf $F$ (up to a shift) supported on the corresponding theta divisor $\Theta$ in $J$. Moreover, the restriction of this sheaf to the nonsingular part $\Theta^{ns} \subset \Theta$ is a line bundle. Because the theta divisor is nonsingular in codimension 1 (this was proved in the previous chapter), $F$ can be recovered from this line bundle by taking the push-forward with respect to the open embedding $\Theta^{ns} \hookrightarrow \Theta$. On the other hand, we can characterize all line bundles $\mathcal{M}$ on $\Theta^{ns}$ appearing above only in terms of the pair $(J, \Theta)$. In fact, we prove that such $\mathcal{M}$ has two properties: (i) $\mathcal{M} \otimes \nu^*\mathcal{M} \simeq \omega_{\Theta^{ns}}$ where $\nu : \Theta \to \Theta$ is the canonical involution; (ii) the class of $\mathcal{M}$ generates the cokernel of the map $\mathrm{Pic}(J) \to \mathrm{Pic}(\Theta^{ns})$. Thus, to recover the curve $C$ from $(J, \Theta)$ one has to pick a line bundle $\mathcal{M}$, as above, extend it to $\Theta$ by taking the push-forward, and then apply the Fourier–Mukai transform. The result will be supported on $C$ (embedded into $J$).

## 21.1. Recovering the Curve from the Theta Divisor

Let $A$ be an abelian variety, $D \subset A$ an effective divisor inducing the principal polarization on $A$. Let us set

$$Z(A, D) := \mathrm{coker}(\mathrm{Pic}(A) \to \mathrm{Pic}(D^{ns})),$$

where $D^{ns} \subset D$ is the nonsingular part of $D$, the homomorphism $\mathrm{Pic}(A) \to \mathrm{Pic}(D^{ns})$ is induced by the embedding $D^{ns} \hookrightarrow A$. If $D' \subset A$ is another effective divisor inducing the same polarization then there exists a unique point $x \in A$ such that $D' = D + x$. It follows that there is a natural transitive system of isomorphisms between groups $Z(A, D)$ for all $D$ inducing the given principal polarization. Now assume in addition that $D$ is symmetric. Then the restriction of the map $[-1]_A : A \to A$ to $D$ induces an involution $\nu : D \to D$.

If $D'$ is another symmetric divisor inducing the principal polarization then $D' = D + x$ for the unique point $x \in A_2$. The corresponding isomorphism $D \simeq D'$ commutes with involutions on $D$ and $D'$. Let $P(A, D) \subset \mathrm{Pic}(D^{ns})$ be the set of isomorphism classes of line bundles $\mathcal{M}$ on $D^{ns}$, such that

(i) $\mathcal{M} \otimes \nu^*\mathcal{M} \simeq \omega_{D^{ns}}$;

(ii) the group $Z(A, D)$ is generated by the image of $\mathcal{M}$.

Clearly, if $D'$ is another symmetric divisor inducing the same polarization then $P(A, D') = P(A, D)$ under the natural isomorphism between $\mathrm{Pic}((D')^{ns})$ and $\mathrm{Pic}(D^{ns})$. Thus, when the polarization on $A$ is fixed we can denote $P(A, D)$ (resp., $Z(A, D)$) simply by $P(A)$ (resp., $Z(A)$).

**Remark.** Since the tangent bundle to $A$ is trivial, one has $\omega_{D^{ns}} = \mathcal{O}_J(D)|_{D^{ns}}$. Thus, if $P(A, D)$ is nonempty then the group $Z(A, D)$ is cyclic and $\nu$ acts on $Z(A, D)$ as $-1$.

Now we can formulate the recipe for recovering the curve from its Jacobian.

**Theorem 21.1.** *Let $J$ be the Jacobian of a curve $C$ of genus $g \geq 2$. Then the set $P(J)$ is nonempty. Pick any theta divisor $\Theta_L \subset J$ (where $L \in \mathrm{Pic}^{g-1}(C)$) and any element $\mathcal{M} \in P(J, \Theta_L)$. Then the Fourier transform $\mathcal{S}(j_*^{ns}\mathcal{M})$, where $j^{ns} : \Theta_L^{ns} \hookrightarrow J$ is the natural embedding, has form $\mathcal{F}[1 - g]$, where $\mathcal{F}$ is a coherent sheaf on $J$. The support of $\mathcal{F}$ is isomorphic to $C$ (in fact, $\mathcal{F}$ is a line bundle of degree $g - 1$ on $C$).*

## 21.2. Computation of $Z(J)$

Recall that all theta divisors in the Jacobian are translations of the natural divisor $\Theta \subset J^{g-1}$. It is more convenient to work with $J^{g-1}$ and this divisor. Note that we have the canonical involution $\nu : \Theta \to \Theta$ corresponding to the map $L \mapsto \omega_C \otimes L^{-1}$. This involution is compatible with involutions on symmetric theta divisors $\Theta_L \subset J$ used in the previous section, via the natural isomorphism $\Theta \simeq \Theta_L$. Thus, we have natural identification of the sets $P(J, \Theta_L)$ (resp., $Z(J, \Theta_L)$) with the set $P(J^{g-1}, \Theta)$ (resp., $Z(J^{g-1}, \Theta)$) defined in the same way in terms of the pair $\Theta \subset J^{g-1}$ and the canonical involution $\nu$ on $\Theta$.

We have a canonical identification of $\mathrm{Pic}^0(J^{g-1})$ with $\mathrm{Pic}^0(J) = \hat{J}$ induced by any standard isomorphism $J \to J^{g-1}$ (given by some line bundle of degree $g - 1$ on $C$). Thus, we can consider the Fourier transform as an equivalence between derived categories of coherent sheaves on $\hat{J}$ and $J^{g-1}$. To prove Theorem 21.1, it suffices to show that for every $\mathcal{M} \in P(J^{g-1}, \Theta)$

the Fourier transform of the sheaf $j^{ns}_*\mathcal{M}$, where $j^{ns} : \Theta^{ns} \hookrightarrow J^{g-1}$ is the natural embedding, has the form specified in Theorem 21.1.

We start by calculating the group $Z(J) = Z(J^{g-1}, \Theta)$. This is not difficult, since we have a smooth compactification for $\Theta^{ns}$. Namely, we can identify $\Theta^{ns}$ with an open subset of $\mathrm{Sym}^{g-1} C$ consisting of effective divisors $D$ of degree $g-1$, such that $h^0(D) = 1$ (indeed, $\Theta^{ns} = W_{g-1} \setminus W^1_{g-1}$ by Corollary 20.3; on the other hand, the morphism $\sigma^{g-1} : \mathrm{Sym}^{g-1} C \to W_{g-1} = \Theta$ is an isomorphism over $\Theta^{ns}$ by Corollary 19.3). Let $Q \subset \mathrm{Sym}^{g-1} C$ be the complementary closed subset consisting of $D \in \mathrm{Sym}^{g-1} C$, such that $h^0(D) > 1$.

**Lemma 21.2.** *Assume that $g \geq 3$. If $C$ is not hyperelliptic then $Q$ has codimension $\geq 2$. If $C$ is hyperelliptic and $\tau : C \to C$ is the hyperelliptic involution then $Q$ coincides with the divisor in* $\mathrm{Sym}^{g-1} C$ *consisting of $D$, such that $D = x + \tau(x) + D'$ for some $x \in C$, $D' \geq 0$.*

*Proof.* When $C$ is not hyperelliptic, the assertion follows from Theorem 20.4. When $C$ is hyperelliptic, it follows from the fact that every divisor $D$ on $C$ with $h^0(D) > 1$ is the sum of a divisor of the form $x + \tau(x)$ with an effective divisor. □

Lemma 21.2 implies that for a nonhyperelliptic curve $C$ one has a canonical isomorphism $\mathrm{Pic}(\Theta^{ns}) \simeq \mathrm{Pic}(\mathrm{Sym}^{g-1} C)$, while for hyperelliptic $C$ of genus $g \geq 3$ the group $\mathrm{Pic}(\Theta^{ns})$ is the quotient of $\mathrm{Pic}(\mathrm{Sym}^{g-1} C)$ by the subgroup generated by the classes of the irreducible components of $Q$.

**Lemma 21.3.** *Assume that $C$ is a hyperelliptic curve of genus $g \geq 3$. Then $Q$ is an irreducible divisor and* $\deg(Q) = -2$*, where* $\deg : \mathrm{Pic}(\mathrm{Sym}^{g-1} C) \to \mathbb{Z}$ *is the homomorphism introduced in Theorem 19.7.*

*Proof.* Let $\tau : C \to C$ be the hyperelliptic involution. For every $d > 1$ let us denote by $Q_d \subset \mathrm{Sym}^d C$ the reduced effective divisor consisting of $D \in \mathrm{Sym}^d C$, such that $D$ contains a pair of points in hyperelliptic involution. By the definition, $Q = Q_{g-1}$. Note that $Q_2 \simeq \mathbb{P}^1$, while $Q_d$ for $d > 2$ is just the image of $Q_2 \times \mathrm{Sym}^{d-2} C$ under the natural map $\mathrm{Sym}^2 C \times \mathrm{Sym}^{d-2} C \to \mathrm{Sym}^d C$, so $Q_d$ is irreducible for every $d \geq 2$. It is easy to check that

$$s_p^*\mathcal{O}(Q_d) \simeq \mathcal{O}\big(Q_{d-1} + R^{d-1}_{\tau(p)}\big) \tag{21.2.1}$$

for every $p \in C$. Consider the embedding $a : Q_2 \hookrightarrow \mathrm{Sym}^d C$ given by $D \mapsto D + D_0$ where $D_0$ is a fixed divisor of degree $d-2$. Then using

(21.2.1) and the relation (19.5.4) from Exercise 1(c) of Chapter 19, we obtain that $a^*\mathcal{O}(R^d_y) \simeq \mathcal{O}_{\mathbb{P}^1}(1)$, while $a^*\mathcal{O}(Q_d) \simeq \mathcal{O}_{\mathbb{P}^1}(Q_2 \cdot Q_2 + d - 2)$. Note that $Q_2 \cdot Q_2 = 1 - g$, so we have $a^*\mathcal{O}(Q_d) \simeq \mathcal{O}_{\mathbb{P}^1}(d - g - 1)$. In particular, $a^*\mathcal{O}(Q_{g-1}) \simeq \mathcal{O}_{\mathbb{P}^1}(-2)$ as required. □

These calculations lead to the following result.

**Proposition 21.4.** *If $C$ is not hyperelliptic then $Z(J) \simeq \mathbb{Z}$. If $C$ is hyperelliptic of genus $g \geq 3$ then $Z(J) \simeq \mathbb{Z}/2\mathbb{Z}$. In both cases the restriction homomorphism* $\text{Pic}\left(J^{g-1}\right) \to \text{Pic}(\Theta^{ns})$ *is injective.*

## 21.3. Proof of Theorem 21.1

Let $A$ be an abelian variety, and let $D \subset A$ be a symmetric effective divisor inducing a principal polarization on $A$. We denote by $\widetilde{A}$ the semidirect product of $\hat{A}$ and $\mathbb{Z}/2\mathbb{Z}$, where $\pm 1 \in \mathbb{Z}/2\mathbb{Z}$ acts on $\hat{A}$ by $[\pm 1]_{\hat{A}}$. The set $P(A, D)$ is equipped with the natural action of $\widetilde{A}(k)$. Namely, an element $\xi \in \hat{A}(k)$ acts by the tensor product with the restriction to $D^{ns}$ of the line bundle $\mathcal{P}_\xi$, while the action of $\mathbb{Z}/2\mathbb{Z}$ is induced by the involution $\nu : D^{ns} \to D^{ns}$. Since the bundles $\mathcal{P}_\xi$ are translation-invariant, this action is compatible with our identifications between the sets $P(A, D)$ for different $D$.

**Proposition 21.5.** *Assume that the restriction homomorphism* $\text{Pic}(A) \to \text{Pic}(D^{ns})$ *is injective. Then the action of $\widetilde{A}$ on $P(A, D)$ is transitive.*

*Proof.* Let $j_{D^{ns}} : D^{ns} \to A$ be the natural embedding. Take $\mathcal{M}, \mathcal{M}' \in P(A, D)$. Since $\mathcal{M}$ generates $Z(A, D)$ and $\nu^*\mathcal{M} \equiv \mathcal{M}^{-1} \bmod j^*_{D^{ns}} \text{Pic}(A)$, we have either $\mathcal{M} \equiv \mathcal{M}' \bmod j^*_{D^{ns}}(\text{Pic}(A))$ or $\nu^*\mathcal{M} \equiv \mathcal{M}' \bmod j^*_{D^{ns}}(\text{Pic}(A))$. Replacing $\mathcal{M}'$ by $\nu^*\mathcal{M}'$ if necessary, we can assume that $\mathcal{M}^{-1} \cdot \mathcal{M}' \simeq j^*_{D^{ns}} M$ for some $M \in \text{Pic}(A)$. Now from injectivity of the restriction homomorphism $\text{Pic}(A) \to \text{Pic}(D^{ns})$ we deduce that $[-1]^*_A M = M^{-1}$. Hence, $\phi_M = \phi_{[-1]^*M} = -\phi_M$, which implies that that $M \in \text{Pic}^0(A)$. □

The injectivity assumption of the previous proposition holds in the case of Jacobians if $g \geq 3$, as follows from Proposition 21.4. Also, in this case instead of looking at symmetric divisors in $J$ we can deal with the similarly defined action of $\widetilde{J}$ on $P(J^{g-1}, \Theta)$. Proposition 21.5 implies that this action is transitive (for $g \geq 3$). In the case $g = 2$ one can show directly that the action of $\widetilde{J}$ on $P(J^1, \Theta)$ is transitive (see Exercise 1 at the end of this chapter). Therefore, it suffices to prove the existence of at least one element in $P(J^{g-1}, \Theta)$ for which the assertion of Theorem 21.1 is true.

Let $\mathcal{P}(g-1)$ be the Poincaré line bundle on $C \times J^{g-1}$ normalized at some point $p \in C$. Let us consider the functor $\Phi_{\mathcal{P}(g-1)} : D^b(C) \to D^b(J^{g-1})$. Theorem 21.1 is an immediate consequence of the following result.

**Proposition 21.6.** *One has $\Phi_{\mathcal{P}(g-1)}(\mathcal{O}_C)[1] = j^{ns}_* \mathcal{M}$ for some element $\mathcal{M} \in P(J^{g-1}, \Theta)$.*

*Proof.* Assume first that $g \geq 3$. Clearly the object $F := \Phi_{\mathcal{P}(g-1)}(\mathcal{O}_C)[1] \in D^b(J^{g-1})$ vanishes outside of $\Theta$. Since $F$ is the derived push-forward of a (shifted) line bundle on $C \times J^{g-1}$, we can represent it as the cone of a morphism $f : V_{-1} \to V_0$ of vector bundles on $J^{g-1}$. Since $f$ is an isomorphism outside of $\Theta$, it is injective, so $F = \text{Coker } f$. Moreover, since $\det(f)$ is an equation of $\Theta$ (see Chapter 17), it follows that $F$ is the push-forward by $j : \Theta \hookrightarrow J^{g-1}$ of a coherent sheaf on $\Theta$.

Now we are going to use the cohomological interpretation of depth, (see [62], Exercise III, 3.4) stating that for a finitely generated module $M$ over a Noetherian ring $R$ and for an ideal $I \subset R$ one has $H^i_I(M) = 0$ for $i < \text{depth}_I M$. In particular, if $R$ is smooth then $H^i_I(R) = 0$ for $i < \text{ht } I$. It follows that for every closed subset $Y \subset J^{g-1}$ of codimension $> 2$ and every vector bundle $V$ on $J^{g-1}$, one has $\mathcal{H}^i_Y V = 0$ for $i \leq 2$. From the exact sequence $0 \to V_{-1} \to V_0 \to F \to 0$ we derive that $\mathcal{H}^i_Y F = 0$ for $i = 0, 1$ and $\text{codim } Y > 2$. Since by Corollary 20.5 the codimension in $J^{g-1}$ of the singular locus of $\Theta$ is $> 2$, this shows that $F = j^{ns}_* \mathcal{M}$, where $\mathcal{M} := j^{ns*} F$ (recall that $j^{ns} : \Theta^{ns} \hookrightarrow J^{g-1}$ is the natural locally closed embedding). Note that by the base change of a flat morphism (see Appendix C) we have

$$Lj^{ns*} F \simeq Rp_{2*}(\mathcal{P}(g-1)|_{C \times \Theta^{ns}})[1].$$

Since $h^0(L) = h^1(L) = 1$ for every $L \in \Theta^{ns}$, applying the base change again we deduce that $\text{rk } \mathcal{M}|_L = 1$ for every $L \in \Theta^{ns}$. Since $\Theta$ is reduced, this implies that $\mathcal{M}$ is a line bundle on $\Theta^{ns}$.

It remains to prove that $\mathcal{M} \in P(J^{g-1}, \Theta)$. First, we are going to apply the duality theory (see [61]) to the projection $p_2 : C \times J^{g-1} \to J^{g-1}$ to prove that

$$R\underline{\text{Hom}}(F, \mathcal{O}_{J^{g-1}}) \simeq \nu^* F[-1],$$

where $\nu : J^{g-1} \to J^{g-1}$ is the involution $L \mapsto \omega_C \otimes L^{-1}$. We have

$$R\underline{\mathrm{Hom}}(F, \mathcal{O}_{J^{g-1}}) \simeq R\underline{\mathrm{Hom}}(Rp_{2*}(\mathcal{P}(g-1)), \mathcal{O}_{J^{g-1}})[-1]$$
$$\simeq Rp_{2*}(\mathcal{P}(g-1)^{-1} \otimes p_2^! \mathcal{O}_{J^{g-1}})[-1].$$

Since the morphism $p_2$ is smooth, we have $p_2^! \mathcal{O}_{J^{g-1}} \simeq p_1^* \omega_C[1]$, where $p_1 : C \times J^{g-1} \to C$ is the projection. Therefore,

$$R\underline{\mathrm{Hom}}(F, \mathcal{O}_{J^{g-1}}) \simeq Rp_{2*}(\mathcal{P}(g-1)^{-1} \otimes p_1^* \omega_C).$$

But $\mathcal{P}(g-1)^{-1} \otimes p_1^*\omega_C) \simeq (\mathrm{id}_C \times \nu)^* \mathcal{P}(g-1)$ by the definition of the involution $\nu$. Hence,

$$R\underline{\mathrm{Hom}}(F, \mathcal{O}_{J^{g-1}}) \simeq \nu^* Rp_{2*}(\mathcal{P}(g-1)) \simeq \nu^* F[-1].$$

Applying the functor $Lj^{ns*}$ to this isomorphism, we obtain

$$R\underline{\mathrm{Hom}}(Lj^{ns*}F, \mathcal{O}_{\Theta^{ns}}) \simeq \nu^* Lj^{ns*} F[-1].$$

Since $Lj^{ns*}F$ has locally free cohomology sheaves, this implies that

$$\mathcal{M}^{-1} \simeq \nu^* L^{-1} j^{ns*} F.$$

But

$$L^{-1} j^{ns*} F \simeq L^{-1} j^{ns*} j_*^{ns} \mathcal{M} \simeq \mathcal{M} \otimes \mathcal{O}_{\Theta^{ns}}(-\Theta).$$

Therefore, $\nu^* \mathcal{M}^{-1} \simeq \mathcal{M}(-\Theta)$, which is condition (i) from the definition of $P(J^{g-1}, \Theta)$.

Let us check condition (ii). Consider the universal divisor $\mathcal{D}_{g-1} \subset C \times \mathrm{Sym}^{g-1} C$. Then the pull-back of the line bundle $\mathcal{P}(g-1)$ on $C \times J^{g-1}$ by the morphism $\mathrm{id} \times \sigma^{g-1} : C \times \mathrm{Sym}^{g-1} C \to C \times J^{g-1}$ is isomorphic to $\mathcal{O}(\mathcal{D}_{g-1} - C \times R_p)$, where $p$ is the point over which $\mathcal{P}(g-1)$ is trivialized. It follows that the line bundle $\mathcal{M}^{-1}$ on $\Theta^{ns}$ is isomorphic to $\alpha^* p_{2*}(\mathcal{O}(\mathcal{D}_{g-1}))(-R_p)$, where $p_2 : C \times \mathrm{Sym}^{g-1} C \to \mathrm{Sym}^{g-1} C$ is the projection, $\alpha : \Theta^{ns} \to \mathrm{Sym}^{g-1} C$ is the embedding. Therefore, $\mathcal{M}^{-1} \simeq \alpha^*(\mathcal{O}_{\mathrm{Sym}^{g-1} C}(-R_p))$, which generates $Z(J)$, so we are done. □

## 21.4. Remarks

From the Lefschetz theorem for Picard groups (see Appendix C) one can easily derive that the restriction map $j^* : \mathrm{Pic}(A) \to \mathrm{Pic}(D)$ is an isomorphism for an arbitrary principally polarized abelian variety $(A, D)$ of dimension $g \geq 4$ (where $D$ is an effective divisor corresponding to the principal polarization). Furthermore, if the dimension of the singular locus Sing $D$ is $<g-4$ then $D$

is locally factorial as follows from [57], exposé 11, (3.14) ch. [56]. Hence, in this case $\mathrm{Pic}(D) = \mathrm{Pic}(D^{ns})$ (because the notions of Cartier divisors and Weil divisors on $D$ coincide) and $Z(A) = 0$. Note that for a Jacobian the dimension of $\mathrm{Sing}\,\Theta$ is $\geq g - 4$. Moreover, Andreotti and Mayer proved in [1] that the closure of the locus of Jacobians constitutes an irreducible component of the locus $N_{g-4}$ of principally polarized abelian varieties with $\dim \mathrm{Sing}\, D \geq g-4$. In [11] Beauville established that for $g = 4$ the locus $N_0$ has two irreducible components. He also proved (assuming that the characteristic is zero) that a generic point of $N_0$, which is not contained in the closure of the locus of Jacobians, corresponds to an abelian variety $A$ with $\mathrm{Sing}\, D$ consisting of one ordinary double point (see [11], 7.5). It follows that the corresponding group $Z(A)$ is isomorphic to $\mathbb{Z}$ and the involution acts on $Z(A)$ as identity. Therefore, the set $P(A)$ in this case is empty. The natural question is whether in higher dimensions one still has $P(A) = \emptyset$ for principally polarized abelian varieties that are not in the closure of the locus of Jacobians.

## Exercises

1. Let $C$ be a curve of genus 2. Show that the involution $\nu$ on the theta divisor $C \simeq \Theta \subset J^1$ coincides with the hyperelliptic involution of $C$. Prove that in this case $\widetilde{J}$ acts transitively on $Z(J^1, \Theta)$.
2. Let $C$ be a curve of genus 2.
   (a) Prove that $P(J^1, \Theta)$ coincides with the set of isomorphism classes of line bundles of degree 1 on $\Theta \simeq C$.
   (b) Let $p \in C$ be a point fixed by the hyperelliptic involution, $i_p : C \hookrightarrow J$ be the corresponding embedding. Identifying $\hat{J}$ with $J$ by means of the standard principal polarization of $J$, we can consider the Fourier transform $\mathcal{S}$ as an autoequivalence of $D^b(J)$. Show that one has

$$\mathcal{S}(i_{p*}\mathcal{O}_C(p))[1] \simeq i_{p*}(\mathcal{O}_C(p)).$$

# 22

# Deligne's Symbol, Determinant Bundles, and Strange Duality

This chapter should be considered as a survey, so we only sketch some proofs in it.

In Sections 22.1 to 22.3, we present the construction and properties of Deligne's symbol $\langle L, M \rangle$, which is a line bundle on the base $S$ associated with a pair of line bundles $L$ and $M$ on a relative curve $C \to S$. The main result about this symbol is the isomorphism (22.3.1) relating it with the determinant line bundles. Then in Section 22.4 we review the strange duality conjecture about generalized theta divisors on the moduli spaces of vector bundles on a curve. These generalized theta divisors are defined by the same determinantal construction as the usual theta divisors (see Section 17.1). Let $U(r, r(g-1))$ (resp. $SU(k)$) be the moduli space of semistable vector bundles of rank $r$ (resp. $k$) and degree $r(g-1)$ (respectively, trivial determinant). Applying the isomorphism (22.3.1) one can easily prove that the pull-back of the (generalized) theta divisor under the tensor product map $U(r, r(g-1)) \times SU(k) \to U(kr, kr(g-1))$ is the external tensor product $\mathcal{O}_{U(r,r(g-1))}(k\Theta) \boxtimes \mathcal{O}_{SU(k)}(r\Theta)$. Therefore, the pull-back of the (generalized) theta function induces a map $H^0(SU(k), \mathcal{O}(r\Theta))^* \to H^0(U(r, r(g-1)), \mathcal{O}(k\Theta))$. The conjecture is that this map is an isomorphism. Using the Fourier–Mukai transform, we reformulate this conjecture in such a way that the roles of $k$ and $r$ become symmetric. Namely, we denote by $F_{r,k}$ the vector bundle on $J$, obtained as the push-forward of $\mathcal{O}(k\Theta)$ with respect to the map $\det : U(r, r(g-1)) \to J$. Then there is a natural morphism $F_{k,r}^{\vee} \to \mathcal{S}(F_{r,k})$ and the conjecture states that it is an isomorphism. From this reformulation one can easily see that the conjecture holds for $(r, k)$ if and only if it holds for $(k, r)$.

## 22.1. Virtual Vector Bundles

A *Picard category* is a (non-empty) category $\mathcal{C}$, such that every morphism in $\mathcal{C}$ is an isomorphism, equipped with a functor $+ : \mathcal{C} \times \mathcal{C} \to \mathcal{C}$ satisfying the associativity constraint, such that for every object $A \in \mathcal{C}$ the functors $X \mapsto A + X$ and $X \mapsto X + A$ are autoequivalences. These axioms imply the existence of the neutral object 0, and of the opposite object $-X$ to every object $X$ equipped with an isomorphism $X + (-X) \simeq 0$. A *commutative Picard category* is a Picard category equipped with a commutativity constraint compatible with the associativity. In such a category one defines $A - B = A + (-B)$ for a pair of object $A, B$. An isomorphism $\phi_1 : A \simeq B + X$ induces an isomorphism $A - B \simeq X$, however, "adding $B$" to this isomorphism we obtain an isomorphism $\phi_2 : A \simeq B + X$ that can differ from $\phi_1$.

**Examples.**

1. The category $\underline{\text{Pic}}(S)$ of line bundles and their isomorphisms on a scheme $S$ has a natural structure of Picard category.
2. Consider the category $\mathcal{P}(S)$ of graded line bundles on a scheme $S$. Its objects are pairs $(L, a)$, where $L$ is a line bundle on $S$, $a$ is an integer-valued locally constant function on $S$ (we think of $L$ as being placed in degree $a$). The operation $+$ is defined by $(L, a) + (L', a') = (L \otimes L', a + a')$. It is equipped with the obvious associativity constraint and the commutativity constraint $v \otimes w \mapsto (-1)^{\deg(v) \cdot \deg(w)} w \otimes v$. With these data, $\mathcal{P}(S)$ is a commutative Picard category.

Let $\mathcal{A}$ be an exact category (a full subcategory of an abelian category stable under extensions). Then one can associate to $\mathcal{A}$ the commutative Picard category $V(\mathcal{A})$ of virtual objects of $\mathcal{A}$ equipped with a functor $(\mathcal{A}, \text{is}) \to V(\mathcal{A}) : A \mapsto [A]$, where $(\mathcal{A}, \text{is})$ is the category with the same objects as $\mathcal{A}$ and isomorphisms in $\mathcal{A}$ as morphisms. To explain the idea of the definition of $V(\mathcal{A})$, let us consider for every Picard category $\mathcal{C}$ functors $[\;] : (\mathcal{A}, \text{is}) \to \mathcal{C}$ equipped with the following additional data:

(a) for every exact triple $A' \to A \to A''$ in $\mathcal{A}$, a functorial isomorphism $[A] \simeq [A'] + [A'']$;

(b) an isomorphism $[0] \simeq 0$;

subject to the axioms:

(c) for an isomorphism $\phi : A \to B$ consider the exact triangle $0 \to A \to B$ (resp. $A \to B \to 0$), then the induced isomorphism $[A] \to [0] + [B] \to [B]$ (resp. $[B] \to [A] + [0] \to [A]$) coincides with $[\phi]$ (resp. $[\phi^{-1}]$);

(d) for an admissible filtration $0 \subset A \subset B \subset C$ (i.e., a filtration with $B/A, C/B \in \mathcal{A}$), the following diagram of isomorphisms is commutative:

$$\begin{array}{ccc} [C] & \longrightarrow & [A]+[C/A] \\ \big\downarrow & & \big\downarrow \\ [B]+[C/B] & \longrightarrow & [A]+[B/A]+[C/B] \end{array} \tag{22.1.1}$$

By the definition, the functor $[\,]:(\mathcal{A},\mathrm{is})\to V(\mathcal{A})$ extends to the universal system of the above type, i.e., for every Picard category $\mathcal{C}$ equipped with a functor $(\mathcal{A},\mathrm{is})\to\mathcal{C}$ and data (a) and (b) subject to (c) and (d), there exists a unique functor of Picard categories $V(\mathcal{A})\to\mathcal{C}$ compatible with data (a) and (b).

The functor $[\,]:(\mathcal{A},\mathrm{is})\to V(\mathcal{A})$ extends to the functor $[\,]:(D^b(\mathcal{A}),\mathrm{is})\to V(\mathcal{A})$, where $D^b(\mathcal{A})$ is the derived category of $\mathcal{A}$, i.e., the quotient of the category of bounded complexes over $\mathcal{A}$ by the subcategory of acyclic complexes (=complexes obtained by successive extensions from shifted exact triples in $\mathcal{A}$).

Let $\mathcal{C}_1,\mathcal{C}_2$ be Picard categories. A *homofunctor* $F:\mathcal{C}_1\to\mathcal{C}_2$ is a functor $F$ such that $F(0)=0$, equipped with an isomorphism of functors $\mathcal{C}_1\times\mathcal{C}_1\to\mathcal{C}_2$

$$F(A_1+_{\mathcal{C}_1}A_2)\simeq F(A_1)+_{\mathcal{C}_2}F(A_2)$$

compatible with associativity constraints.

For a scheme $S$ we denote $\underline{K}(S)=V(\mathrm{Vect}(S))$, where $\mathrm{Vect}(S)$ is the category of vector bundles on $S$. We call objects of $\underline{K}(S)$ *virtual vector bundles* on $S$. For a morphism of schemes $f:X\to S$ one has a canonical homofunctor $f^*:\underline{K}(S)\to\underline{K}(X)$. If $f$ is proper and flat and $X$ is quasi-projective, then one can also define a homofunctor $f_*:\underline{K}(X)\to\underline{K}(S)$ compatible with the derived push-forward functor $D^b(X)\to D^b(S)$. The map $V\mapsto(\det V,\mathrm{rk}\,V)$ extends to a homofunctor $\det:\underline{K}(S)\to\mathcal{P}(S)$.

## 22.2. Deligne's Symbol

Recall that the Weil reciprocity law asserts that for every pair of rational functions $f$ and $g$ on a smooth projective curve $C$ such that the divisors of $f$ and $g$ are disjoint, one has

$$f(\mathrm{div}(g))=g(\mathrm{div}(f)), \tag{22.2.1}$$

where $\mathrm{div}(f)$ denotes the divisor of $f$, for a divisor $D=\sum n_i p_i$ disjoint from $\mathrm{div}(f)$ we set $f(D)=\prod f(p_i)^{n_i}$. The proof can be easily reduced to the case

$C = \mathbb{P}^1$ as follows. Consider $f$ as a morphism $f : C \to \mathbb{P}^1$. Then we have $\operatorname{div}(f) = f^*(\operatorname{div}(t))$ where $t = t_1/t_0$ is the canonical rational function on $\mathbb{P}^1$, $\operatorname{div}(t) = (0) - (\infty)$. Let $N : k(C)^* \to k(t)^*$ be the norm homomorphism. Then we have $g(\operatorname{div}(f)) = N(g)(\operatorname{div}(t))$ and $f(\operatorname{div}(g)) = t(\operatorname{div}(N(g))$, so the statement follows from the similar assertion for $t$ and $N(g)$.

To a pair of line bundles $L$ and $M$ on a curve $C$, Deligne associates (see [31], Section 6) the one-dimensional vector space $\langle L, M\rangle$ generated by the symbols $\langle l, m\rangle$, where $l$ and $m$ are rational sections of $L$ and $M$, such that the divisors of $l$ and $m$ are disjoint, subject to the relations

$$\langle l, fm\rangle = f(\operatorname{div}(l))\langle l, m\rangle, \tag{22.2.2}$$

$$\langle gl, m\rangle = g(\operatorname{div}(m))\langle l, m\rangle. \tag{22.2.3}$$

These relations are consistent due to Weil reciprocity law. The immediate consequences of the definition are the following isomorphisms

$$\langle L_1 \otimes L_2, M\rangle \simeq \langle L_1, M\rangle \otimes \langle L_2, M\rangle, \tag{22.2.4}$$

$$\langle L, M_1 \otimes M_2\rangle \simeq \langle L, M_1\rangle \otimes \langle L, M_2\rangle. \tag{22.2.5}$$

This construction makes sense over arbitrary base $S$: if $C/S$ is a relative curve, $L$ and $M$ are line bundles on $C$, then one gets the line bundle $\langle L, M\rangle$ on $S$. Namely, one has to work étale locally over $S$ and use the definition $f(D) = N_{D/S}(f)$, where $D/S$ is a relative Cartier divisor, $N_{D/S}$ is the norm homomorphism.

Recall that for a finite flat morphism $g : S' \to S$ there is a canonical functor $N_{S'/S} : \underline{\operatorname{Pic}}(S') \to \underline{\operatorname{Pic}}(S)$ between the categories of line bundles induced by the norm homomorphism (one should think about line bundles as $\mathbb{G}_m$-torsors, then $N_{S'/S}(L)$ is the push-forward of $L$ with respect to the norm homomorphism $N_{S'/S} : g_*\mathcal{O}_{S'}^* \to \mathcal{O}_S^*$). More generally, for a vector bundle $E$ on $S'$ one can define a functor $N_{E/S} : \underline{\operatorname{Pic}}(S') \to \underline{\operatorname{Pic}}(S)$ in the same way, using the homomorphism

$$N_{E/S} : g_*\mathcal{O}_{S'}^* \to \mathcal{O}_S^* : u \mapsto \det(u, g_*E),$$

where $\det(u, g_*E)$ denotes the determinant of the action of $u$ on $g_*E$.

Now let $C/S$ be a relative curve, $D \subset C$ be a relative Cartier divisor. Then for a vector bundle $E$ on $D$ and a line bundle $M$ on $C$, one has a canonical isomorphism

$$\langle \det E, M\rangle \simeq N_{E/S}(M), \tag{22.2.6}$$

where $\det E \in \underline{\operatorname{Pic}}(C)$ is defined using a resolution of $E$ by vector bundles on $C$.

### 22.3. Determinant Bundles

Let $f : C \to S$ be a relative curve. Then for a pair of vector bundles $E_0$ and $E_1$ (resp., $F_0$ and $F_1$) of the same rank on $C$, one has a canonical isomorphism of line bundles on $S$:

$$\langle \det(E_0 - E_1), \det(F_0 - F_1)\rangle \simeq \det f_*((E_0 - E_1) \otimes (F_0 - F_1)). \quad (22.3.1)$$

Here $E_0 - E_1$ and $F_0 - F_1$ are virtual vector bundles of rank 0.

The idea of proof is to trivialize both parts of equation (22.3.1) locally over $S$ and then check that the transition functions are the same. Namely, locally over $S$ one can find a vector bundle $E_2$ of the same rank as $E_i$ and embeddings $u_0 : E_0 \to E_2$ and $u_1 : E_1 \to E_2$. Then by additivity of both parts in $(E_0 - E_1)$, it suffices to prove the statement in the situation when there exists an embedding $u : E_0 \to E_1$. Moreover, we can assume that $\det(u)$ does not vanish on every fiber of $f$, so that $E = \operatorname{coker}(u)$ is a vector bundle on a relative Cartier divisor $D \subset C$. Then in view of (22.2.6), the isomorphism (22.3.1) reduces to

$$N_{E/S} \det(F_0 - F_1) \simeq \det f_*(E \otimes (F_0 - F_1)).$$

One can trivialize both parts locally by choosing an isomorphism $v : F_0|_D \to F_1|_D$ and then check that these trivializations glue into a global isomorphism.

In particular, for every pair of line bundles $L$ and $M$ one has

$$\langle L, M\rangle = \det f_*((L - \mathcal{O}_C) \otimes (M - \mathcal{O}_C)). \quad (22.3.2)$$

Let $C_0$ be a curve over $k$, $J$ be its Jacobian, $\mathcal{P}_{C_0}$ be a Poincaré line bundle on $C_0 \times J$. Applying the above isomorphism to $S = J \times J$, $C = C_0 \times J \times J$ we obtain that $\langle p_{12}^*\mathcal{P}_{C_0}, p_{23}^*\mathcal{P}_{C_0}\rangle$ coincides with the biextension $\mathcal{B}$ on $J \times J$ associated with the line bundle $\det p_{2*}(\mathcal{P}_{C_0}) = \det \mathcal{S}(\mathcal{O}_C)$ on $J$ (see Section 17.4). Thus, the biextension on $J \times J$ given by the Deligne symbol $\langle L, M\rangle$ is the inverse of the biextension corresponding to the principal polarization of $J$.

### 22.4. Generalized Theta Divisors and Strange Duality Conjecture

Let $C$ be a curve of genus $g$. Let us denote by $U(r, d)$ the moduli space of semistable vector bundles of rank $r$ and degree $d$ on $C$, and by $SU(r)$ the moduli space of semistable bundles of rank $r$ and trivial determinant. These spaces can be constructed using the Geometric Invariant Theory (see [99]). We will only mention that points of $U(r, d)$ (resp., $SU(r)$) are in bijection with equivalence classes of semistable bundles of rank $r$ and degree $d$ (resp., with trivial determinant), where the equivalence relation is the following: $E \sim F$ if $E$ and $F$ have isomorphic associated graded bundles with respect to some filtrations whose succesive quotients are stable bundles of the same slope (see Lemma 14.4, (ii)). In particular, $U(r, d)$ (resp., $SU(r)$) has an open subset whose points are in bijection with stable bundles. On the moduli space $U(r, r(g-1))$ one has a natural divisor $\Theta_r$ supported on the set of vector bundles $E$, such that $h^0(E) \neq 0$. As in the case $r = 1$, the fiber of the corresponding line bundle at $E$ is canonically isomorphic to $\det R\Gamma(E)^{-1}$. For every line bundle $L$ of degree $g-1$ one can define the (generalized) theta divisor $\Theta_L$ in $U(r, 0)$ as the preimage of $\Theta_r$ under the morphism $E \mapsto E \otimes L$. It is known (see, e.g., [36]) that the restriction of $\Theta_L$ to $SU(r)$ (which we denote also by $\Theta_L$) generates the Picard group of $SU(r)$. Note that for $F \in U(r, 0)$ there is a canonical isomorphism

$$\mathcal{O}(\Theta_L)|_F \simeq \det R\Gamma(F \otimes L)^{-1} \simeq \det R\Gamma(F)^{-1} \otimes \det R\Gamma(L - \mathcal{O})^{-r}.$$

Now let us consider the morphism

$$t : U(r, r(g-1)) \times U(k, 0) \to U(kr, kr(g-1)) : (E, F) \mapsto E \otimes F.$$

The pull-back $t^*\mathcal{O}(\Theta_{kr})$ can be computed using the canonical isomorphism

$$\begin{aligned} \det R\Gamma(E \otimes F) \simeq \det R\Gamma(E)^k \otimes \det R\Gamma(F)^r \\ \otimes \det R\Gamma(\mathcal{O})^{-kr} \otimes \langle \det E, \det F \rangle, \end{aligned} \tag{22.4.1}$$

which follows from (22.3.1). In particular, we have an isomorphism

$$t^*\mathcal{O}(\Theta_{kr})|_{U(r,r(g-1)) \times SU(k)} \simeq \mathcal{O}_{U(r,r(g-1))}(k\Theta_r) \boxtimes \mathcal{O}_{SU(k)}(r\Theta_L).$$

Thus, the pull-back of the canonical section of $\mathcal{O}(\Theta_{kr})$ can be considered as an element of $H^0(U(r, r(g-1)), \mathcal{O}(k\Theta_r)) \otimes H^0(SU(k), \mathcal{O}(r\Theta_L))$, or equivalently as a map

$$H^0(SU(k), \mathcal{O}(r\Theta_L))^* \to H^0(U(r, r(g-1)), \mathcal{O}(k\Theta_r)). \tag{22.4.2}$$

The *strange duality conjecture* states that this map is an isomorphism. There is a generalization of this map for other values of ranks and degrees (see [35]). It is shown in [35] that the dimensions of both vector spaces (given by the Verlinde formula) are the same. Also, in [14] the strange duality conjecture is reformulated as the following geometric statement: the linear system $|r\Theta_L|$ on $SU(k)$ is spanned by the divisors $t^{-1}(\Theta_{kr})|_{F \times SU(k)}$, where $F$ varies in $U(r, 0)$.

We want to rewrite this duality in a more symmetric way. Namely, consider the map $\det : U(r, d) \to J^d = J^d(C)$. Let us fix a line bundle $L$ of degree

$g-1$. For $r, k > 0$ let us denote by $F_{r,k}$ the push-forward of $\mathcal{O}(k\Theta_r)$ by the map $\det : U(r, r(g-1)) \to J$, where we identify $J^{r(g-1)}$ with $J$ using $L^r$. It is known that higher cohomology groups of $\mathcal{O}(k\Theta_r)$ on fibers of det vanish, so $F_{r,k}$ is a vector bundle on $J$. Using the isomorphism $U(k, 0) \simeq U(k, k(g-1)) : F \mapsto F \otimes L$, we can consider the theta divisor $\Theta_L$ in $U(k, 0)$. Then (22.4.1) can be rewritten as follows:

$$t^*\mathcal{O}(\Theta_{kr}) \simeq \mathcal{O}(\Theta_L) \boxtimes \mathcal{O}(\Theta_k) \otimes (\det \times \det)^*\mathcal{P} \otimes \det R\Gamma(L)^{kr},$$

where $\mathcal{P} := \mathcal{B}^{-1}$ is the biextension on $J \times J$ corresponding to the principal polarization of $J$ (so we have $\mathcal{P}_{L,M} = \langle L, M\rangle^{-1}$). Hence, the pull-back of the canonical section of $\Theta_{kr}$ induces the canonical section

$$s_{r,k} \in H^0(J \times J, F_{r,k} \boxtimes F_{k,r} \otimes \mathcal{P}) \otimes \det R\Gamma(L)^{kr}.$$

Note that since the isomorphisms (22.4.1) for the pairs $(E, F)$ and $(F, E)$ coincide up to sign, we have $s_{k,r} = \pm s_{r,k}$. The section $s_{r,k}$ induces a homomorphism

$$p_2^* F_{k,r}^{\vee} \to p_1^* F_{r,k} \otimes \mathcal{P} \otimes \det R\Gamma(L)^{kr},$$

which is the same as a homomorphism

$$F_{k,r}^{\vee} \to \mathcal{S}(F_{r,k}) \otimes \det R\Gamma(L)^{kr}, \tag{22.4.3}$$

where $\mathcal{S} : \mathcal{D}^b(J) \to \mathcal{D}^b(J)$ is the Fourier transform defined using $\mathcal{P}$. The induced homomorphism on fibers at 0 coincides with (22.4.2). We claim that the strange duality conjecture is equivalent to the assertion that (22.4.3) is an isomorphism. Indeed, it is easy to see that for every $\xi \in J$ one has

$$t_{r\xi}^* F_{r,k} \simeq F_{r,k} \otimes \mathcal{P}_{k\xi} \otimes \det R\Gamma(\mathcal{O} - \xi)^{kr}.$$

In particular, $\mathcal{S}(F_{r,k})$ is a vector bundle satisfying

$$t_{k\xi}^* \mathcal{S}(F_{r,k}) \simeq \mathcal{S}(F_{r,k}) \otimes \mathcal{P}_{-r\xi} \otimes \det R\Gamma(\xi - \mathcal{O})^{kr}.$$

There is a similar isomorphism with $\mathcal{S}(F_{r,k})$ replaced by $F_{k,r}$ and the homomorphism (22.4.3) commutes with these isomorphisms. This immediately implies that the homomorphism (22.4.3) is an isomorphism if and only if its restriction to fibers at 0 is an isomorphism (this was first observed by M. Popa [117]).

**Theorem 22.1.** *If the strange duality conjecture holds for a pair $(r, k)$, then it holds for $(k, r)$.*

*Proof.* The maps (22.4.3) for $(r, k)$ and for $(k, r)$ are both obtained from the same section $s_{r,k}$ (up to a sign) using duality and adjointness of $(p_i^*, p_{i*})$ for two different projections $p_1, p_2$ of $J \times J$ to $J$. Now our assertion follows from the following general statement. Let $F_1, F_2$ be vector bundles on $J$, $s$ a global section of $(F_1 \boxtimes F_2) \otimes \mathcal{P}$ on $J \times J$. Then $s$ induces two morphisms

$$l_s : F_1^\vee \to \mathcal{S}(F_2),\ r_s : F_2^\vee \to \mathcal{S}(F_1).$$

In this situation $l_s$ is an isomorphism if and only if $r_s$ is an isomorphism. To prove this, first we construct a canonical isomorphism

$$R\underline{\mathrm{Hom}}(\mathcal{S}(F_1), \mathcal{O}_J) \simeq \mathcal{S}^{-1}(F_1^\vee).$$

Indeed, using the Grothendieck duality we find a canonical isomorphism

$$\begin{aligned} R\underline{\mathrm{Hom}}(\mathcal{S}(F_1), \mathcal{O}_J) &= Rp_{2*}R\underline{\mathrm{Hom}}(p_1^*F_1 \otimes \mathcal{P}, p_2^!\mathcal{O}_J) \\ &\simeq Rp_{2*}(p_1^*R\underline{\mathrm{Hom}}(F_1, \omega_J) \otimes \mathcal{P}^{-1})[g] \\ &\simeq \mathcal{S}^{-1}(R\underline{\mathrm{Hom}}(F_1, \mathcal{O}_J)). \end{aligned}$$

Now one can check that under this isomorphism, $r_s$ corresponds to the dual morphism to $\mathcal{S}^{-1}(l_s)$. □

It is easy to see that the conjecture holds in the case $k = 1, r \geq 1$. Hence, by Fourier transform and duality we get that the conjecture holds also for $r = 1$ and $k \geq 1$. The first proof of this fact was obtained in [15]. It used the technique of representing generic bundles of rank $k$ on $C$ as push-forwards of line bundles on a $k$-sheeted covering of $C$.

## Exercises

1. Prove that for a line bundle $L$ on a relative curve $C/S$, one has an isomorphism

$$(\det f_*(L - \mathcal{O}_C))^2 = \langle L, L \otimes \omega_{C/S}^{-1} \rangle,$$

where $\omega_{C/S}$ is the relative canonical bundle. This isomorphism should be considered as a categorification of the codimension-1 part of the Riemann-Roch theorem for the morphism $f$. [*Hint*: Apply (22.3.2) to $M = \omega_{C/S} \otimes L^{-1}$ and use relative Serre duality.]

2. Let $f : C_1 \to C_2$ be a finite morphism of curves.
   (a) Construct the canonical isomorphism

$$\langle N_{C_1/C_2}(L), M\rangle \simeq \langle L, f^*M\rangle,$$

   where $L$ is a line bundle on $C_1$, $M$ is a line bundle on $C_2$.
   (b) Define the homomorphism $\mathrm{Nm}_f : J(C_1) \to J(C_2)$ inducing the usual norm homomorphism on $k$-points. Show that it is dual to the pull-back homomorphism $J(C_2) \to J(C_1)$ under the standard self-duality of Jacobians.

## Appendix C. Some Results from Algebraic Geometry

Here is a collection of some theorems that we use which are not contained in Hartshorne [61], [62] or Griffiths-Harris [52]. All schemes below are assumed to be Noetherian and of finite Krull dimension.

**Descent.** *Let $f : X \to Y$ be a flat surjective morphism. Then the category of quasicoherent sheaves on $Y$ is equivalent to the category of pairs $(\mathcal{F}, \alpha)$, where $\mathcal{F}$ is a quasicoherent sheaf on $X$, $\alpha : p_1^*\mathcal{F} \to p_2^*\mathcal{F}$ is an isomorphism on $X \times_Y X$, satisfying the condition*

$$p_{13}^*(\alpha) = p_{23}^*(\alpha) \circ p_{12}^*(\alpha),$$

*where $p_{ij} : X \times_Y X \times_Y X \to X \times_Y X$ are projections.*

The pair $(\mathcal{F}, \alpha)$ as above is called the *descent data*. (For the proof see [56], exp. VIII or [100], ch. VII.) If $f$ is finite of degree $d$ which is invertible in $\mathcal{O}(Y)$, then the similar statement holds for derived categories of coherent sheaves (see [107], Appendix (the assumption on the degree was erroneously omitted in *loc. cit.*)).

**Proper Base Change.** *Let $f : X \to Y$ be a proper morphism, where $Y =$ Spec$(A)$ is an affine scheme, $\mathcal{F}$ be a coherent sheaf on $X$, flat over $Y$. Then there exists a finite complex $K^\bullet$ (with $K^n = 0$ for $n < 0$) of finitely generated projective $A$-modules and an isomorphism of functors*

$$H^p(X \times_Y \operatorname{Spec} B, \mathcal{F} \otimes_A B) \simeq H^p(K^\bullet \otimes_A B), \ \ p \geq 0,$$

*on the category of $A$-algebras $B$.*

For the proof, see [95], II.5 (see also [55], 6.10 and 7.7). This proof shows that the complex $K^\bullet$ in fact represents $Rf_*(\mathcal{F})$. Also, if $R^n f_*(\mathcal{F}) = 0$ for $n > g$, then the complex $K^\bullet$ can be chosen in such a way that $K^n = 0$ for $n > g$. Indeed, we can replace arbitrary complex $K^\bullet$ as above by its truncation $\tau_{\leq g} K^\bullet$, defined by

$$\tau_{\leq g} K^n = \begin{cases} K^n, & n < g, \\ \ker(K^g \to K^{g+1}), & n = g, \\ 0, & n > g. \end{cases}$$

Because $H^n(K^\bullet) = 0$ for $n > g$, one can easily see that the terms of $\tau_{\leq g} K^\bullet$ are finitely generated projective $A$-modules. Also, since the natural embedding $\tau_{\leq g} K^\bullet \to K^\bullet$ is a quasi-isomorphism, Lemma II.5.2 of [95] implies that we can replace $K^\bullet$ by $\tau_{\leq g} K^\bullet$ in the statement of the theorem.

**Base Change of a Flat Morphism.** *Let $f : X \to Y$ be a flat morphism and let $u : Y' \to Y$ be a morphism. Consider the cartesian square*

$$\begin{array}{ccc} X' & \xrightarrow{v} & X \\ \downarrow{\scriptstyle f'} & & \downarrow{\scriptstyle f} \\ Y' & \xrightarrow{u} & Y \end{array} \tag{22.4.4}$$

*where $X' = Y' \times_Y X$. Then there is a natural isomorphism of functors*

$$Lu^* \circ Rf_* \simeq Rf'_* \circ Lv^*$$

*from the category $D^-_{qc}(X)$ to $D^-_{qc}(Y')$.*

Here is a sketch of a proof. First of all, there is a canonical morphism of functors

$$Lu^* \circ Rf_* \to Rf'_* \circ Lv^* \tag{22.4.5}$$

constructed using adjoint pairs $(Lu^*, Ru_*)$ and $(Lv^*, Rv_*)$ and the isomorphism $Ru_* \circ Rf'_* \simeq Rf_* \circ Rv_*$. The fact that it is an isomorphism can be checked locally in $Y'$, so we can assume that $Y'$ and $Y$ are affine. Then it suffices to prove that $Ru_*$ applied to the base change arrow (22.4.5) is an isomorphism. By the projection formula,

$$Ru_*Lu^*Rf_*(F) \simeq Ru_*(\mathcal{O}_{Y'}) \otimes^{\mathbb{L}} Rf_*(F),$$

$$Ru_*Rf'_*Lv^*(F) \simeq Rf_*Rv_*Lv^*(F) \simeq Rf_*(Rv_*(\mathcal{O}_{X'}) \otimes^{\mathbb{L}} F).$$

By the flat base change we have $Rv_*\mathcal{O}_{X'} \simeq f^*Ru_*(\mathcal{O}_{Y'})$, so applying the projection formula again, we get the result.

In the case when $f$ is smooth this isomorphism can be also derived from the flat base change, the projection formula and the Grothendieck duality (see Lemma 1.3 of [21]). The reader can find a much more general statement in the notes [85] (see Proposition 3.10.3).

**Lefschetz Theorem for Picard Groups.** *Let $X$ be a smooth projective scheme, $Y \subset X$ be the zero locus of a section of an ample line bundle $\mathcal{O}_X(1)$. Assume that* $\dim Y \geq 3$ *(resp.* $\dim Y \geq 2$*) and $H^i(Y, \mathcal{O}_Y(-n)) = 0$ for $n > 0$ and $i = 1, 2$ (resp., $i = 1$). Then the restriction map* $\mathrm{Pic}(X) \to \mathrm{Pic}(Y)$ *is an isomorphism (resp., injective).*

For the proof, see [57], Exposé 11 (3.12) and Exposé 12 (3.4).

**Orlov's Theorem.** *Let $F$ be an exact functor from $D^b(X)$ to $D^b(Y)$, where $X$ and $Y$ are smooth projective varieties over a field. Assume that $F$ is full and faithful and has the right adjoint functor. Then there exists a unique (up to isomorphism) object $E \in D^b(X \times Y)$ such that $F \simeq \Phi_E$.*

The proof is given in [104].

# Bibliographical Notes and Further Reading

*Chapter 1.* The classification of holomorphic line bundles on a complex torus is a standard topic in complex geometry and in the theory of theta functions, see, e.g., [18], [52], [64], [69], [95], [131]. Our presentation is close to that of [95] with slight variations in the abstract nonsense part.

*Chapter 2.* An excellent exposition of the theory of Heisenberg groups including the complete proof of the Stone-von Neumann theorem can be found in Mumford's lectures [97]. The particular case of real Heisenberg groups is also presented in [84].

*Chapter 3.* We tried to use the minimal amount of data when defining theta series. The usual definition (see, e.g., [18]) starts with an isotropic decomposition $\Gamma = \Gamma_1 \oplus \Gamma_2$, while we use only a Lagrangian sublattice in $\Gamma$ and the quadratic map $\alpha : \Gamma \to U(1)$. One can recover the standard definition by taking $\alpha = \alpha_0(\Gamma_1, \Gamma_2)$ and $\Gamma_1$ or $\Gamma_2$ as a Lagrangian sublattice. Our proof of the Lefschetz theorem essentially follows [95]. The proof of Theorem 3.9 is borrowed from [81] (see also [133], IV.6). The identity of Exercise 7(c) was discovered by L. Kronecker (see [77]) and rediscovered in [135] and [111].

*Chapter 4.* The intertwining operators corresponding to Lagrangian subspaces and lattices in the real Heisenberg groups are considered thoroughly in [84]. It was also observed in *loc. cit.* that the study of these operators leads to the evaluation of some Gauss sums. Our derivation of the Turaev-van der Blij formula in Appendix B is a natural development of this idea. Note that this formula plays an important role in proving the reciprocity relations for Gauss sums ([33], [127]). It is interesting to note that earlier derivations of such reciprocity used theta functions (see [74]). We did not pursue another important application of the study of intertwining operators for $\mathcal{H}(V)$ – the construction of the Weil representation of the double covering of the symplectic group (see [84] or [132]).

*Chapter 5.* For other derivations of the classical functional equation for theta functions, see [84] or [97]. The distinguishing feature of our approach is that it does not use an explicit set of generators of the relevant subgroup of $Sp_{2g}(\mathbb{Z})$. Moret-Bailly found a nice interpretation of the appearance of 8th roots of unity in the functional equation: it corresponds to triviality of the 8th power of certain natural line bundle on the moduli space of abelian varieties with some additional data (see [90]). The action of the complex multiplication on theta functions is discussed in [108] and in [125]. For a nice exposition of the functional equation and for a generalization of the theory of theta functions to noncommutative tori, see [121] (noncommutative theta functions are also discussed in [87]).

*Chapter 6.* Most of the material in this chapter is based on the paper [5]. Here are some more references on homological mirror symmetry: the case of elliptic curve is (partially) treated in [115] and [112] (see also [75]); for the case of higher-dimensional abelian varieties, see [44] and [73]. The homological mirror conjecture has also inspired the study of autoequivalences of derived categories of coherent sheaves on Calabi-Yau manifolds, most notably the paper [123], where analogues of Dehn twists are constructed in this context.

*Chapter 7.* Another proof of Theorem 7.2 (using Hodge theory) can be found in [18] and [69]. In the algebraic context the computation of the cohomology of nondegenerate line bundles on abelian varieties is also given in [95].

*Chapter 8.* This chapter is a condensed exposition of the similar material in Mumford's book [95].

*Chapter 9.* Our construction of the dual abelian variety essentially follows [95]. The only novelty we introduce is the use of the Fourier–Mukai transform in the proof of Theorem 9.4 in Chapter 11. The procedure of taking the quotient of a scheme by the action of a finite group scheme is also discussed in [34].

*Chapter 10.* The notion of biextension is due to Grothendieck (see [58]). For the relation between biextensions and theta functions, see Breen's book [24]. A valuable source on commutative group schemes is Oort's book [103].

*Chapter 11.* The Fourier–Mukai transform appeared in the paper of S. Mukai [91]. The relative version of this transform for abelian schemes is also quite useful (see, e.g., [92], [109]). Most of the results of this chapter are borrowed from [91]. The formula for the cohomological Fourier–Mukai transform can be found in [92], Proposition 1.17. The action of the Fourier–Mukai transform on the Chow group was studied by Beauville in [12] and [13]. The general formalism of functors between derived categories of coherent sheaves has been studied and applied in [21], [22], [26], [27], [86], [104]. The action of the $SL_2(\mathbb{Z})$ on the derived categories of abelian varieties and its generalizations are studied in [106] and [110]. Other generalizations of the Fourier–Mukai transform have to do with differential operators on abelian varieties, see [82], [119]. The Fourier–Mukai transform can also be applied to the study of some natural torsion line bundles on the moduli space of abelian varieties, see [109].

*Chapter 12.* The Mumford group is discussed at length in [95] (see also [98]). The relative version of the theory of Heisenberg groups is developed in [89] (see also [110]). The proof of the quartic Riemann's theta identity can be found in practically every book on theta functions. Among algebraic versions of this identity (valid in arbitrary characteristic) let us mention the adelic version from [97]. Our approach is closer to the one adopted in [98]. The identity of Exercise 6(c) for the Kronecker function is an example of the associative version of the classical Yang-Baxter equation (see [113]).

*Chapter 13.* Theorem 13.7 is not stated explicitly in [95], but it follows easily from Theorem 23.3 of *loc. cit.* (see also discussion after Theorem 20.2 of *loc. cit.*). A generalization of this theorem to abelian schemes can be found in [32] (Proposition 1.2). The quadratic form associated with a symmetric line bundle is studied in [98]. Theorem 13.5 is well known to the specialists, however, we were not able to locate it in the literature.

*Chapter 14.* Classification of vector bundles on elliptic curves over $\mathbb{C}$ was obtained by Atiyah [7]. The case of arbitrary characteristic is due to Oda [102]. Our method is closer to that of S. Kuleshov in [78]. For the connection of vector bundles (and more generally principal $G$-bundles) on elliptic curves with loop groups, see [9].

*Chapter 15.* The equivalences between derived categories discussed in this chapter (and their generalization to the case of twisted coherent sheaves) were constructed in [107]. D. Orlov in [105] gave a different construction and proved (basing also on the results of [104]) that these are all exact equivalences that can occur between derived categories of abelian varieties. In [110], the theory of intertwining functors presented in this chapter is generalized to abelian schemes and the corresponding analogue of Weil representation is studied. The relation between indexes of nondegenerate line bundles established in Proposition 15.8 seems to be new.

*Chapter 16.* Our discussion of symmetric powers is inspired by [6], Exp. XVII, 6.3 (however, we avoid using Grothendieck representability theorem). Similar constructions can also be found in [65]. The first algebraic construction of the Jacobian of a curve is due to A. Weil [130] and is based on the study of rational group laws. The ideas of Weil's book have much wider applications (see [23]). For another construction of the Jacobian, see [67].

*Chapter 17.* The study of the principal polarization of the Jacobian and Riemann's theorem (Theorem 17.4) are standard topics in the theory of Riemann surfaces and theta functions, see, e.g., [41], [52]. The connection between determinant bundles and theta functions is discussed in [25]. Theta characteristics are studied in Mumford's paper [94]. Besides the theorems proven in Section 17.6, it contains an algebraic proof of the fact that the parity is preserved under deformations of the pair consisting of a curve and a theta-characteristic on it. A different (analytic) proof of this result can be found in Atiyah's paper [8].

*Chapter 18.* We follow closely [114], where the Cauchy-Szegö kernel is interpreted as a triple Massey product between coherent sheaves on a curve. Other proofs of Fay's trisecant identity can be found in [42], [118], [68] (arbitrary characteristic), [18], [40], [97] and [116]. The role of the trisecant identity in characterization of Jacobians among principally polarized abelian varieties is discussed in [4]. A generalization of Fay's trisecant identity to relative curves is considered in [19] and [37].

*Chapter 19.* Our computation of the Picard group of $\mathrm{Sym}^n C$ essentially follows [16]; for an alternative approach, see [29]. The computation of the Chern classes of the bundles $E_d$ is due to Schwarzenberger [122]. The formula (19.5.3) is due to Mattuck [87]. These matters are also briefly discussed in Fulton's book [46], Example 4.3.3. More information on the geometry of the morphism $\mathrm{Sym}^n C \rightarrow J^n(C)$ can be found in Kempf's book [67].

*Chapter 20.* We mostly follow Kempf's paper [66] (additional details can be found in [67]). For analytic proofs of the Riemann-Kempf singularity theorem, see [41], [52], and [96]. An excellent source on special divisors is the book [3].

*Chapter 21.* We essentially follow [16] with slight variations. A nice exposition of various approaches to Torelli theorem and to the Schottky problem can be found in Mumford's lectures [96].

*Chapter 22.* The symbol $\langle L, M\rangle$ is defined and studied thoroughly in [6], Exp. XVIII. In [31], it is applied to the problem of "categorification" of the Riemann-Roch formula. The strange duality conjecture is discussed in [14] and [35]. The interpretation in terms of Fourier–Mukai transform can be found in [117]. However, it seems that the equivalence of the strange duality conjecture for $(r, k)$ and for $(k, r)$ (see Theorem 22.1) was not noticed previously.

# References


[1] A. Andreotti, A. Mayer, On period relations for abelian integrals on algebraic curves, *Ann. Scuola Norm. Sup. Pisa* 21 (1967), 189–238.

[2] P. Appell, Sur les fonctions périodiques de deux variables, *Journ. de Math.* Sér. IV, 7 (1891) 157–219.

[3] E. Arbarello, M. Cornalba, P. A. Griffiths, J. Harris, *Geometry of Algebraic Curves*, vol. 1. Springer-Verlag, New York, 1985.

[4] E. Arbarello, Fay's trisecant formula and a characterization of Jacobian varieties, in *Algebraic Geometry*, Bowdoin (Brunswick, Me, 1985), 49–61; American Mathematical Society (AMS), Proc. Sympos. Pure Math. 46, part 1, Providence, RI, 1987.

[5] D. Arinkin, A. Polishchuk, Fukaya category and Fourier transform, in *Winter School on Mirror Symmetry, Vector Bundles and Lagrangian Submanifolds* (Cambridge, MA, 1999), 261–274. AMS, Providence, RI, 2001.

[6] M. Artin, A. Grothendieck, J.-L. Verdier, *Théorie de topos et cohomologie étale de schémas (SGA 4)*, tome 3. *Lecture Notes in Math.* 305, Springer-Verlag, Berlin, New York, 1973.

[7] M. F. Atiyah, Vector bundles over an elliptic curve, *Proc. Lond. Math. Soc.* VII (1957), 414–452.

[8] M. F. Atiyah, Riemann surfaces and spin structures, *Ann. Sci. École Norm. Sup.* 4 (1971), 47–62.

[9] V. Baranovsky, V. Ginzburg, Conjugacy classes in loop groups and $G$-bundles on elliptic curves, *Int. Math. Res. Notices* 15 (1996), 733–751.

[10] V. Batyrev, Dual polyhedra and mirror symmetry for Calabi-Yau hypersurfaces in toric varieties, *J. Alg. Geom.* 3 (1994), 493–535.

[11] A. Beauville, Prym varieties and the Schottky problem, *Inventiones Math.* 41 (1977), 149–196.

[12] A. Beauville, Quelques remarques sur la transformation de Fourier dans l'anneau de Chow d'une variété abélienne, *Algebraic Geometry* (Tokyo/Kyoto 1982); *Lecture Notes in Math.* 1016, 238–260. Springer-Verlag, Berlin, 1983.

[13] A. Beauville, Sur l'anneau de Chow d'une variété abélienne, *Math. Ann.* 273 (1986), 647–651.

[14] A. Beauville, Vector bundles on curves and generalized theta functions: recent results and open problems, in "*Current Topics in Complex Algebraic Geometry*" (Berkeley, CA, 1992/1993), 17–33; *Math. Sci. Res. Inst. Publ.*, 28, Cambridge University Press, Cambridge, UK, 1995.

[15] A. Beauville, M. S. Narasimhan, S. Ramanan, Spectral curves and the generalised theta divisor, *J. Reine Angew. Math.* 398 (1989), 169–179.

[16] A. Beilinson, A. Polishchuk, Torelli theorem via Fourier transform, in *Moduli of Abelian Varieties*, C. Faber, et al., eds., 127–132. Birkhäuser, Basel, 2001.

[17] P. Berthelot, A. Grothendieck, L. Illusie, *Théorie des intersections et théorème de Riemann-Roch (SGA 6). Lecture Notes in Math.* 225. Springer-Verlag, Berlin, New York, 1971.

[18] Ch. Birkenhake, H. Lange, *Complex Abelian Varieties.* Springer-Verlag, 1992.

[19] I. Biswas, A. K. Raina, Relative curves, theta divisor, and Deligne pairing, *Int. Math. Res. Notices* 9 (1995), 425–435.

[20] F. van der Blij, An invariant of quadratic forms modulo 8, *Indag. Math.* 21 (1959), 291–293.

[21] A. I. Bondal, D. O. Orlov, Semiorthogonal Decomposition for Algebraic Varieties, preprint alg-geom./9506012.

[22] A. I. Bondal, D. O. Orlov, Reconstruction of a variety from the derived category and groups of autoequivalences, *Compositio Math.* 125 (2001), 327–344.

[23] S. Bosch, W. Lutkebohmert, M. Raynaud, *Néron Models.* Springer-Verlag, Berlin, 1990.

[24] L. Breen, *Functions theta et théorème du cube. Lecture Notes in Math.* 980. Springer-Verlag, Berlin, 1983.

[25] L. Breen, The cube structure on the determinant bundle, in *Theta functions – Bowdoin 1987*, part 1, 663–673. AMS, Providence, RI, 1989.

[26] T. Bridgeland, Equivalences of triangulated categories and Fourier-Mukai transforms, *Bull. Lond. Math. Soc.* 31 (1999), 25–34.

[27] T. Bridgeland, Flops and Derived Categories, *Invent. Math.* 147 (2002), 613–632.

[28] J.-L. Brylinski, P. Deligne, Central extensions of reductive groups by $K_2$, Publ. Math. IHES 94 (2001), 5–85.

[29] A. Collino, The rational equivalence ring of symmetric products of curves, *Illinois J. Math.* 19 (1975), 567–583.

[30] B. Conrad, *Grothendieck duality and base change.* Springer-Verlag, Berlin, 2000.

[31] P. Deligne, Le déterminant de la cohomologie, *Current Trends in Arithemetical Algebraic Geometry, 93–177, Contemp. Math.*, no. 67, AMS, Providence, 1987.

[32] P. Deligne, G. Pappas, Singularités des espaces de modules de Hilbert, en les caractéristiques divisant le discriminant, *Compositio Math.* 90 (1994), 59–79.

[33] F. Deloup, Linking forms, reciprocity for Gauss sums and invariants of 3-manifolds, *Trans. AMS* 351 (1999), 1895–1918.

[34] M. Demazure, P. Gabriel, *Groupes Algébriques*, tome I. North-Holland Publishing Co., Amsterdam, 1970.

[35] R. Donagi, L. Tu, Theta functions for SL(n) versus GL(n), *Math. Res. Letters* 1 (1994).

[36] J. M. Drezet, M. S. Narasimhan, Groupe de Picard des variétés de modules de fibrés semi-stables sur les courbes algébriques, *Invent. Math.* 97 (1989), 53–94.

[37] F. Ducrot, Fibré déterminant et courbes relatives, *Bull. Soc. Math. France* 118 (1990), 311–361.

[38] G. Faltings, C.-L. Chai, *Degeneration of Abelian Varieties.* Springer-Verlag, Berlin, 1990.

[39] H. M. Farkas, On Fay's tri-secant formula, *J. Analyse Math.* 44 (1984/85), 205–217.
[40] H. M. Farkas, I. Kra, *Riemann surfaces*. 2nd ed., Springer-Verlag, New York, 1992.
[41] H. M. Farkas, H. E. Rauch, *Theta Functions with Applications to Riemann Surfaces*. Williams and Wilkins, Baltimore, 1974.
[42] J. D. Fay, *Theta functions on Riemann surfaces*. *Lecture Notes in Math.* 352. Springer-Verlag, Berlin, New York, 1973.
[43] K. Fukaya, Morse homotopy, $A^\infty$-category, and Floer homologies, *Proceedings of GARC Workshop on Geometry and Topology '93* (Seoul, 1993), 1–102.
[44] K. Fukaya, Mirror symmetry of abelian variety and multi theta functions, *J. Algebraic Geom.* 11 (2002), 393–512.
[45] K. Fukaya, Y.-G. Oh, H. Ohta, K. Ono, Lagrangian intersection Floer theory – anomaly and obstruction, 2000, see http://www.kusm.kyoto-u.ac.jp/~fukaya/fukaya.html.
[46] W. Fulton, *Intersection Theory*. Springer-Verlag, Berlin, 1984.
[47] P. Gabriel, Sur les catégories abéliennes localement noethériennes et leurs applications aux algèbres étudiées par Dieudonné, Sém. J.-P. Serre, 1960.
[48] S. Gelfand, Yu. Manin, *Methods of Homological Algebra*. Springer-Verlag, Berlin, 1996.
[49] J. Giraud, *Cohomologie non-abélienne*. Springer-Verlag, Berlin, New York, 1971.
[50] A. Givental, Equivariant Gromov-Witten invariants, *Int. Math. Res. Notices* 13 (1996), 613–663.
[51] V. Golyshev, V. Lunts, D. Orlov, Mirror symmetry for abelian varieties, *J. Algebraic Geom.* 10 (2001), 433–496.
[52] P. Griffiths, J. Harris, *Principles of Algebraic Geometry*. Wiley-Interscience, New York, 1978.
[53] M. Gross, Topological mirror symmetry, *Invent. Math.* 144 (2001), 75–137.
[54] A. Grothendieck, Sur quelques points d'algèbre homologique, *Tohoku Math. J.* (2) 9 (1957), 119–221.
[55] A. Grothendieck, J. Dieudonné, *Eléments de géométrie algébrique. III. Etude cohomologique des faisceaux cohérents*, *Publ. Math. IHES* 11 (1961), 17 (1963).
[56] A. Grothendieck, *Revetements Étale et Groupe Fondamental (SGA 1)*. Springer-Verlag, Berlin, New York, 1971.
[57] A. Grothendieck, *Cohomologie Locale des Faisceaux Coherents et Theoremes de Lefschetz Locaux et Globaux* (*SGA* 2). Amsterdam, North-Holland Publishing Co., 1969.
[58] A. Grothendieck, *Groupes de Monodromie en Géométrie Algébrique. I (SGA 7 I)*, *Lecture Notes in Math.* 288, Springer-Verlag, Berlin, New York, 1972.
[59] R. C. Gunning, Some identities for Abelian integrals, *Amer. J. Math.* 108 (1986), 39–74.
[60] G. Harder, M. S. Narasimhan, On the cohomology groups of moduli spaces of vector bundles on curves, *Math. Ann.* 212 (1974/75), 215–248.
[61] R. Hartshorne, *Residues and Duality*. *Lecture Notes in Mathematics* 20. Springer-Verlag, Berlin, 1966.

[62] R. Hartshorne, *Algebraic Geometry*. Springer-Verlag, Berlin, 1977.

[63] G. Humbert, Théorie générale des surface hyperelliptiques, *Journ. de Math.* Sér. IV, 9 (1893) 29–170, 361–475.

[64] J. Igusa, *Theta Functions*. Springer-Verlag, New York, Heidelberg, 1972.

[65] B. Iversen, Linear determinants with applications to the Picard scheme of a family of algebraic curves, *Lecture Notes in Math.* 174, Springer-Verlag, Berlin, New York, 1970.

[66] G. Kempf, On the geometry of a theorem of Riemann, *Ann. of Math.* 98 (1973), 178–185.

[67] G. R. Kempf, *Abelian Integrals*, Universidad Nacional Autónoma de México, México, 1983.

[68] G. R. Kempf, Fay's trisecant formula, Algebraic geometry and complex analysis (Pátzcuaro, 1987), 99–106. *Lecture Notes in Math.* 1414. Springer-Verlag, Berlin, 1989.

[69] G. Kempf, *Complex Abelian Varieties and Theta Functions*. Springer-Verlag, Berlin, 1991.

[70] M. Kontsevich, Homological algebra of mirror symmetry, in *Proceedings of ICM* (Zürich, 1994), 120–139. Birkhäuser, Basel, 1995.

[71] M. Kontsevich, Enumeration of rational curves via torus actions, in *The Moduli Spaces of curves (Texel Island, 1994)*, R. Dijkgraad et al., eds., 335–368. Birkhäuser, 1995.

[72] M. Kontsevich, Yu. Manin, Gromov-Witten classes, quantum cohomology and enumerative geometry, *Comm. Math. Phys.* 164 (1994), 525–562.

[73] M. Kontsevich, Y. Soibelman, Homological mirror symmetry and torus fibration, *Symplectic Geometry and Mirror Symmetry* (Seoul 2000), 203–263, World Sci. Publishing, River Edge, NJ, 2001.

[74] A. Krazer, *Zur Theorie der mehrfachen Gausschen Summen*, 181. H. Weber Festschrift, Leipzig, 1912.

[75] B. Kreussler, Homological mirror symmetry in dimension one, in *Advances in Algebraic Geometry Motivated by Physics* (Lowell, MA, 2000), E. Previato, ed., 179–198. AMS, Providence, RI, 2001.

[76] I. M. Krichever, Algebro-geometric construction of the Zaharov-Shabat equations and their periodic solutions. *Soviet Math. Dokl.* 17 (1976), 394–397; Integration of nonlinear equations by the methods of nonlinear geometry, *Funk. Anal. i Pril*, 11 (1977), 15–31.

[77] L. Kronecker, Zur theorie der elliptischen functionen (1881), in *Leopold Kronecker's Werke*, vol. IV, 311–318. Chelsea Pub. Co., New York, 1968.

[78] S. A. Kuleshov, Construction of bundles on an elliptic curve, in *Helices and Vector Bundles*. Cambridge University Press, 1990.

[79] S. Lang, *Abelian Varieties*. Interscience Publishers, London, 1959.

[80] S. Lang, *Algebra*. Addison-Wesley, Reading, MA, 1965.

[81] S. Lang, *Elliptic functions*. Addison-Wesley, Reading, MA, 1973.

[82] G. Laumon, Transformation de Fourier generalisée, Preprint Alg.-Geom./9603004.

[83] C. Leung, S.-T. Yau, E. Zaslow, From special Lagrangian to Hermitian-Yang-Mills via Fourier-Mukai Transform, *Adv. Theor. Math. Phys.* 4 (2000), 1319–1341.

[84] G. Lion, M. Vergne, *The Weil Representation, Maslov Index, and Theta Series*, Birkhäuser, Boston, 1980.

[85] J. Lipman, Notes on derived categories and derived functors, see http://www.math.purdue.edu/~lipman/.

[86] A. Maciocia, Generalized Fourier-Mukai transforms, *J. Reine Angew. Math.* 480 (1996), 197–211.

[87] Yu. I. Manin, Mirror symmetry and quantization of abelian varieties, in *Moduli of abelian varieties (Texel Island, 1999)*, C. Faber et al., eds., 231–254, Birkhäuser, Basel, 2001.

[88] A. P. Mattuck, Symmetric products and Jacobians, *Amer. J. Math.* 83 (1961), 189–206.

[89] L. Moret-Bailly, *Pinceaux de variétés abéliennes*, *Astérisque* 129 (1985).

[90] L. Moret-Bailly, Sur l'équation fonctionnelle de la fonction theta de Riemann, *Compositio Math.* 75 (1990), 203–217.

[91] S. Mukai, Duality between D(X) and D($\hat{X}$) with its application to Picard sheaves, *Nagoya Math. J.* 81 (1981), 153–175.

[92] S. Mukai, Fourier functor and its application to the moduli of bundles on an abelian variety, *Adv. Studies in Pure Math.* 10 (1987), 515–550.

[93] S. Mukai, Abelian varieties and spin representation, preprint of Warwick University (1998) (English translation from *Proceedings of the Symposium "Hodge theory and Algebraic Geometry,"* 110–135. Sapporo, 1994.)

[94] D. Mumford, Theta characteristics of an algebraic curve, *Ann. Sci. École Norm. Sup.* 4 (1971), 181–192

[95] D. Mumford, *Abelian Varieties.* Oxford University Press, London, 1974.

[96] D. Mumford, *Curves and Their Jacobians.* University of Michigan Press, Ann Arbor, 1975.

[97] D. Mumford, et al. *Tata Lectures on Theta I-III.* Birkhauser, Boston, 1982–1991.

[98] D. Mumford, On the equations defining abelian varieties, *Invent. Math.* 1 (1966), 287–354, 3 (1967), 75–135, 215–244.

[99] D. Mumford, J. Fogarty, F. Kirwan, *Geometric Invariant Theory.* Springer-Verlag, Berlin, 1994.

[100] J. Murre, *Lectures on an Introduction to Grothendieck's Theory of the Fundamental Group.* Lecture Notes. Tata Institute of Fundamental Research, Bombay, 1967.

[101] J. von Neumann, Die Eindeutigkeit der Schrödingerschen Operatoren, *Ann. Mat. Pura Appl.* 104 (4) (1931), 570–578.

[102] T. Oda, Vector bundles on an elliptic curve, *Nagoya Math. J.* 43 (1971), 41–72.

[103] F. Oort, *Commutative group schemes, Lecture Notes in Math.* 15, Springer-Verlag, 1966.

[104] D. O. Orlov, Equivalences of derived categories and K3 surfaces, *Journal Math. Sci.*, New York, 84 (1997), 1361–1381.

[105] D. O. Orlov, *On equivalences of derived categories of coherent sheaves on abelian varieties*, preprint math. AG/9712017.

[106] A. Polishchuk, A remark on the Fourier-Mukai transform, *Math. Research Letters* 2 (1995), 193–202.

[107] A. Polishchuk, Symplectic biextensions and a generalization of the Fourier-Mukai transform, *Math. Research Lett.* 3 (1996), 813–828.

[108] A. Polishchuk, Theta identities with complex multiplication, *Duke Math. J.* 96 (1999), 377–400.

[109] A. Polishchuk, Determinant bundles for abelian schemes, *Compositio Math.* 121 (2000), 221–245.

[110] A. Polishchuk, *Analogue of Weil representation for abelian schemes*, Journal für die reine und angewandte Mathematik 543 (2002), 1–37.

[111] A. Polishchuk, Massey and Fukaya products on elliptic curves, *Adv. Theor. Math. Phys.* 4 (2000), 1187–1207.

[112] A. Polishchuk, $A_\infty$-structures on an elliptic curve, preprint math. AG/0001048.

[113] A. Polishchuk, Classical Yang-Baxter equation and the $A_\infty$-constraint, *Advances in Math.* 168 (2002), 56–95.

[114] A. Polishchuk, Triple Massey products on curves, Fay's trisecant identity and tangents to the canonical embedding, preprint math. AG/0107194.

[115] A. Polishchuk, E. Zaslow, Categorical mirror symmetry in the elliptic curve, in *Winter School on Mirror Symmetry, Vector Bundles and Lagrangian Submanifolds* (Cambridge, MA, 1999), 275–295. AMS, Providence, RI, 2001.

[116] C. Poor, Fay's trisecant formula and cross-ratios, *Proc. AMS* 114 (1992), 667–671.

[117] M. Popa, Verlinde bundles and generalized theta linear series, preprint math. AG/0002017.

[118] A. K. Raina, Chiral fermions on a Riemann surface and the trisecant identity, *Mathematical physics* (Islamabad, 1989), 326–338. World Scientific Publishing, River Edge, NJ, 1990.

[119] M. Rothstein, Sheaves with connection on abelian varieties, *Duke Math. Journal* 84 (1996), 565–598.

[120] W.-D. Ruan, Lagrangian torus fibrations and mirror symmetry of Calabi-Yau manifolds, preprint math. DG/0104010.

[121] A. Schwarz, Theta functions on noncommutative tori, *Lett. Math. Phys.* 58 (2001), 81–90.

[122] R. L. E. Schwarzenberger, Jacobians and symmetric products, *Illinois J. Math.* 7 (1963), 257–268.

[123] P. Seidel, R. Thomas, Braid group actions on derived categories of coherent sheaves, *Duke Math. J.* 108 (2001), 37–108.

[124] J.-P. Serre, Géométrie algébrique et géométrie analytique, *Ann. Inst. Fourier* 6 (1956), 1–42.

[125] G. Shimura, Theta functions with complex multiplication, *Duke Math. J.* 43 (1976), 673–696.

[126] A. Strominger, S.-T. Yau, E. Zaslow, Mirror symmetry is T-duality, *Nucl. Phys. B* 479 (1996), 243–259.

[127] V. Turaev, Reciprocity for Gauss sums on finite abelian groups, *Math. Proc. Camb. Phil. Soc.* 124 (1998), 205–214.

[128] J.-L. Verdier, *Des catégories dérivées des catégories abéliennes*, *Asterisque* 239 (1996).

[129] C. Weibel, *An Introduction to Homological Algebra*, Cambridge University Press, Cambridge, 1994.

[130] A. Weil, *Variétés abélienne et courbes algébrique, Hermann, and Cie.*, Paris, 1948.

[131] A. Weil, *Introduction à l'étude des variétés kählériennes*, Hermann, Paris, 1958.

[132] A. Weil, Sur certains groupes d'opérateurs unitaires, *Acta Math.* 111 (1964), 143–211.

[133] A. Weil, *Elliptic Functions According to Eisenstein and Kronecker.* Springer-Verlag, Berlin, New York, 1976.

[134] S.-T. Yau, ed., *Essays on Mirror Manifolds*, International Press, Hong Kong, 1992.

[135] D. Zagier, Periods of modular forms and Jacobi theta functions, *Invent. Math.* 104 (1991), 449–465.

[136] Yu. Zarhin, Endomorphisms of abelian varieties and points of finite order in characteristic p., *Math. Notes of the Academy of Sciences of the USSR* 21 (1977), 734–744.

[137] I. Zharkov, Torus fibrations of Calabi-Yau hypersurfaces in toric varieties, *Duke Math. J.* 101 (2000), 237–257.

[138] I. Zharkov, Theta-functions for indefinite polarizations, preprint math. AG/0011112.

# Index